Geography for Nongeographers

FRANK R. SPELLMAN

Government Institutes
An imprint of
THE SCARECROW PRESS, INC.
Lanham • Toronto • Plymouth, UK
2010

Published by Government Institutes
An imprint of The Scarecrow Press, Inc.
A wholly owned subsidiary of The Rowman & Littlefield Publishing Group, Inc.
4501 Forbes Boulevard, Suite 200, Lanham, Maryland 20706
http://www.govinstpress.com

Estover Road, Plymouth PL6 7PY, United Kingdom

The reader should not rely on this publication to address specific questions that apply to a particular set of facts. The author and the publisher make no representation or warranty, express or implied, as to the completeness, correctness, or utility of the information in this publication. In addition, the author and the publisher assume no liability of any kind whatsoever resulting from the use of or reliance upon the contents of this book.

British Library Cataloguing in Publication Information Available

Library of Congress Cataloging-in-Publication Data

Spellman, Frank R.
Geography for nongeographers / Frank R. Spellman.
p. cm.
ISBN 978-1-60590-686-7 (pbk. : alk. paper)—ISBN 978-1-60590-687-4 (electronic)
1. Geography. I. Title.
G128.S655 2010
910′.02—dc22

2009046974

™ The paper used in this publication meets the minimum requirements of American National Standard for Information Sciences—Permanence of Paper for Printed Library Materials, ANSI/NISO Z39.48-1992.

Printed in the United States of America

For
JoAnn Garnett-Chapman: The Ultimate Traveler

Contents

Preface

Why is this book on geography for the nongeographer? Answer: This book is for people, nonspecialists, who want to learn about geography without taking a formal course. It can serve as a classroom supplement, tutorial aid, self-teaching guide, or home-schooling text.

Based on personal experience, I have found that many professionals do have some background in geography (don't we all?), but many of these folks need to stay current (remember that former country X is country Y today and country Z tomorrow). We live in a dynamic, ever-changing world and society. Moreover, many practitioners are specialists (e.g., engineers, environmental scientists, environmental health professionals, medical professionals, epidemiologists, water quality technicians, toxicologists, environmental scientists, occupational health and safety professionals, industrial hygienists, teachers, and so forth). Herein lies the problem—that is, specialization. My view is that practitioners of any blend, flavor, or type should absolutely know the ins and outs of their field of specialization but in doing so should also be generalists with a wide range of knowledge—not just specialists whose knowledge may be too narrowly focused. Again, based on personal experience, my students, for example, who have generalized their education (spread out their exposure to include disciplines in several aspects of general studies), versus those who narrowly specialize, are afforded many more opportunities to broaden their view of the world and increase their chances for upward mobility in employment ventures. They have had a much better opportunity to ascend to upper management positions because with on-the-job experience and their generalized education—their holistic view of the world and life around them—they become well-rounded individuals, ready to undertake just about any challenge they may face. Is this not the standard characteristic of a successful person?

Geography for Nongeographers fills the gap between general introductory geography texts and the more advanced environmental geography books used in graduate courses. *Geography for Nongeographers* fills this gap by surveying and covering the basics of geography. This book is a nontechnical survey for those with little background in geography—presented in reader-friendly written style.

Geography is a multidisciplinary field that incorporates aspects of biology, chemistry, physics, ecology, geology, meteorology, pedology, sociology, and many other fields. Books on the subject are typically geared toward professionals in these fields. This makes undertaking a study of geography daunting to those without this specific background. However, this complexity also indicates geography's broad scope of impact. Because geography affects us (sometimes in profound ways such as volcanism, earthquakes, border wars, social/cultural norms, and so forth), it is important to understand some basic concepts of the discipline.

Along with basic geographical principles, the text provides a clear, concise presentation of the consequences of the physical interactions with the environment we inhabit. Even if you are not tied to a desk, the book provides you, the nongeographer, with the jargon, concepts, and key concerns of geography and geography in action. This book is compiled in an accessible, user-friendly format, unique in that it explains scientific concepts in the most basic way possible. Moreover, the text is packed with information students need to get ready for SAT exams.

Each chapter ends with a chapter review test to help evaluate mastery of the concepts presented. Before going on to the next chapter, you should take the review test, compare your answers to the key provided in the appendix, and review the pertinent information for any problems you missed. If you miss many items, review the whole chapter.

Again, this text is accessible to those who have no experience with geography—it is an entry-level, nonfiction science book especially designed for lay people—nonspecialists. If you work through the text systematically, you will acquire an understanding of and skill in geographical principles—adding a critical component to your professional knowledge.

I

THE BASICS

Topography and various landforms have a lot to do with where people live on Earth. Obviously, the scene shown here is not conducive to fulltime human habitation.

CHAPTER 1

Introduction

> As a young man, my fondest dream was to become a geographer. However, while working in the customs office I thought deeply about the matter and concluded it was too difficult a subject. With some reluctance I then turned to physics as a substitute.
>
> —Attributed to Albert Einstein (later debunked)

> Geography, to many, remains a baffling subject, seemingly simplistic yet intimidating.
>
> —Keith M. Bell (2007)

For many of us, the only thing we remember about the geography we were taught in school is a few of those capitals of countries and states throughout the world and in the United States that we had to memorize. Also, there were those maps—maps everywhere: maps pinned to every classroom wall and the maps in textbooks—and that globe of the world that stood in the corner of the classroom. Those maps and globes caused many of us to gaze at them and to let our minds escape in wonderment to so many mysterious, far-off places.

Geography that is all about maps and the capitals of states and countries is how many of us then (the challenge of facing a blank outline map of the United States with the charge to name the states and capitals) and today would sum up what geography is all about. Intuitively, of course, we know that geography is about much more; it is just so difficult, beyond the maps, foreign countries, and capital cities, to define. This book explains what geography is all about. One thing is certain: though we spend some time discussing maps and mention a capital or two, geography is much more than just maps and capitals. You will learn this as you make your way through this book.

What Is Geography?

While the word geography is derived from Greek (*geo* referring to Earth and *graphy* meaning picture or writing) and literally means "to write about the Earth," the subject of geography, as mentioned, is much more than the names of capitals and maps. Geography is a science, an all-encompassing science that seeks to bridge and understand, through an understanding of location and place, the world's physical and human features. Geography teaches us where things are and how they got there; it looks at the spatial connection between people, places, and the Earth.

Some have called geography the "mother of all sciences." This title is well suited and fitting because all other disciplines took root from its existence. Geography is not only an analytic tool but also a spatial science (dealing with people, landscapes, money, and other infinite uses) that deals with many aspects of social science (e.g., history, psychology, and anthropology), physical science (e.g., geology and weather and climate), and technical science (e.g., geographic information system [GIS], geodesy, and remote sensing; Bell 2007).

Did You Know?

Most people have an idea what a geologist does but don't have a clue of what a geographer does. While geography is commonly divided into human geography and physical geography, the difference between physical geography and geology is often confusing. We distinguish the difference by pointing out that geographers study the surface of the Earth, its features, and why they are where they are. Geologists look deeper into the Earth and study its rocks . . . and also study every child's favorite, the dinosaurs.

Another view of geography states, "The more I work in the social-studies field the more convinced I become that Geography is the foundation of all. When I call it the 'queenly science,' I do not visualize a bright-eyed young woman recently a princess but rather an elderly, somewhat beat-up dowager, knowing in the way of power" (p. 760). The author of this statement, James A. Michener (1970), who is also the author of best-selling historical novels such as *The Source, Hawaii, Centennial, Alaska, Texas, Caribbean, Chesapeake, Caravans, The Drifters,* and several others, knew a few things about geography and several other subjects. Using his wide-ranging geographic knowledge of Earth, Michener simply makes the point that geography is a parent to biology, geology, chemistry, physics, history, and economics. In light of this, I feel that geographers are the sons and daughters of the geographers who as scientists study the relationship between people and their environments.

Michener (1970, 761) made another important point about geography, one that I feel is germane to the purpose of this text. Consider the following: "With growing emphasis on ecology and related problems of the environment, geography will undoubtedly grow in importance and relevance. I wish that the teaching of it were going to improve commensurately; most of the geography courses I have known were rather poorly taught and repelled the general student like me." Based on personal experience, I have found that Michener's sentiments are all too true, even at the college level.

Looking back, the state of training and the relevance of geography have not always been problematic. Consider the following by John Rennie Short (2004): There was a time when

> formal geography texts constitute[d] most of the body of schoolbooks in the eighteenth and early nineteenth century in the United States. Texts on history, for example, did not appear until the 1880s. These early geography texts were compendiums of knowledge, widely used in schools and the more literate homes. The early geography texts were the encyclopedia of their day. (p. 3)

It is not my intent to make this book an encyclopedia of today; it is my intent, however, to ensure that this book is relevant and informative for every day.

Mother of All Sciences

You know, when we use the stock phrase the "mother of all . . ." in our public discourse and popular culture, we need to be a bit careful. Remember it was Saddam Hussein who promised in a speech the "mother of all battles" if the U.S.-led coalition forces attempted to evict his army of occupation from Kuwait in 1991. The truth be told, the coalition-led invasion turned out for Hussein to be the mother of all slaughters, the mother of all debacles, the mother of all embarrassments; General Patton, who said in World War II that running through the enemy will be like crap running through a goose, would have been proud, because Hussein's army was our goose. Today, it is not uncommon to see or hear the "mother of all" phrase used in terms of the current financial crisis gripping the globe—the mother of all bailouts. I am sure you have heard this common discourse—and things being reality, who can argue against that? One thing is certain; if you are going to label something as the "mother of all" then you better be able to back it up.

The point is that when we say geography is the mother of all sciences, we have to be careful with such usage. Some might question the validity of this statement; they might be somewhat baffled. However, the professional geographer would know this statement is true, and he or she would have little difficulty in defending its accuracy. The professional geographer would simply point out that nearly every other type of scientific profession or pursuit, be it ecology, geology, population study, demographics, agriculture, literacy, wildfire, or environmental management, is based on the geographic location of the place under study. Geography plays a role in nearly every decision we make. Choosing sites, targeting market sectors, planning distribution networks, responding to emergencies, or redrawing city, county, state, or country boundaries—all of these problems involve questions in geography. Moreover, with the developing technology of GIS, which integrates software, and data for capturing, managing, analyzing, and displaying forms of geographically referenced material, geography has been propelled into the digital age—and for the geographer and the rest of us, this is a good thing.

The professional geographer is a naturalist, an observer of the main facets of nature; he or she is one who recognizes and understands the richness and variety of nature, one who can recognize the contours of landscapes, correlate satellite images with the area, read maps, and interpret landscape. Most of all, a geographer is a preserver of nature—the ultimate mother of us all.

The Big Picture

Unlike the important but narrowly focused science of geology—the science that deals with and answers many questions about planet Earth—geography is holistic in its approach; it presents the big picture, one that has environmental events and human actions intertwined.

All of the disciplines emanate from the core of geographic sciences. No one made this point clearer than the renowned American geologist/geographer, Nevin Fenneman. Fenneman (1865–1945) was noted for his work on the physiography (i.e., study of processes and patterns in the natural environment) of the United States (Short 2004). He opined that there is in geography a central core that is pure geography and nothing else, but there is much beyond this core that is none the less geography, though it belongs also to overlapping sciences. Figure 1.1 is derived from Fenneman's work, and it demonstrates his contention that "the seeds are in the core, and the core is regional geography, and this is why the subject propagates itself and maintains a separate existence" (Bell 2007).

History of Geography

One of the main foundational factors adding to the credence of geography being classified as the mother of all sciences is its history. Geography has been around for more than 2,200 years, dating back to the Greeks when Eratosthenes supposedly coined the term from the words, as pointed out earlier, *geo* (Earth) and *graphein* (to write)—literally meaning *description of the Earth*. As geography developed, many other scientific disciplines developed later and became increasingly specialized.

As mentioned, the name geography is attributed to Eratosthenes, but it is Anaximander of Miletus (c. 610 BC–c. 545 BC) who is considered to be the true founder of geography. Anaximander is credited with the early measurement of latitude and with the prediction of eclipses. From Europe to Africa to Asia, Eratosthenes, the Chinese, and Arabs amassed and synthesized incredible amounts of information about the Earth. Building on the body of knowledge of early geographers, Immanuel Kant (1724–1804) distinguished between the two ways of classifying things: spatially (having the nature of space) and the category of time (temporal). Modern geography evolved from the works of German geographers Alexander von Humboldt and Carl Ritter. They moved geography away from the study of tomes of data to the study of regions for the ultimate understanding of Earth. Table 1.1 lists notable geographers.

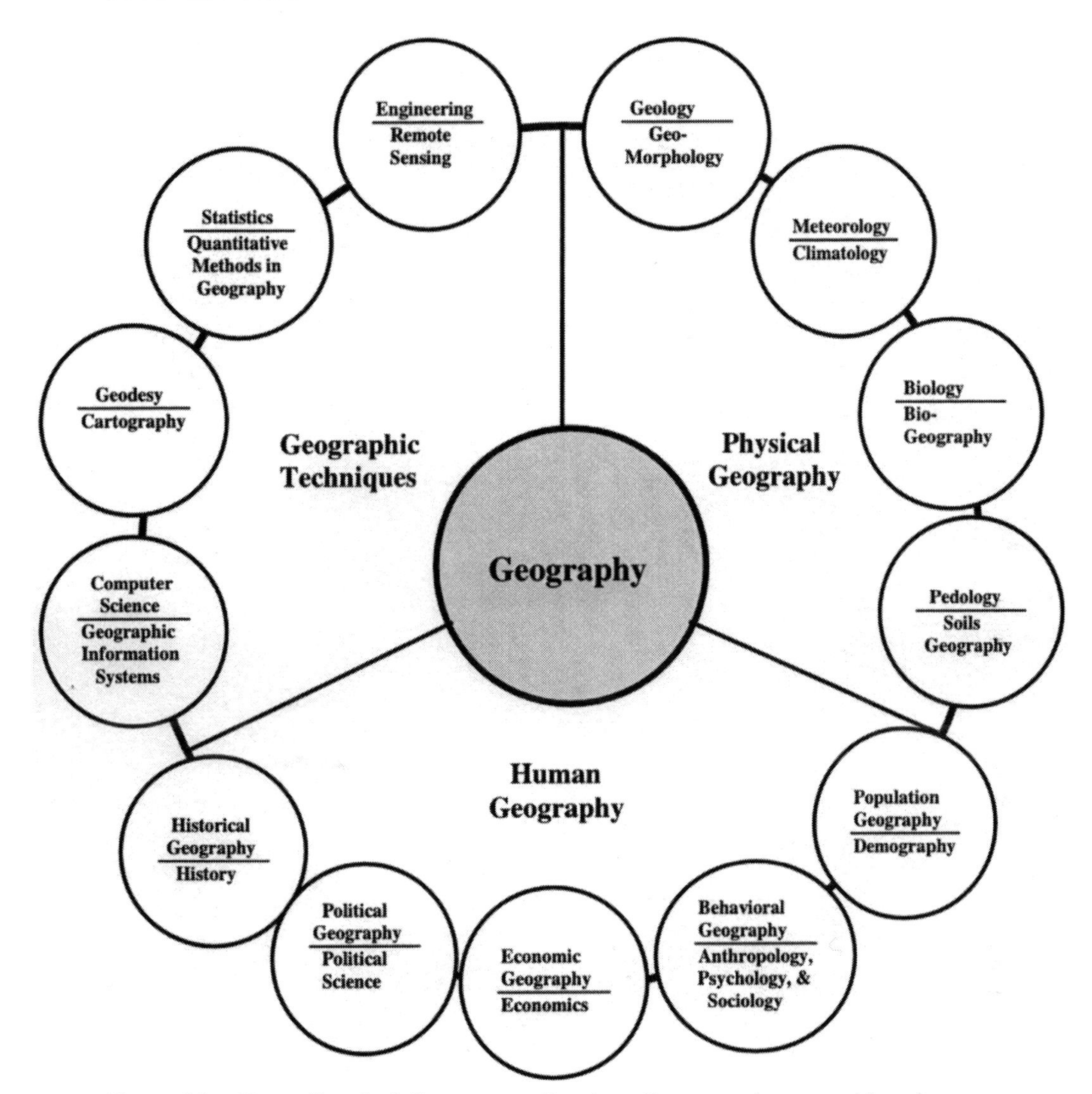

Figure 1.1. Shows the disciplines emanating from the core of geographic sciences. Adaptation from Bell (2007).

Branches of Geography

Earlier, in figure 1.1, it was pointed out that geography is a merged/mingled or integrative science and a synthesizer (combiner of parts) of knowledge. Each specialized field in geography, as shown in figure 1.1, overlaps other related branches of science.

Speaking of branches, there are two major branches of geography:

- Physical geography—study of the physical features and changes on the Earth's surface; focuses on geography as an Earth science.

Table 1.1. Notable Geographers

Geographer	*Period*	*Major Accomplishment(s)*
Eratosthenes	(276 BC–194 BC)	Calculated the size of the Earth
Ptolemy	(c. 90–c.168)	Compiled Greek and Roman knowledge into the book *Geographica*
Gerardus Mercator	(1512–1594)	Produced the Mercator projection
Alexander von Humboldt	(1769–1859)	Considered father of modern geography
Carl Ritter	(1779–1859)	Also considered father of modern geography
Arnold Henry Guyot	(1807–1884)	Noted glaciologist
William Rorris Davis	(1850–1934)	Father of American geography; developer of the cycle of erosion
Paul Vidal de la Blache	(1845–1918)	Wrote the principles of human geography
Sir Halford John Mackinder	(1861–1947)	Founder of Geographical Association
Walter Christaller	(1893–1969)	Human geographer and inventor of central place theory
Yi-Fu Tuan	(1930–)	Started humanistic geography
David Harvey	(1935–)	Author of theories on spatial and urban geography
Edward Soja	(1941–)	Noted for his work on regional development, planning, and governance
Michael Frank Goodchild	(1944–)	GIS scholar
Doreen Massey	(1944–)	Key scholar in the space and places of globalization and its pluralities
Nigel Thrift	(1949–)	Originator of nonrepresentational theory

- Cultural geography—study of humans and their ideas, patterns, and processes that impact and shape human ideas and actions on the Earth.

Did You Know?

In regard to culture, in the context of this text, it is defined as the way of life that distinguishes a group of people (i.e., their religions, language, customs, beliefs, knowledge, law, lifestyles, foods, music, etc.).

Other branches of geography include the following:

- biophysical geography—study of the natural environment and the interrelationships of all the living things in that environment;
- topography—shapes of the land and the bodies of water in a given location;
- political geography—study of the political organization of areas;
- social geography—study of groups of people and the interrelationships among groups and communities;
- economic geography—study of resources and resource use;
- historical geography—study of the ways in which the relationships between people and their environments have changed over time;
- urban geography—focuses on the cities;
- cartography—art and science of mapmaking;
- environmental geography—describes the spatial aspects of interactions between humans and the natural world;
- geomatics—involves the use of traditional spatial techniques used in cartography and topography and their application to computers; and
- regional geography—studies the regions of all sizes across the Earth.

Tools of Geography

As with every science there are certain tools used by practitioners in the specific scientific field. Geography is no different. Probably the most important tools of geography are listed below.

- Globes and Maps
- GIS
- Remote Sensing
- Observations
- Surveys
- Mathematical Models

GLOBES AND MAPS

Because they are the most useful models of the Earth, globes and maps are the two basic tools of geographers. Globes and maps are not reality. Globe and mapmakers look at the surface of Earth and then decide what to place on the globe or the map. The art and science of making globes and maps is called cartography. Again, globes and maps are not reality; they are subjective and not objective. We can get globes and maps to say or depict anything we want. So, to define globes and maps we can say that they are three-dimensional models (globes) or graphical representation of features (qualitative and quantitative) of the surface of the Earth.

Globes of Earth are often used in geography because they accurately represent the shape of the Earth, the shapes of landmasses and bodies of water, parallels and meridians, direction, and distance—globes are the best tools to show the shortest distance between two places. The problem with globes is that they can be not only big and bulky but also expensive. Another problem with using a globe is that only one half of the globe can be viewed at one time.

Although maps are flat representations of the Earth, they are easier to use than globes. For example, it is easier to carry a map just about anywhere; they are portable. A map provides an easy-to-use reference and can show the Earth's entire surface or just a particular part. Maps also show more detail such as a wide range of topics including physical and cultural features of the Earth. Again, the problem with maps is that they are not reality and can have distortions (inaccuracies) because they are flat and not three-dimensional representations.

Map Key Terms

The following key terms and definitions, taken from the U.S. Geological Survey (USGS; 2006), are useful in any discussion of maps and globes.

- *Aspect*—individual azimuthal map projections are divided into three aspects: the polar aspect, which is tangent at the pole; the equatorial aspect, which is tangent at the equator; and the oblique aspect, which is tangent anywhere else.
- *Azimuth*—the angle measured in degrees between base line radiating from a center point and another line radiating from the same point. Normally, the base line points north, and degrees are measured clockwise from the base line.
- *Conformality*—a map projection is conformal when at any point the scale is the same in every direction. Therefore, meridians and parallels intersect at right angles, and the shapes of very small areas and angles with very short sides are preserved. The size of most areas, however, is distorted.
- *Developable surface*—a simple geometric form capable of being flattened without stretching. Many map projections can then be grouped by a particular developable surface: cylinder, cone, or plane.
- *Equal areas*—a map projection is equal area if every part, as well as the whole, has the same area as the corresponding part on the Earth, at the same reduced scale. No flat map can be both equal area and conformal.
- *Equidistant*—show true distances only from the center of the projection or along a special set of lines. For example, an Azimuthal Equidistant map centered at Washington shows the correct distance between Washington and any other point on the projection. It shows the correct distance between Washington and San Diego and between Washington and Seattle. But it does not show the correct distance between San Diego and Seattle. No flat map can be both equidistant and equal area.
- *Graticule*—the spherical coordinate system based on lines of altitude and longitude.
- *Great Circle*—a circle formed on the surface of a sphere by a plane that passes through the center of the sphere. The equator, each meridian, and each other full circumference of the Earth forms a great circle. The arc of the great circle shows the shortest distance between points on the surface of the Earth.
- *Linear scale*—the relation between a distance on a map and the corresponding dis-

tance on the Earth. Scale varies from place to place on every map. The degree of variation depends on the projection used in making the map.

- *Map projection*—a systematic representation of a round body such as the Earth or a flat (plane) surface. Each map projection has specific properties that make it useful for specific purposes.
- *Rhumb line*—a line on the surface of the Earth cutting all meridians at the same angle. A rhumb line shows true direction. Parallels and meridians, which also maintain constant true directions, may be considered special cases of the rhumb line. A rhumb line is a straight line on a Mercator projection. A straight rhumb line does not show the shorter distance between points on the equator or on the same meridian.

Parts of Maps

Maps consist of parts that can be used to read the maps to analyze the physical and human landscapes of the world. Map parts are described below.

- Title—identifies map and contents.
- Legend (key)—explains the meaning of colors and symbols used on the map and may include the key to elevation (distance above or below sea level).
- Direction indicator—identifies direction or orientation on a map. Usually direction is provided on a map by a single arrow labeled "N" that points north. Other maps are printed with a compass rose (directional indicator) symbol that indicates direction on a map with arms that point to the cardinal and intermediate directions. The cardinal directions are north, south, east, and west, and the intermediate directions are northeast, southeast, northwest, and southwest (see figure 1.2).
- Map scales—provide information used to measure distances on maps. Different scales are used on different maps and are necessary for developing map representations because the size of a map in relation to the size of the real world is different; scale is shown by giving the ratio between distances on the map and actual distances on Earth. Areas can be represented using a variety of scales. The amount of detail shown on a map is dependent on the scale used.
  - Written or statement scale—uses a statement or phrase to relate a distance on the map to the distance it represents on Earth (i.e., one inch equals four miles; English units in United States).
  - Representative fraction (RF)—uses fractions or ratios to relate distance on a map to distant on the Earth (i.e., 1:250,000 or 1/250,000).
  - Graphic or bar scale—a short line that represents the number of miles or kilometers on the Earth's surface equal to a distance on the map (divided into equal parts labeled with miles or kilometers).
- Grid—used to locate places on a map. Global grids are formed by the crossing of parallels (line of latitude) and meridians (lines of longitude). A letter or number coordinate grid system is used to locate places on maps of smaller places like state maps, city maps, and highway maps.

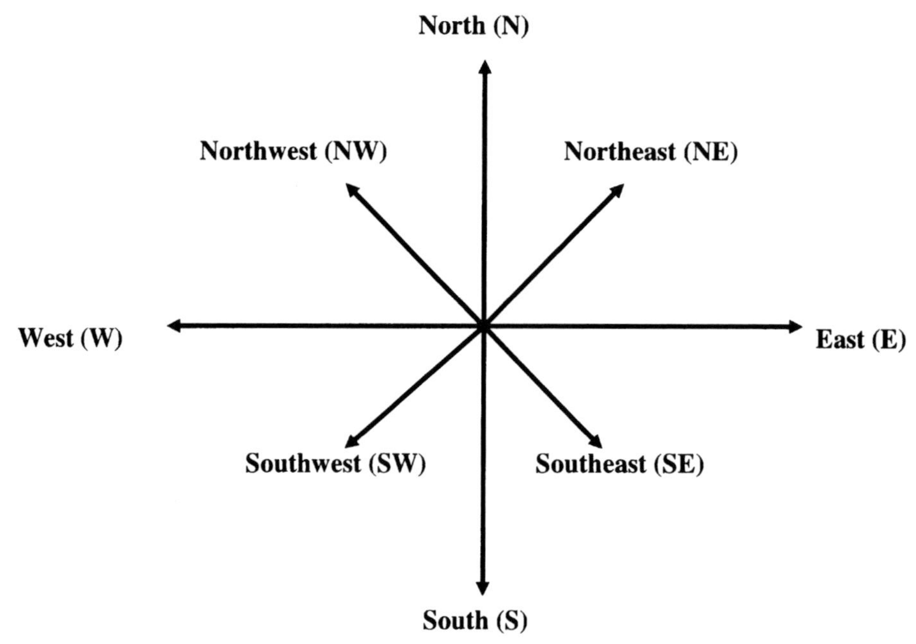

Figure 1.2. Map direction finder; compass rose.

Types of Maps

Maps vary according to their purpose. All the different kinds of maps can be put into two broad categories based on the type of information they show.

- General map—this type of map displays qualitative information, mainly showing a place or area size or where things are located and their qualities—nominal and cardinal data.
- Thematic map—a general purpose map that shows specific quantitative data or information, often on a single theme or topic. Examples of thematic maps include population, economic resource, language, ethnicity, climate, highway, and vegetation maps.

Map Projections (USGS 2006)

Globes provide the most accurate depiction of surface features on Earth; they are the only true representation of distance, area, direction, and proximity. However, you can't fold up a globe and put it into your pocket. Moreover, if you need to locate a particular city street in a specific city, you would need a very large globe. You could, of course, make a huge globe and then just cut the sections you need from the globe. I think you get my point. Globes are accurate but not hands-on practical or easy to transport from place to place. This is the reason flat maps were developed—three-dimensional Earth is projected onto a two-dimensional map. You can fold up a flat map and carry it almost anywhere. However, even though these flat maps are convenient and do display spatial information, they give a distorted view. At present, there is no flat map that does not have some type of distortion. Simply, when a map projection is used to portray all or part of the Earth on a flat surface, it can't be done without some distortion.

Every map projection has its own set of advantages and disadvantages. There is no "best" projection. Some projection types are conical, planar, or cylindrical. Choice of projection depends on the ultimate use of the map.

- Mercator projection—a cylindrical projection that preserves shape (see figure 1.3). It is used for navigation or maps of equatorial regions. Distances are true only along equator. This projection is not perspective, equal area, or equidistant.
- Transverse Mercator projection—used by USGS for many quadrangle maps; also used for mapping large areas that are mainly north-south in extent (see figure 1.4). It is mathematically projected on cylinder tangent to a meridian.
- Miller cylindrical projection—used to represent the entire Earth in a rectangular frame (see figure 1.5). Popular for world maps. Map is not equal area equidistant, conformal, or perspective.
- Robinson projection—uses tabular coordinates rather than mathematical formulas to make the world "look right." It is not conformal, equal area, equidistant, or perspective.
- Orthographic projection—used for perspective views of the Earth, Moon, and other planets (see figure 1.6). The Earth appears as it would on a photograph from deep space. Map is perspective but not conformal or equal area.
- Gnomonic projection (meaning sundial-like or great circles in straight lines)—used along with the Mercator by some navigators to find the shortest path between two points (see figure 1.7). It is considered to be the oldest projection and is geometrically projected on a plane. Point of projection is the center of a globe.
- Lambert conformal conic projection—used by USGS for many 7.5- and 15-minute

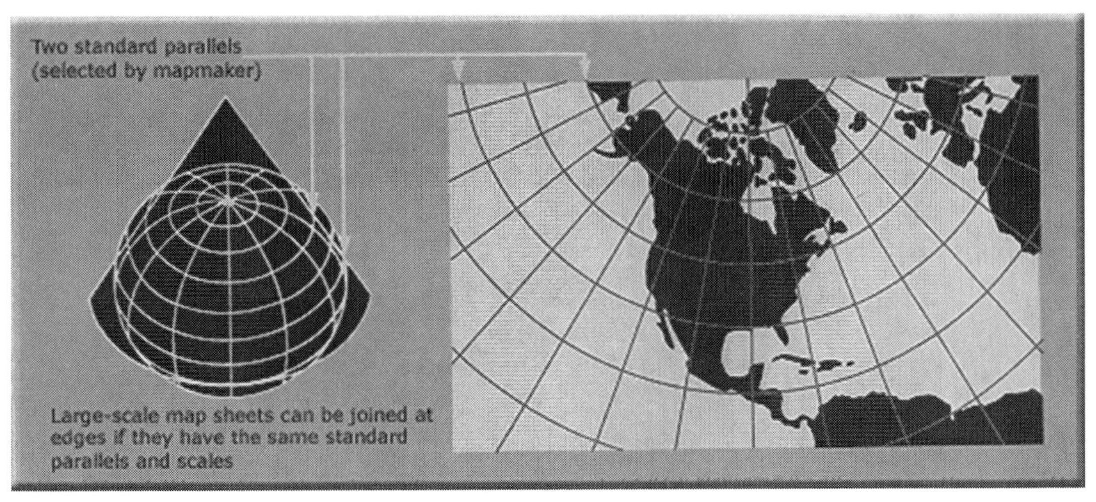

Figure 1.3. Mercator projection. Source: U.S. Geological Survey.

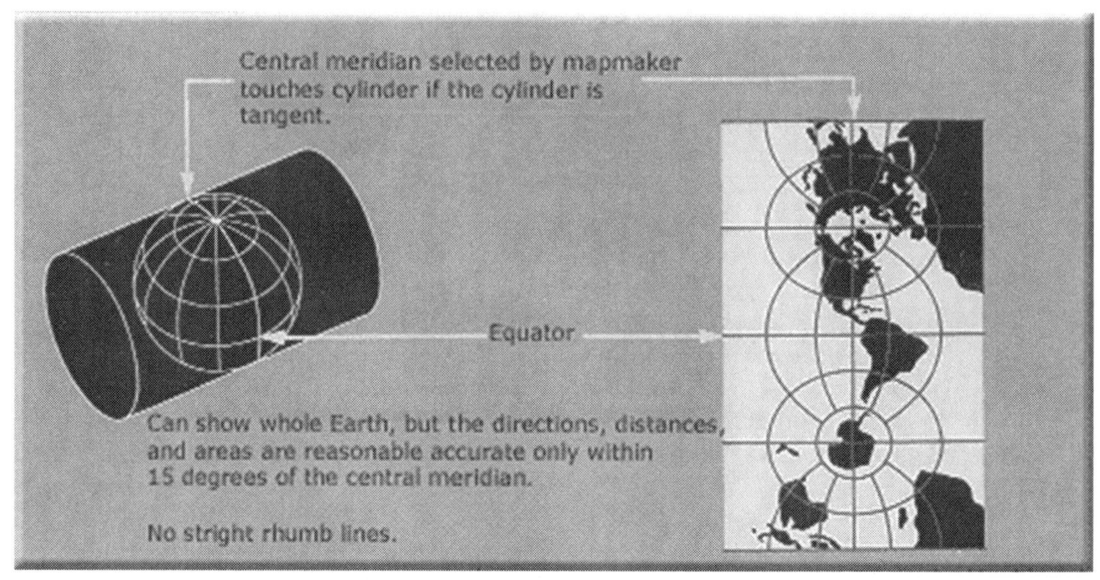

Figure 1.4. Transverse Mercator projection. Source: U.S. Geological Survey.

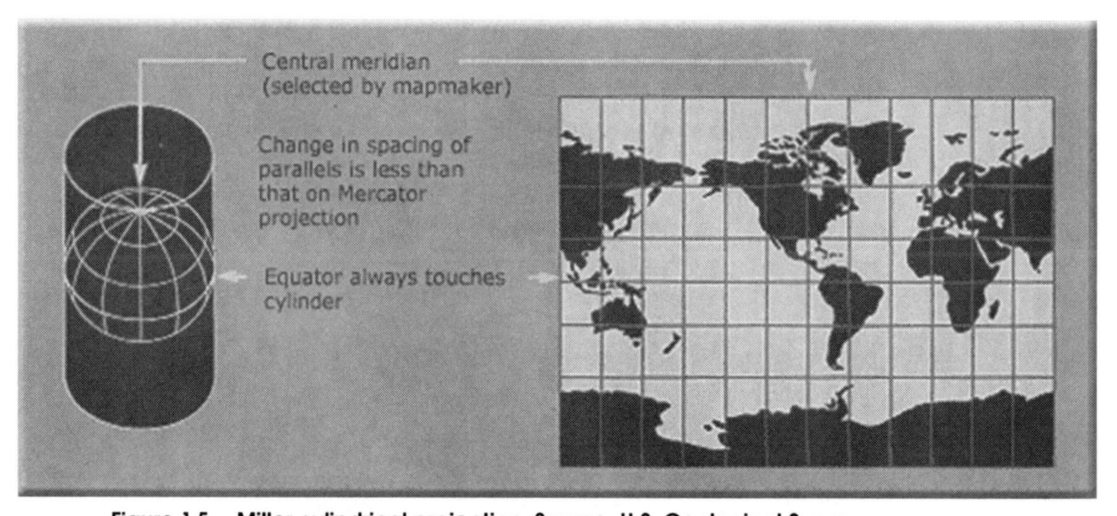

Figure 1.5. Miller cylindrical projection. Source: U.S. Geological Survey.

topographic maps and for State Base Map series (see figure 1.8). It is also used to show a country or region that is mainly east-west in extent. This conic projection is mathematically projected on a cone conceptually secant at two standard parallels.

Geographic Information Systems (USGS 2007)

Geographic Information Systems (GIS) is the newest development on the geographical frontier. GIS unites databases and maps through the power of the computer. In

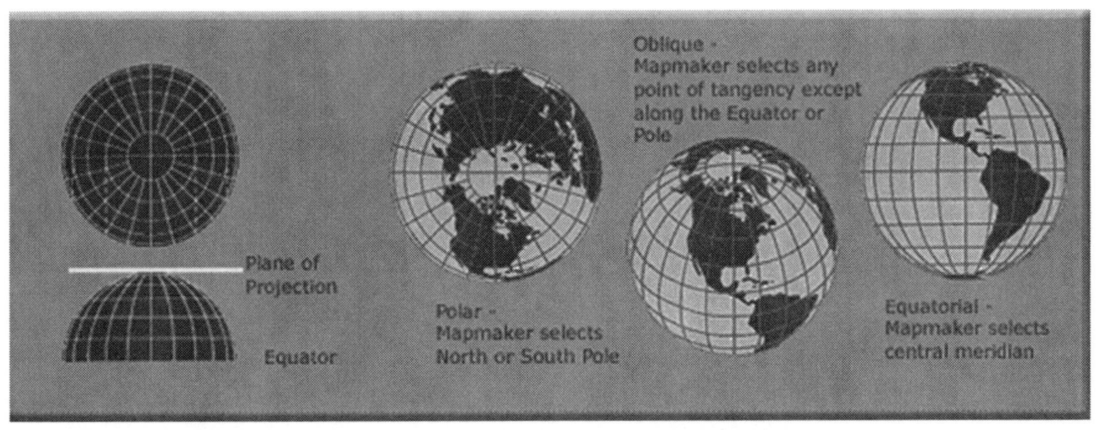

Figure 1.6. Orthographic projection. Source: U.S. Geological Survey.

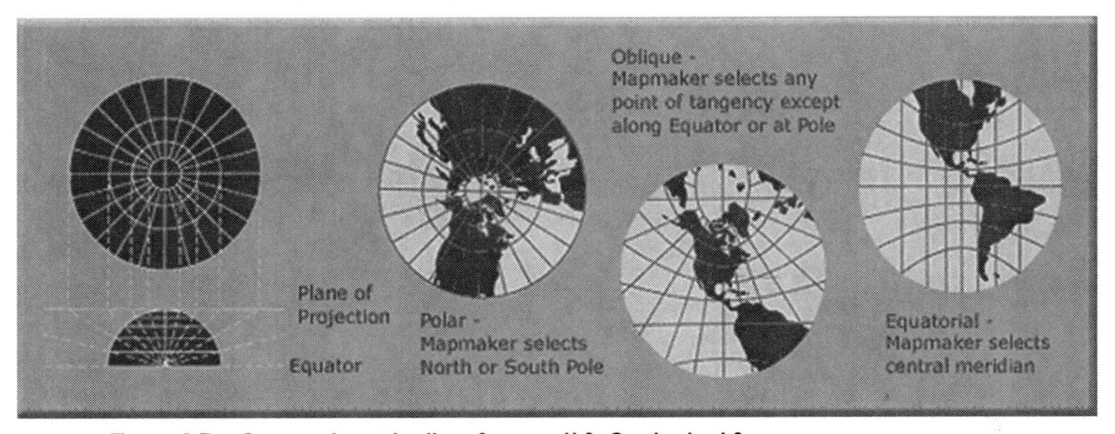

Figure 1.7. Gnomonic projection. Source: U.S. Geological Survey.

addition, the computer aids in drawing maps rapidly and without the use of conventional drawing materials (pens and ink). Moreover, GIS can be used for scientific investigations, resource management, and development planning. For example, GIS might be used to find wetlands that need protection from pollution. GIS has revolutionized many of the tasks geographers do, from mapmaking to spatial analysis.

So, exactly what is GIS and how does it work? GIS is a computer system that weds mapping, database, procedures, and operating personnel for spatial analysis by capturing, storing, analyzing, and displaying geographically referenced information, that is, data identified according to location. The power of GIS comes from the ability to relate different information in a spatial context and to reach a conclusion about this relationship. Most of the information we have about our world contains a location reference, placing that information at some point on the globe. When rainfall information is collected, it is important to know where the rainfall is located. This is done by using a location reference system, such as longitude and latitude, and perhaps elevation. Comparing the rainfall information with other information, such as the

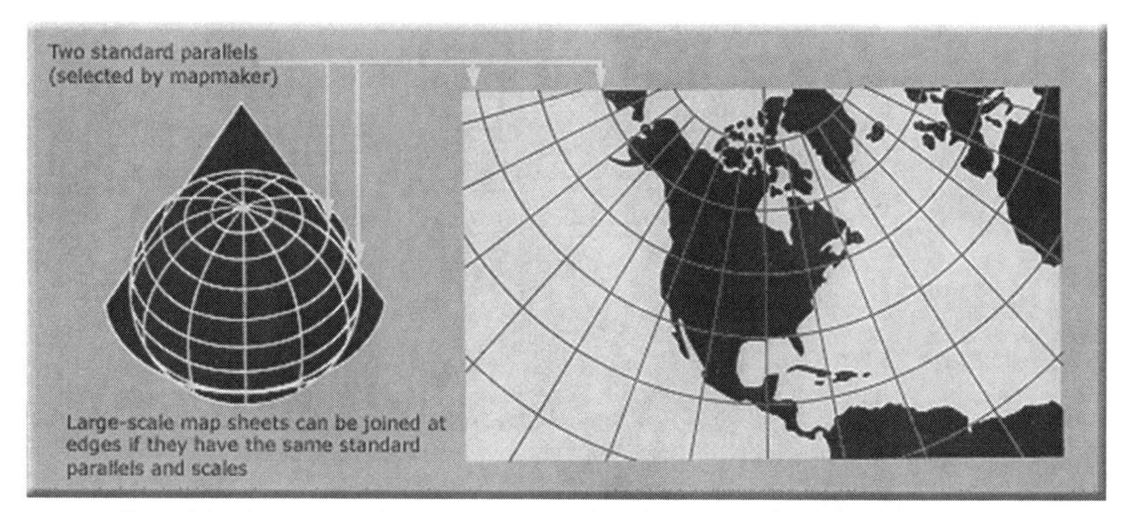

Figure 1.8. Lambert conformal conic projection. Source: U.S. Geological Survey.

location of marshes across the landscape, may show that certain marshes receive little rainfall. This fact may indicate that these marshes are likely to dry up, and this inference can help us make the most appropriate decisions about how humans should interact with the marsh. A GIS, therefore, can reveal important new information that leads to better decision making.

Many computer databases that can be directly entered into a GIS are being produced by federal, state, tribal, and local governments; private companies; academia; and nonprofit organizations. Different kinds of data in map form can be entered into a GIS (see, e.g., figure 1.9).

REMOTE SENSING

This is the process of detecting and monitoring the physical characteristics of an area by measuring its reflected and emitted radiation at a distance from the targeted area. Remote sensing is used in this book to refer to methods that are solely or primarily deployed through air or space. Included in this concept are studies of biological populations using remote imaging techniques. Related methods that are used most frequently on the ground (e.g., photography), whether underwater or from airplanes or satellites, are not included in the term "remote sensing" (USGS 2009).

OBSERVATIONS

Geographers use questionnaires or participant observers to investigate spatial phenomena.

Figure 1.9. Digital line graph of rivers. Source: U.S. Geological Survey.

SURVEYS

Geographers conduct several annual or ongoing large-scale topographical surveys.

MATHEMATICAL MODELS

Geographers use quantifiable observations, trends, and so forth, to predict or understand complex spatial relationships. These models could describe typical or human phenomena (e.g., pollution models, watershed and flood models, etc.).

The Five Themes of Geography

All sciences evolve with time. Geography and its formal presentation to students of geography evolved from William Pattison's (1963) four traditions of geography, which included spatial tradition, area studies tradition, man-land tradition, and Earth science tradition, to Edward Taaffe's (1973) spatial view in context. The spatial view in context held three views: the man-land view, the areas study view, and the spatial view—Earth science is considered to be subsumed in the other three traditions. In 1984, Salvatore Natoli introduced *Guidelines for Geographic Education and the Fundamental Themes in Geography* (popularly known as *The Five Themes of Geography*). The five themes answer five important questions that can help organize information about places. These themes are as follows:

- Location (what is the location of a place?)
  - Relative location
  - Absolute location
- Place (what is the character of a place?)
  - Human characteristics
  - Physical characteristics
- Human-environmental interactions (how do people interact with the natural environment of a place?)
  - Humans adapt to the environment
  - Humans modify the environment
  - Humans depend on the environment
- Movement (how do people, goods, and ideas move between places?)
  - People
  - Goods
  - Ideas
- Regions (how are place similar to, and different from, other places?)
  - Formal
  - Functional
  - Vernacular (perceptual)

The five themes became the framework upon which the content of geography could be taught in the K–12 classrooms until the national geography standards were published in 1994. Because the five themes and six standards are wedded, the themes remain a valuable tool for learning a geographic perspective. The standards, discussed later, work to strengthen instructional planning.

LOCATION

An important theme in geography is *location.* Every place on Earth can be described in both absolute and relative terms. An *absolute location* is usually found by using imaginary lines (latitude and longitude—global location) that mark positions on the surface of the Earth or a street location (local location). The global grid formed by lines of latitude and longitude crossing each other can be used to name the precise or absolute location of any place on Earth. Every place has only one absolute location that never changes.

Did You Know?

Imaginary lines that mark positions on the surface of the Earth include the equator, which circles the globe halfway between the north and south poles. It divides the world into two hemispheres, northern and southern. Lines of latitude (parallels) are imaginary lines that run parallel to the equator. Latitude is the distance a place is north or south of the equator. Latitude is measured in degrees from 0 to 90 north, and from 0 to 90 south. Lines of longitude (meridians) are imaginary lines that run north and south from pole to pole. Measuring longitude is the distance a place is east or west of the prime meridian (an imaginary line that runs through Greenwich, England, at 0 longitude; it divides the Earth into two hemispheres: the Eastern and Western Hemispheres).

Relative locations are those where the position of a place is in relation to another place. Every place has many relative locations, which can change over time. Relative locations are described by landmarks, time, direction, or distance from one place to another and may associate one place with another.

PLACE

Another important geographic theme is *place*. Each place on the Earth has its own physical features. It can be described in terms of its land, water, weather, soil, altitude, and plant and animal life. Each place also has its human or cultural characteristics such as government, language, ethnic groups, art, literature, tools, laws, customs, lifestyles, population, economy, types of agriculture, and so forth. The image people have of a place is based on their experiences, both intellectual and emotional.

HUMAN/ENVIRONMENTAL INTERACTION

An important theme in geography is human and environmental interaction. How do humans and the environment affect each other? There are three key concepts to human/environmental interaction: humans adapt to the environment; humans modify the environment; and humans depend on the environment. All places and interactions have pluses and minuses for humans and the environment (e.g., clearing a forest for farming produces food but also destroys trees; burning coal provides energy but also pollutes the air). One person's advantage may be another person's disadvantage. People adapt to their surroundings by using different types of clothing, food, and shelter. People change their physical surroundings. Some of these changes are intentional, and others have been accidental (i.e., tunnels, clearing of forests, planting of crops, and use of land and natural resources).

MOVEMENT

Another important geographic theme is movement. This theme includes the movement and exchange of people, goods, services, and ideas (types of transportation, written materials, telecommunications, travel, imports, exports, etc.). We interact with each other through travel, trade, cyberspace, and political events. We live in a global village and global economy. Movement may also include natural movements of the Earth's physical features: water cycle, ocean currents, wind, volcanic eruptions, mass wasting, animal migration, and so on.

REGIONS

Yet another important theme in geography is regions. Regions, groups of places with at least one common characteristic, are important because they make the study of geography manageable. Geographers often divide the world into regions, or areas (they may be any size), based on certain physical features, such as land type, landform, water bodies, climate, or plant and animal life. They also divide the world into regions based on certain human characteristics, such as the way people are governed, religion, industry, agriculture, or the kind of language they speak. The boundaries of regions may overlap. There are three types of regions:

- Formal regions—areas in which a certain characteristic is found through the area; they are defined by governmental or administrative boundaries (i.e., United States, Canada, or Seattle).
- Functional regions—consist of one central place and the surrounding places affected by it (i.e., Tennessee Valley Authority, United Airlines landing area, Andes Mountains, the Amazon drainage basin, or a military route service area). If the function ceases to exist, the region no longer exists.
- Vernacular (native) regions—are areas loosely defined by people's perception (i.e., the Middle East, the South, the Pacific Northwest, east of the mountains in Washington State, etc.).

Key Points of Geographical Mapping and Location

- Every map projection has some degree of distortion because a curved surface cannot be represented on a flat surface without distorting curvature.
- Regions have in common a relative location, spatial extent, and boundaries.
- A street address is an example of absolute location.
- Spatial interaction is the movement of people and things between places.
- Situation is the relative location of a place or activity.
- Connectivity describes the paths and ways in which different places are linked.

- Dispersion is the amount of spread of an item over an area.
- Density is the quantity of an item within a unit area.
- Latitude is a measure of distance north and south of the equator.
- Latitude lines are always parallel to each other.
- On a globe, lines of latitude intersect meridians of longitude at right angles.
- An isometric map is a type of quantitative thematic map.
- The size and location of a place described by its local physical characteristics is called its site.
- A contour is an isoline showing points of equal elevation.
- All meridians are one-half the length of the equator.
- The scale on the surface of the globe is the same in every direction.
- Density, dispersion, and pattern are elements common to all spatial distributions.
- Perceptual regions reflect personal or popular impressions of territory and spatial divisions.
- An isoline on a map connects points of equal value to the mapped item.
- The four main properties of maps are area, shape, distance, and direction.
- The north and south poles, equator, and prime meridian are all key reference points in the grid system.
- The map scale defines the relationship between the size of an Earth feature and its size on the map.

National Geography Standards (NASA 2009)

It was mentioned earlier that the science of geography is an evolving science, basically a work in progress. Nowhere is this more apparent than in the current trend to blend the 1994 National Geography Standards with the five themes. In reality, these geography standards are a framework of benchmarks against which the context of various geography courses can be measured in the United States. The eighteen standards listed in table 1.2 have been categorized into six essential elements.

The Bottom Line

This chapter clearly demonstrates that geography is not just listing the capitals of the United States or filling in the names of countries or states on a blank map. Geography is the understanding of the spatial relationships of phenomena on the face of the Earth. The themes of location, place, human and environmental interaction, movement, and regions are important to the study of geography.

Chapter Review Questions

1.1. Why are location and place important geographic themes?
1.2. How are regions important to geographers?
1.3. Why do geographers use other fields of study?

Table 1.2. The Six Essential Elements of the National Geography Standards

The World in Spatial Terms

The geographically informed person knows and understands the following:

Standard 1: How to use maps and other geographic representations, tools, and technologies to acquire, process, and report information from a spatial perspective.
Standard 2: How to use mental maps to organize information about people, places, and environments in a spatial context.
Standard 3: How to analyze the spatial organization of people, places, and environments on Earth's surface.

Places and Regions

The geographically informed person knows and understands the following:

Standard 4: The physical and human characteristics of places.
Standard 5: That people create regions to interpret Earth's complexity.
Standard 6: How culture and experience influence people's perceptions of places and regions.

Physical Systems

The geographically informed person knows and understands the following:

Standard 7: The physical processes that shape the patterns of Earth's surface.
Standard 8: The characteristics and spatial distribution of ecosystems on Earth's surface.

Human Systems

The geographically informed person knows and understands the following:

Standard 9: The characteristics, distribution, and migration of human populations on Earth's surface.
Standard 10: The characteristics, distribution, and complexity of Earth's cultural mosaics.
Standard 11: The patterns and networks of economic interdependence on Earth's surface.
Standard 12: The processes, patterns, and functions of human settlement.
Standard 13: How the forces of cooperation and conflict among people influence the division and control of Earth's surface.

Environment and Society

The geographically informed person knows and understands the following:

Standard 14: How human actions modify the physical environment.
Standard 15: How physical systems affect human systems.
Standard 16: The changes that occur in the meaning, use, distribution, and importance of resources.

The Uses of Geography

The geographically informed person knows and understands the following:

Standard 17: How to apply geography to interpret the past.
Standard 18: How to apply geography to interpret the present and plan for the future.

1.4. _________ geography is concerned with physical changes on the Earth's surface.
1.5. _________ geography focuses on cities.
1.6. _________ and _________ are not reality.
1.7. _________ is the spherical coordinate system based on lines of latitude and longitude.
1.8. _________ is a line on the surface of the Earth cutting all meridians at the same angle.
1.9. A _________ _________ is also called a compass rose.
1.10. A _________ _________ projection is used to represent the entire Earth in a rectangular frame.

References and Recommended Reading

Bell, K. M. 2007. What is geography? http://74.6.239.67/search/cache?ei=UTF-8&p=geography+mother +of+all+sciences&icp+1&w+geograph (accessed May 12, 2009; site no longer posted).

Bergman, J. 2005. Rocks and the rock cycle. www.windows.ucar.edu/tour/ling=/earth/geology/rocks_intro.html (accessed May 23, 2008).

Blue, J. 2007. Descriptor terms. USGS *Gazetteer of Planetary Nomenclature.* http://planetarynames.wr.usgs.gov/jsp/append5.jsp (accessed November 11, 2009).

Campbell, N. A. 2004. *Biology: Concepts and connections,* 4th CD-ROM ed. Menlo Park, Calif.: Benjamin-Cummings Publishing.

Huxley, T. H. 1876. *Science and education, volume III: Collected essays.* New York: D. Appleton.

Larsson, K. A. 1993. Prediction of the pollen season with a cumulated activity method. *Grana* 32:111–14.

Michener, J. A. 1970. The mature social studies teacher. *Social Education* (November): 760–67.

NASA. 2009. *National Geography Standards.* http://edmail.gsfc.NASA.gov.inv99project.site/pages/geo.stand.htm (accessed July 9, 2009).

Natoli, S. J. 1984. *Guidelines for geographic education and the fundamental themes in geography.* New York: Association of American Geographers.

Pattison, W. D. 1963. *The four traditions of geography.* Columbus, Ohio: Annual Convention for National Council of Geography Education.

Press, R., and F. Siever. 2001. *Earth,* 3rd ed. New York: W. H. Freeman.

Short, J. R. 2004. *Representing the Republic: Mapping the United States, 1600–1900.* Chicago: Reaktion Press.

Spellman, F. R. 1998. *Environmental science and technology: Concepts and applications.* Lancaster, Pa.: Technomic Publishing.

———. 2008. *The Science of air: Concepts and applications.* Boca Raton, Fla.: CRC Press.

Spellman, F. R., and N. E. Whiting. 2006. *Environmental science and technology,* 2nd ed. Lanham, Md.: Government Institutes.

Taaffe, E. J. 1973. *Geography of transportation,* 2nd ed. Englewood Cliffs, N.J.: Prentice Hall.

U.S. Geological Survey (USGS). 2006. Map projections. http://egsc.usgs.gov/isb/pubs/MapProjections/projections.html (accessed May 15, 2009).

———. 2007. Geographic information systems. http//egsc.usgs.gov/isb/pubs/gis_poster (accessed May 16, 2009).

———. 2009. Remote sensing. http://usgs.gov/science.php?term=981 (accessed May 1, 2009).

II

PHYSICAL GEOGRAPHY

Bryce Canyon, Utah. Photograph shows a classic example of the result of weathering and erosion working in tandem to carve colorful Claron limestones into thousands of spires (hoodoos), fins, arches, and mazes. Weathering processes include the chemical and/or physical breakdown of rock materials. Erosion involves the removal or transportation of materials by agents such as running water, ice, wind, etc. Photo by Frank R. Spellman.

CHAPTER 2

Landforms

> Geography is a synoptic discipline that synthesizes findings of other sciences through the concept of Raum (area or space).
>
> —Immanuel Kant, 1780

Based on personal experience, when introducing students to the geological and geographical aspects of environmental science initially, there is some confusion as to the exact difference between geology and geography. This is partially the case because, while there are several differences between the two sciences, they are also wed in many respects. To save on time and to avoid confusion, the best way to differentiate between the two sciences is to simply point out that for the purpose of this book geology is defined as the science that deals with the natural structure of Earth and geography deals with the human-drawn or human-made national borders and lines on Earth. One thing is certain: certain aspects of geology, such as Earth's internal forces involved in the building and development of mountains, continental plains, and coastal basins, are relevant to physical geography. Moreover, rock type and structure are important as variables that influence the effectiveness of rain, wind, and weathering processes on landforms.

We have all seen the photographs of the Earth taken from outer space; they show the surface of the Earth as being far from uniform. Earth's surface is covered with natural features called landforms. These features are classified by type in order to describe them—mountains, valleys, plains, and so forth. The features' names help us to locate specific places. Along with classification by type, landforms can also be classified and organized by the genetic processes that create them.

Genetic Landforming Processes

Genetic landforming processes work across the globe's seven large landmasses, the continents. These processes include Aeolian landforms, coastal and oceanic landforms, erosion landforms, fluvial landforms, mountain and glacial landforms, slope landforms, and volcanic landforms. Genetic landforming processes and individual landforms are described in the following text; many of the individual landforms are described in greater detail later in the text. Keep in mind that many of the landforms listed and described below are produced by more than one landforming process. For example, landforms produced by erosion and weathering usually occur in both coastal and fluvial environments. However, to eliminate redundancy, each landform is only listed and described once under a specific process.

AEOLIAN LANDFORMS

An Aeolian (derived from the Greek Aeolus—god of the winds) landform is a surface feature produced on Earth (and other planets, e.g., Mars) by the erosive or constructive action of the wind.

These wind-constructed landforms are rarely preserved on the surface of the Earth except in arid regions where, in most cases, there is a lack of moving water to erase them. Aeolian landforms are listed and described below (BAR 2009).

- barchan—an arc-shaped sand ridge (dune), comprising well-sorted sand.
- blowout—sandy depressions in a sand dune ecosystem caused by the removal of sediments by wind.
- desert pavement—a desert surface that is covered with closely packed, interlocking angular or rounded rock fragments of pebble and cobble size.
- desert varnish—dark coating found on exposed rock surfaces in arid environments.
- dune—hill of sand built by Aeolian processes.
- erg—large, relatively flat area of desert covered with windswept sand with little or no vegetative cover (NASA 2009).
- loess—a homogeneous, typically nonstratified, porous, friable, slightly coherent, often calcareous, fine-grained, silty, pale yellow or buff, windblown sediment (SEMP 2009).
- medanos—in South America, medanos refers to continental dunes, whereas *dunes* refer to dunes of coastal regions.
- playa—a dry or ephemeral lakebed, generally extending to the shore, or a remnant of an endorheic lake (a lake that does not flow into the sea).
- sandhill—an ecological community type found in many parts of the world.
- ventifact—rocks that have been abraded, pitted, etched, grooved, or polished by wind-driven sand or ice crystals (Segerstrom 1962).
- yardang—a wind-abraded ridge found in a desert environment.

COASTAL AND OCEANIC LANDFORMS

The main agents responsible for formation of coastal landforms are deposition and erosion, which are the result of waves, tides, and currents. Coastal and oceanic landforms are also heavily dependent on the type of rock present (i.e., the hardness or softness of rock—how resistant/nonresistant it is to erosion). Many coastal and oceanic landforms are listed and described below.

- abyssal fan—underwater structures that look like deltas formed at the end of many large rivers.

- abyssal plain—flat or very gently sloping areas of the deep ocean basin floor.
- arch—a natural formation where a rock arch forms, with a natural passageway underneath. They commonly form where cliffs are subject to erosion from the sea, rivers, or weathering; the processes find weaknesses in rocks and work on them, making them bigger until they break through.
- archipelago—a chain or cluster of islands that are formed tectonically (i.e., by forces and movements within the Earth).
- atoll—an annular reef enclosing a lagoon in which there are no promontories other than reefs and islets composed of reef detritus; in an exclusively morphological sense, a ring-shaped ribbon reef enclosing a lagoon in the center (McNeil 1954; Fairbridge 1950).
- ayre—a body of water divided from the sea by a narrow bar of land.
- barrier bar—a long and narrow (linear) landform within or extending into a body of water, typically composed of sand, silt, or small pebbles.
- bay—an area of water bordered by land on three sides.
- beach—landform along the shoreline of a body of water (Bascom 1980). A raised beach is an emergent coastal landform, a wave-cut platform raised above the shore line (Johnson and Libbey 1997).
- beach cusps—shoreline formations made up of various grades of sediment in an arc pattern.
- beach ridge—a wave-swept or wave-deposited ridge running parallel to a shoreline.
- bight—a bend or curve in the line between land and water, or a large, shallow bay.
- channel—the physical confine of a river, slough, or ocean strait consisting of a bed and banks.
- cliff—a significant, vertical, or near vertical erosion-formed rock exposure.
- coast—where the land meets the sea.
- continental shelf—the undersea extended perimeter of each continent and associated coastal plain.
- coral reef—sea or ocean structures produced by living organisms.
- cove—a circular or oval coastal inlet with a narrow entrance.
- dune—hill of sand built by Aeolian processes.
- estuary—a semienclosed coastal body of water with one or more rivers or streams flowing into it, and with a free connection to the open sea.
- firth—various coastal waters in Scotland.
- fjord—glacier cut U-shaped drowned valley.
- headlands—coastline feature surrounded by water on three sides.
- inlet—a narrow body of water between islands or leading inland from a larger body of water, often leading to an enclosed body of water, such as a bay, lagoon, or marsh (Bruun and Mehta 1978).
- isthmus—the narrow strip of land connecting two larger areas of land.
- lagoon—body of comparatively shallow salt or brackish water separated from the deeper sea by a shallow or exposed sandbank, coral ref, or similar feature.
- machair—refers to a fertile, low-lying grassy plain found on some of the northwest coastlines of Ireland and Scotland (Angus 1997).

- mid-ocean ridge—an underwater mountain range, typically having a valley known as a rift running along its spine, formed by plate tectonics.
- ocean basin—large geologic basin covered by seawater and well below sea level.
- oceanic trench—narrow topographic depressions of the sea floor that can be hemispheric in length.
- ocean plateau—large, relatively flat submarine region that rises well above the level of the ambient seabed.
- peninsula—a piece of land that is nearly surrounded by water but connected to mainland via an isthmus.
- ria—a drowned river valley.
- river delta—a landform that is created at the mouth of a river where that river flows into an ocean, sea, estuary, lake, reservoir, or another river.
- salt marsh—a type of marsh that is transitional intertidal between land and salty or brackish water.
- sea cave—type of cave formed primarily by the wave action of the sea.
- shoal—a sandbar that is somewhat linear within or extending into a body of water, typically composed of sand, silt, or small pebbles.
- sound—a large sea or ocean inlet larger than a bay, deeper than a bight, and wider than a fjord, or it may identify a narrow sea or ocean channel between two bodies of land.
- spit—a deposition landform off coasts; one end is connected to land and the other far end juts into open water.
- stack—a landform consisting of a steep and often vertical column or columns of rock in the sea near a coast.
- straight—a narrow, navigable channel of water that connects two large navigable bodies of water.
- surge channel—a narrow inlet on a rocky shoreline.
- tombolo (Italian for mound)—a deposition landform in which an island is attached to the mainland by a narrow piece of land such as a spit or bar.
- volcanic arc—a chain of volcanic islands or mountains formed by plate tectonics as an oceanic tectonic plate subducts under another tectonic plate and produces magma.
- wave-cut platform—the narrow, flat area often seen at the base of a sea cliff or along a large lakeshore caused by the action of the waves.

EROSION LANDFORMS

The following landforms are the result of either erosion or weathering or a combination of both.

- butte—an isolated hill with steep, often vertical sides and a small, relatively flat top (smaller than mesas, plateaus, and tables).
- canyon—a deep valley between cliffs, often carved from the landscape by a river.

- cuesta—a ridge formed by gently tilted sedimentary rock strata in a homoclinal structure (Monkhouse 1978).
- dissected plateau—a plateau area that has been uplifted, then severely eroded so that the relief is sharp.
- eolianite—any rock formed by the lithification of sediment deposited by Aeolian processes (i.e., the wind).
- gulch—a deep V-shaped valley formed by erosion.
- gully—a landform created by running water eroding sharply into soil, typically on a hillside.
- hogback—a homoclinal ridge, formed from a monocline, composed of steeply tilted strata rock protruding from the surrounding area.
- hoodoo—a tall, thin spire of rock that protrudes from the bottom of an arid drainage basin or badland.
- lavaka—a type of erosional feature common in Madagascar.
- limestone pavement—a natural karst landform consisting of a flat, incised surface of exposed limestone that resembles an artificial pavement.
- malpais—a landform characterized by eroded rocks of volcanic origin in an arid environment.
- mesa—an elevated area of land with a flat top and sides that are usually steep cliffs.
- pediment—a gently inclined erosional surface carved into bedrock.
- peneplain—the final stage in fluvial or stream erosion.
- potrero—a long mesa that at one end slopes upward to higher terrain.
- tea table—a rock formation that is a remnant of new strata that have eroded away.
- tepui—a tabletop mountain (mesa) found only in the Guayana highlands of South America.
- valley—a depression with predominant extent in one direction.

FLUVIAL LANDFORMS

Fluvial landforms result from erosion by water flowing on land surfaces. Fluvial landforms include the following (Spellman 1996):

- ait—a small island in a river.
- alluvial fan—a fan-shaped deposit formed where a fast-flowing stream flattens, slows, and spreads, typically at the exit of a canyon onto a flatter plain.
- anabranch—a section of a river or stream that diverts from the main channel or stem of a watercourse and rejoins the main stem downstream.
- arroyo—usually dry creek bed or gulch that temporarily (or seasonally) fills with water after a heavy rain.
- bayou—a small, slow-moving stream or creek, or a lake or pool that lies in an abandoned channel of a stream.
- braided river—one of a number of channel types and has a channel that consists of a network of small channels separated by small and often temporary islands called braid bars.

- Carolina Bay—elliptical depressions concentrated along the Atlantic seaboard within coastal Delaware, Maryland, New Jersey, Virginia, Georgia, North Carolina, South Carolina, and north-central Florida.
- drainage basin—an extent of land where water from rain or snowmelt drains downhill into a body of water, such as a river, lake, reservoir, estuary, wetland, sea, or ocean.
- exhumed river channel—a ridge of sandstone that remains when the softer floodplain mudstone is eroded away.
- gully—landform created by running water eroding sharply into soil, typically on a hillside.
- lacustrine plain—a plain that originally formed in a lacustrine environment, that is, as the bed of a lake from which the water has disappeared, by natural drainage, evaporation, or other geophysical processes.
- lake—a terrain feature, a body of liquid on the surface of a world that is localized to the bottom of a basin and moves slowly if it moves at all.
- levee—a natural or artificial slope or wall to regulate water levels.
- meander—a bend in a sinuous watercourse.
- oasis—an isolated area of vegetation in a desert, typically surrounding a spring or similar water source.
- oxbow lake—a U-shaped body of water formed when a wide meander from the mainstem of a river is cut off to create a lake.
- pond—a body of water smaller than a lake.
- proglacial lake—a lake formed either by the damming action of a moraine or ice dam during the retreat of a melting glacier or by meltwater trapped against an ice sheet due to isostatic depression of the crust around the ice.
- rapid—a section of a river where the riverbed has a relatively steep gradient causing an increase in water velocity and turbulence.
- riffle—a shallow stretch of a river or stream, where the current is below the average stream velocity and where the water forms small rippled waves as result.
- rock-cut basin—cylindrical depressions cut into stream or riverbeds, often filled with water.
- spring—any natural occurrence where water from below the surface of the Earth flows onto the surface of the Earth and is thus where the aquifer surface meets the ground surface.
- stream—a flowing body of water with a current, confined within a bed and stream banks.
- stream terrace—relict feature, such as a floodplain, from periods when a stream was flowing at a higher elevation and has downcut to a lower elevation.
- swamp—a wetland featuring temporary or permanent inundation of large areas of land by shallow bodies of water.
- wadi—Arabic term traditionally referring to a valley.
- waterfall—a body of water resulting from water, often in the form of a stream, flowing over an erosion-resistant rock formation that forms a nickpoint, or sudden break in elevation.

MOUNTAIN AND GLACIAL LANDFORMS

Mountain and glacial landforms include the following:

- arête—a thin, almost knifelike, ridge of rock that is typically formed when two glaciers erode parallel U-shaped valleys.
- cirque—an amphitheaterlike valley or valley head, formed at the head of a glacier by erosion.
- crevasse—a huge crack formed by two glaciers colliding.
- dirt cone—a feature of a glacier in which dirt that has fallen into a hollow in the ice forms a coating that insulates the ice below.
- drumlin—an elongated, whale-shaped hill formed by glacial action.
- esker—a long, winding ridge of stratified sand and gravel.
- glacier—a large, slow-moving mass of ice, formed from compacted layers of snow, that slowly deforms and flows in response to gravity and high pressure.
- glacier cave—a cave formed within the ice of a glacier.
- glacier foreland—region between the current leading edge of the glacier and the moraines of latest maximum.
- hill—a landform that extends above the surrounding terrain, in a limited area.
- kame—an irregularly shaped hill or mound composed of sand, gravel, and till that accumulates in a depression on a retreating glacier and is then deposited on the land surface with further melting of the glacier.
- kame delta—a glacial landform made by a stream flowing through glacial ice and depositing material upon entering a lake or pond at the end or terminus of the glacier.
- kettle—a shallow, sediment-filled body of water formed by retreating glaciers or draining floodwaters.
- monadnock—an isolated rock hill, knob, ridge, or small mountain that rises abruptly from a gently sloping or virtually level surround plain.
- moraine—any glacially formed accumulation of unconsolidated glacial debris (soil and rock) that can occur in the currently glaciated and formerly glaciated regions, such as those areas acted upon by a past ice age.
- moulin—a narrow, tubular chute, hole, or crevasse through which water enters a glacier from the surface.
- nunatak—an exposed, often rocky element of a ridge, mountain, or peak not covered with ice or snow within an ice field or glacier.
- outwash fan—a fan-shaped body of sediments deposited by braided streams from a melting glacier.
- pingo—a mound of earth-covered ice found in the Arctic and subarctic that can reach up to 70 meters (230 feet) in height and up to 600 meters (2,000 feet) in diameter.
- pyramidal peak (glacial horn)—a mountaintop that has been modified by the action of ice during glaciation and frost weathering.
- rift valley—linear-shaped lowland between highlands or mountain ranges created by the action of a geologic rift or fault.

- sandur—a glacial outwash plain formed of sediments deposited by meltwater at the terminus of a glacier.
- side valley—refers to a valley whose brook or river is confluent to a greater one.
- stream terrace—a relict feature from periods when a stream was flowing at a higher elevation and has downcut to a lower elevation.
- tunnel valley—a deep but narrow valley with a 'U'-shaped cross-section and frequently a 'U'-shaped plain that is usually found filled with glacial till.
- valley—a depression with predominant extent in one direction.

SLOPE LANDFORMS

Generally, when discussing slope, we are primarily concerned with hillslopes—that is, the slopes connecting hilltops with river channels in valley bottoms. Slopes include slopes created by river sediments, rainwash, and rockfall (talus). Although discussed in detail later in the text, it should be pointed out that slope has a great influence on mass movement and wasting. Slope not only influences the evolution of landforms but also gives geologists important information about the formation of landforms. Slope landforms include the following:

- alas—a steep-sided depression formed by the melting of permafrost; it may contain a lake.
- defile—a narrow pass or gorge between mountains or hills.
- dell—a small wooded valley.
- escarpment—a transition zone between different provinces that involves a sharp, steep elevation differential, characterized by a cliff or steep slope.
- glen—a valley, typically one that is long, deep, and often glacially U-shaped.
- gully—a landform created by running water eroding sharply into soil, typically on a hillside.
- hill—a landform that extends above the surrounding terrain, in a limited area.
- knoll—a small, natural hill.
- mountain pass—a saddle point in between two areas of higher elevation.
- ravine—a very small valley, which is often the product of stream-cutting erosion.
- ridge—a geological feature that consists of a continuous, elevational crest for some distance.
- rock shelter—a shallow cavelike opening at the base of a bluff or cliff.
- scree—an accumulation of broken rock fragments at the base of mountain cliffs.
- vale—a wide river valley, with a particularly wide floodplain or flat valley bottom.

VOLCANIC LANDFORMS

A volcanic landform is characterized by the type of material it is made of. Later processes modify the original landform to other forms. Volcanic landforms include the following:

- caldera—very large, cauldronlike depression that is usually formed by the collapse of land following a volcanic eruption. Calderas are enclosed depressions that collect rainwater and snowmelt, and thus lakes often form within a caldera.
- geyser—a hot spring characterized by intermittent discharge of water ejected turbulently and accompanied by steam.
- lava—molten rock.
- lava spine—an upright cylindrical mass of lava caused by the upward squeezing of pasty lava inside a volcanic vent.
- lava tube—a natural conduit through which lava travels beneath the surface of a lava flow, expelled by a volcano during an eruption.
- maar—a broad, low-relief volcanic crater that is caused by an explosion caused by groundwater coming into contact with hot magma.
- malpais—a landform characterized by eroded rocks of volcanic origin in an arid environment.
- mamelon—a hill formed by eruption of stiff lava.
- mid-ocean ridge—an underwater mountain range, typically having a valley known as a rift running along its spine, formed by plate tectonics.
- oceanic trench—narrow topographic depressions of the sea floor.
- pit crater—a depression formed by a sinking of the ground surface lying above a void or empty chamber.
- pseudocrater—a volcanic landform that resembles a true volcanic crater but differs in that it is not an actual vent from which lava has erupted.
- subglacial mound—a type of subglacial volcano that forms when lava erupts beneath a thick glacier or ice sheet.
- tuya—a distinctive, flat-topped, steep-sided volcano formed when lava erupts through a thick glacier or ice sheet.
- volcanic dam—a natural dam produced by volcanic activity.
- volcanic field—a spot of the Earth's crust that is prone to localized volcanic activity.
- volcanic plateau—a plateau produced by volcanic activity.
- volcanic plug—a volcanic landform created when magma hardens within a vent on an active volcano.

Earth's Geological Processes

In the preceding discussion of landforms, it should be apparent that, without Earth's geological processes, there would be no landforms. Thus, geology—with its interface to geography—is about much more than landforms. Geology is about patterns and processes. Geology is about materials—materials that make up the Earth. The materials that make up the Earth, of course, are mainly rocks (including dust, silt, sand, and soil). Rocks in turn are composed of minerals, and minerals are composed of atoms.

Earth processes are constantly acting upon and within the Earth to change it. Examples of these ongoing processes include formation of rocks; chemical cementation of sand grains together to form rock; construction of landforms such as mountain ranges; and erosion of mountain ranges. These are internal processes that get their energy from the interior of the Earth—most from radioactive decay (nuclear energy). Other examples of ongoing Earth processes include those that are more apparent to us (external processes) because they occur relatively quickly and are visible. These include volcanic eruptions, dust storms, mudflows, and beach erosion. The energy source for these processes is solar and gravitational. It is important to point out that many of these processes are cyclical in nature. The two most important cyclical processes are the hydrologic (water) cycle and the rock cycle.

HYDROLOGICAL CYCLE

Simply, the water cycle describes how water moves through the environment and identifies the links between groundwater, surface water, and the atmosphere (see figure 2.1). As illustrated, water is taken from the Earth's surface to the atmosphere by evaporation from the surface of lakes, rivers, streams, and oceans. This evaporation process occurs when the Sun heats water. The Sun's heat energizes surface molecules, allowing them to break free of the attractive force binding them together and then evaporate and rise as invisible vapor in the atmosphere.

Water vapor is also emitted from plant leaves by a process called *transpiration*. Every day, an actively growing plant transpires five to ten times as much water as it can hold at once. As water vapor rises, it cools and eventually condenses, usually on tiny particles of dust in the air. When it condenses, it becomes a liquid again or turns directly into a solid (ice, hail, or snow). These water particles then collect and form clouds. The atmospheric water formed in clouds eventually falls to Earth as precipitation. The precipitation can contain contaminants from air pollution. The precipitation may fall directly onto surface waters, be intercepted by plants or structures, or fall onto the ground. Most precipitation falls in coastal areas or in high elevations. Some of the water that falls in high elevations becomes runoff water, the water that runs over the ground picking up the sand, silt, and clay from the soil and carries particles to lower elevations to form streams, lakes, and alluvial fertile valleys.

The water we see is known as *surface water*. Surface water can be broken down into five categories: oceans, lakes, rivers and streams, estuaries, and wetlands.

The health of rivers and streams is directly linked to the integrity of habitat (and geology/geography) along the river corridor and in adjacent wetlands. Stream quality will deteriorate if activities damage vegetation along riverbanks and in nearby wetlands. Trees, shrubs, and grasses filter pollutants from runoff and reduce soil erosion. Removal of vegetation also eliminates shade that moderates stream temperature. Stream temperature, in turn, affects the availability of dissolved oxygen in the water column for fish and other aquatic organisms.

ROCK CYCLE

With time and changing conditions, the igneous, sedimentary, and metamorphic rocks of the Earth are subject to alteration by the processes of weathering (erosion),

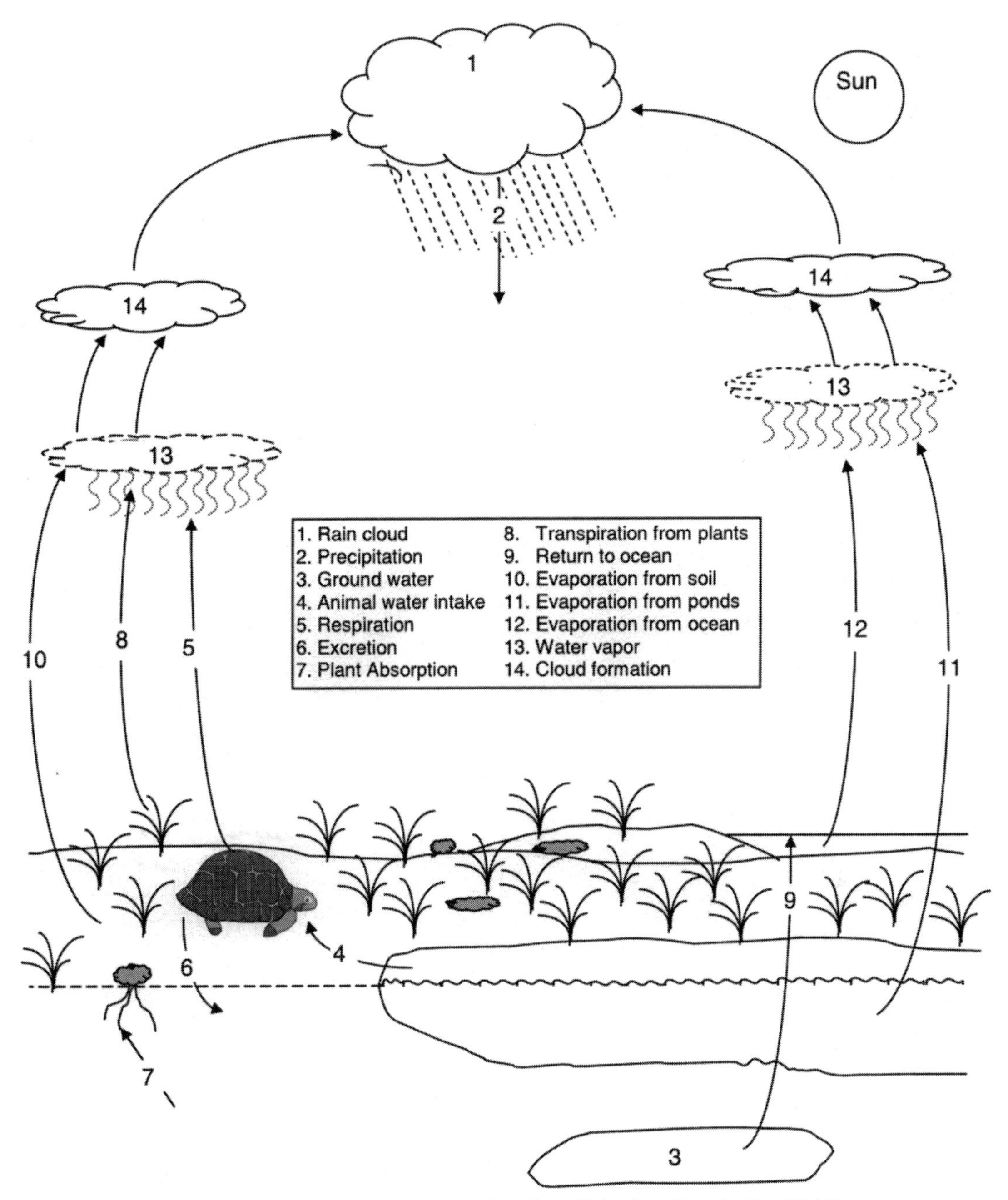

Figure 2.1. Water cycle. Modified from Carolina Biological Supply Co. (1966).

volcanism, and tectonism. Known as the *rock cycle* (see figure 2.2), this series of events (or group of changes) represents a response of Earth materials to various forms of energy. As shown in figure 2.2, most surface rocks started out as igneous rocks (rocks produced by crystallization from a liquid). When igneous rocks are exposed at the surface, they are subject to weathering. Erosion moves particles into rivers and oceans where they are deposited to become sedimentary rocks. Sedimentary rocks can be buried or pushed to deeper levels in the Earth, when changes in pressure and temperature cause them to become metamorphic rocks. At high temperatures, metamorphic rocks may melt to become magmas. Magmas rise to the surface, crystallize to become igneous rocks, and the processes starts over. Keep in mind that the cycles are not always completed, for there can be many short circuits along the way, as indicated in figure 2.2.

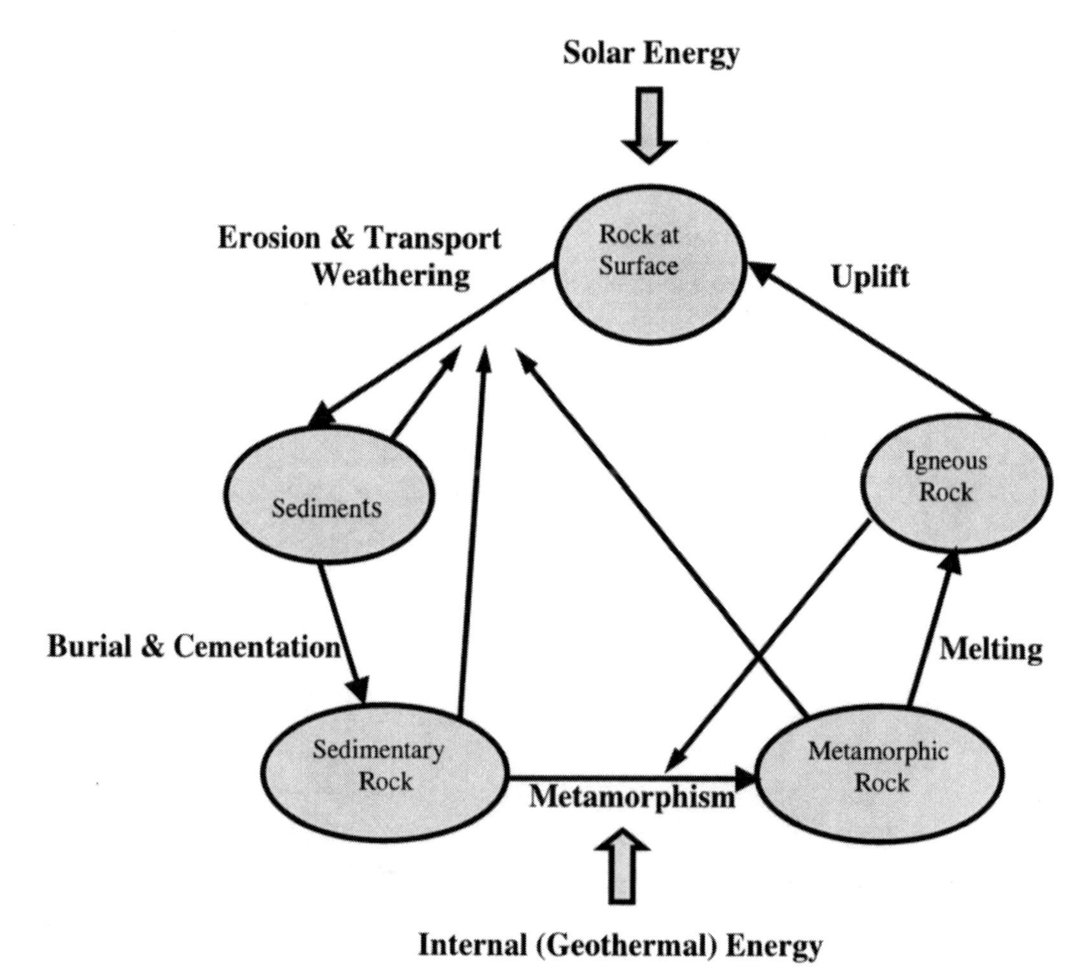

Figure 2.2. Rock Cycle, if uninterrupted, will continue completely around the outer margin of the diagram. However, as shown by the arrows, the cycle may be interrupted or "short-circuited" at various points in its course.

Planet Earth

Earth has a radius of about 6,370 kilometers, although it is about 22 kilometers larger at the equator than at the poles. The circumference of the Earth is about 24,874 miles, and the surface area comprises roughly 197 million square miles. About 29 percent are surface lands; the remaining 71 percent of the Earth's surface is covered by water.

Earth, like the other planets in our solar system, revolves around the Sun within its own orbit and period of revolution. The Earth also rotates on its own axis. The Earth rotates from west to east and makes one complete rotation each day. It is this rotating motion that gives us the alternating periods of daylight and darkness that we know as day and night.

The Earth also precesses (or wobbles) as it rotates on its axis, much as a top wobbles as it spins. This single wobble has to do with the fact that the Earth's axis is tilted at an angle of 23 1/2 degrees. The tilting of the axis is also responsible for the seasons. Over the years, various hypotheses have been put forward in attempts to determine what is the force, or excitation mechanism, that propels the wobble, such as atmospheric phenomena, continental water storage (changes in snow cover, river runoff, lake levels, or reservoir capacity), interaction at the boundary of Earth's core and its surrounding mantle, and earthquakes. In an explanation by the National Aeronautics and Space Administration (NASA; 2000), the principal cause of the wobble is fluctuating pressure on the bottom of the ocean, caused by temperature and salinity changes and wind-driven changes in the circulation of the oceans.

Earth revolves around the Sun in a slightly elliptical orbit approximately once every 365 1/4 days. During this solar year, the Earth travels at a speed of more than sixty thousand miles per hour, and on the average it remains about ninety-three million miles from the Sun.

Principal Divisions of the Earth

The Earth consists of four interconnected "geospheres": the *atmosphere*, a gaseous envelope surround the Earth; the *hydrosphere*, the waters filling the depressions and covering almost 71 percent of the Earth; the *lithosphere*, the solid part of the Earth that underlies the atmosphere and hydrosphere; and the *biosphere*, composed of all living organisms.

ATMOSPHERE

The atmosphere is the body of air that surrounds our planet. But what is air? Air is a mixture of gases that constitutes the Earth's atmosphere. What is the Earth's atmosphere? The atmosphere is that thin shell, veil, or envelope of gases that surrounds Earth like the skin of an apple—thin, very thin—but very, very vital. The approximate composition of dry air is, by volume at sea level, nitrogen, 78 percent; oxygen, 21 percent (necessary for life as we know it); argon, 0.93 percent; and carbon dioxide, 0.03 percent, together with very small amounts of numerous other constituents (see table 2.1). The water vapor content is highly variable and depends on atmospheric conditions. Air is said to be pure when none of the minor constituents is present in sufficient concentration to be injurious to the health of human beings or animals, to damage vegetation, or to cause loss of amenity (e.g., through the presence of dirt, dust, or odors or by diminution of sunshine).

Where does air come from? Genesis 1:2 states that God separated the water environment into the atmosphere and surface waters on the second day of creation. Many scientists state that 4.6 billion years ago a cloud of dust and gases forged the Earth and also created a dense molten core enveloped in cosmic gases. This was the *proto-atmosphere* or *proto-air*, composed mainly of carbon dioxide, hydrogen, ammonia, and carbon monoxide, but did not last long before it was stripped away by a tremendous outburst of charged particles from the Sun. As the outer crust of Earth began to solidify, a new atmosphere began to form from the gases outpouring from gigantic hot springs and volcanoes. This created an atmosphere of air composed of carbon dioxide, nitrogen oxides, hydrogen, sulfur dioxide, and water vapor. As the Earth cooled, water vapor condensed into highly acidic rainfall, which collected to form oceans and lakes.

For much of Earth's early existence (the first half), only trace amounts of free oxygen were present. But then green plants evolved in the oceans, and they began to add oxygen to the atmosphere as a waste gas, and later oxygen increased to about 1 percent of the atmosphere and with time to its present 21 percent.

How do we know for sure about the evolution of air on Earth? Are we guessing, using "voodoo" science? There is no guessing or voodoo involved with the historical geological record. Consider, for example, geological formations that are dated to two billion years ago. In these early sediments, there is a clear and extensive band of red sediment ("red bed" sediments)—sands colored with oxidized (ferric) iron. Previously, ferrous formations had been laid down showing no oxidation. But there is more evidence. We can look at the time frame of 4.5 billion years ago, when carbon dioxide

Table 2.1. Composition of Air in the Earth's Atmosphere

Gas	*Chemical Symbol*	*Volume (%)*
nitrogen	N_2	78.08
oxygen	O_2	20.94
carbon dioxide	CO_2	0.03
argon	Ar	0.093
neon	Ne	0.0018
helium	He	0.0005
krypton	Kr	trace
xenon	Xe	trace
ozone	O_3	0.00006
hydrogen	H_2	0.00005

in the atmosphere was beginning to be lost in sediments. The vast amount of carbon deposited in limestone, oil, and coal indicate that carbon dioxide concentrations must once have been many times greater than today, which stands at only 0.03 percent. The first carbonated deposits appeared about 1.7 billion years ago, the first sulfate deposits about 1 billion years ago. The decreasing carbon dioxide was balanced by an increase in the nitrogen content of the air. The forms of *respiration* practiced advanced from fermentation 4 billion years ago to anaerobic *photosynthesis* 3 billion years ago to aerobic photosynthesis 1.5 billion years ago. The aerobic respiration that is so familiar today only began to appear about five hundred million years ago. The atmosphere itself continues to evolve, but human activities—with their highly polluting effects—have now overtaken nature in determining the changes. And, when you get right down to it, that is one of the overriding themes of this text—human beings and their affect on planet Earth.

The atmosphere is an important geologic agent and is responsible for the processes of weathering that are continually at work on the Earth's surface.

HYDROSPHERE

The hydrosphere includes all the waters of the oceans, lakes, and rivers, as well as groundwater—which exists within the lithosphere. Approximately forty million cubic miles of water cover or reside within the Earth. The oceans contain about 97 percent of all water on Earth. The other 3 percent is freshwater: (1) snow and ice on the surface of Earth contains about 2.25 percent of the water; (2) usable groundwater is approximately 0.3 percent; and (3) surface freshwater is less than 0.5 percent.

In the United States, for example, average rainfall is approximately 2.6 feet (a volume of 5,900 cubic kilometers). Of this amount, approximately 71 percent evaporates (about 4,200 cubic centimeters), and 29 percent goes to streamflow (about 1,700 cubic kilometers).

Beneficial freshwater uses include manufacturing, food production, domestic and public needs, recreation, hydroelectric power production, and flood control. Streamflow withdrawn annually is about 7.5 percent (440 cubic kilometers). Irrigation and industry use almost half of this amount (3.4 percent or 200 cubic kilometers per year). Municipalities use only about 0.6 percent (35 cubic kilometers per year) of this amount.

Historically, in the United States, water usage is increasing (as might be expected). For example, in 1975, 40 billion gallons of freshwater were used. In 1990, the total increased to 455 billion gallons. Projected use in 2002 is about 725 billion gallons.

The primary sources of freshwater include the following:

1. captured and stored rainfall in cisterns and water jars;
2. groundwater from springs, artesian wells, and drilled or dug wells;
3. surface water from lakes, rivers, and streams;
4. desalinized seawater or brackish groundwater; and
5. reclaimed wastewater.

In the United States, current federal drinking water regulations actually define three distinct and separate sources of freshwater. They are surface water, groundwater, and groundwater under the direct influence of surface water (GUDISW). This last classification is the result of the Surface Water Treatment Rule (SWTR). The definition of the conditions that constitute GUDISW, while specific, is not obvious. This classification is discussed in detail later.

Did You Know?

The distance between the East Coast and the West Coast of the United States is approximately 2,800 miles (4,500 kilometers).

LITHOSPHERE

The lithosphere is of prime importance to the geologist and geographer. This, the solid, inorganic, rocky crust portion of the Earth, is composed of rocks and minerals that, in turn, comprise the continental masses and ocean basins. The rocks of the lithosphere are of three basic types: igneous, sedimentary, and metamorphic.

Soil Geology

We use soil for our daily needs, but we do not sufficiently take account of its slow formation and fast loss. Simply, we take soil for granted. It's always been there—with the implied corollary that it will always be there—right? But where does soil come from?

Of course, soil was formed, and in a never-ending process, it is still being formed. However, as mentioned, soil formation is a slow process—one at work over the course of millennia, as mountains are worn away to dust through bare rock succession.

Any activity, human or natural, that exposes rock to air begins the process. Through the agents of physical and chemical weathering, through extremes of heat and cold, through storms and earthquake and entropy, bare rock is gradually worn away. As its exterior structures are exposed and weakened, plant life appears to speed the process along.

Lichens cover the bare rock first, growing on the rock's surface, etching it with mild acids and collecting a thin film of soil that is trapped against the rock and clings. This changes the conditions of growth so much that the lichens can no longer survive and are replaced by mosses.

The mosses establish themselves in the soil trapped and enriched by the lichens and collect even more soil. They hold moisture to the surface of the rock, setting up another change in environmental conditions.

Well-established mosses hold enough soil to allow herbaceous plant seeds to invade the rock. Grasses and small flowering plants move in, sending out fine root systems that hold more soil and moisture, and work their way into minute fissures in the rock's surface. More and more organisms join the increasingly complex community.

Weedy shrubs are the next invaders, with heavier root systems that find their way into every crevice. Each stage of succession affects the decay of the rock's surface and adds its own organic material to the mix. Over the course of time, mountains are worn away, eaten away to soil, as time, plants, weather, and extremes of weather work on them.

The parent material, the rock, becomes smaller and weaker as the years, decades, centuries, and millennia go by, creating the rich, varied, and valuable mineral resource we call soil.

Perhaps no term causes more confusion in communication between various groups of average persons, soil geologists, soil scientists, soil engineers, and Earth scientists than the word *soil.* In simple terms, soil can be defined as the topmost layer of decomposed rock and organic matter that usually contains air, moisture, and nutrients and can therefore support life. Most people would have little difficulty in understanding and accepting this simple definition. Then why are various groups confused on the exact meaning of the word soil? Quite simply, confusion reigns because soil is not simple—it is quite complex. In addition, the term soil has different meanings to different groups (like pollution, the exact definition of soil is a personal judgment call). Let's take a look at how some of these different groups view soil.

Average people seldom give soil a first or second thought. Why should they? Soil isn't that big a deal—that important—it doesn't impact their lives, pay their bills, or feed their bulldog, right?

Not exactly. Not directly.

The average person seldom thinks about soil as soil. He or she may think of soil in terms of dirt but hardly ever as soil. Why is this? Having said the obvious about the confusion between soil and dirt, let's clear up this confusion.

First of all, soil is not dirt. Dirt is misplaced soil—soil where we don't want it, contaminating our hands and fingernails, clothes, and automobiles and tracked in on the floor. Dirt is what we try to clean up and to keep out of our living environments.

Secondly, soil is too special to be called dirt. Why? Because soil is mysterious and, whether we realize it or not, essential to our existence. Because we think of it as common, we relegate soil to an ignoble position. As our usual course of action, we degrade it, abuse it, throw it away, contaminate it, ignore it—we treat it like dirt, and only feces hold a more lowly status than it does. Soil deserves better.

Why?

Again, because soil is not dirt—how can it be? It is not filth, or grime, or squalor. Instead, soil is clay, air, water, sand, loam, organic detritus of former life-forms (including humans), and most important, the amended fabric of Earth itself; if water is Earth's blood, and air is Earth's breath, then soil is its flesh and bone and marrow—simply put, soil is the substance that most life depends on.

Soil scientists (or pedologists) are people interested in soils as a medium for plant growth. Their focus is on the upper meter or so beneath the land surface (this is known as the weathering zone, which contains the organic-rich material that supports plant growth) directly above the unconsolidated parent material. Soil scientists have developed a classification system for soils based on the physical, chemical, and biological properties that can be observed and measured in the soil.

Soils engineers are typically soil specialists who look at soil as a medium that can be excavated using tools. Soils engineers are not concerned with the plant-growing potential of a particular soil but rather are concerned with a particular soil's ability to support a load. They attempt to determine (through examination and testing) a soil's particle size, particle-size distribution, and the plasticity of the soil.

Earth scientists (or geologists) have a view that typically falls between pedologists and soils engineers—they are interested in soils and the weathering processes as past indicators of climatic conditions, and in relation to the geologic formation of useful materials ranging from clay deposits to metallic ores.

Would you like to gain a new understanding of soil? Take yourself out to a plowed farm field somewhere, anywhere. Reach down and pick a handful of soil, and look at it—really look at it closely. What are you holding in your hand? Look at the two descriptions that follow, and you may gain a better understanding of what soil actually is and why it is critically important to us all.

1. A handful of soil is alive, a delicate living organism—as lively as an army of migrating caribou and as fascinating as a flock of egrets. Literally teeming with life of incomparable forms, soil deserves to be classified as an independent ecosystem or, more correctly stated, as many ecosystems.
2. When we pick up a handful of soil, exposing Earth's stark bedrock surface, it should remind us (and maybe startle us) to the realization that without its thin, living soil layer, Earth is a planet as lifeless as our own Moon (Spellman 1998).

Did You Know?

The United States' Coastal Plain and its interior Great Plains vary in altitude by amounts up to 4,000 feet (1,220 meters).

BIOSPHERE

Thanks to the life-giving qualities of air and water, Earth is populated by countless species of plants and animals. This horde of organisms comprises the biosphere. Most of the planet's life is found from three meters below the ground to thirty meters above it and in the top two hundred meters of the oceans and seas.

In regard to the life-forms that make up the biosphere, have you ever asked what life is? What does it mean to be alive? Have you ever tried to define life? If so, how did you define it? If these questions strike you as odd, consider them for a moment (they are almost as difficult as defining the origin of life). Of course, we all have an intuitive sense of what life is, but if you had difficulty, as is probably the case, with answering these questions, you are not alone. These questions are open to debate and have been from the beginning of time. One thing is certain; life is not a simple concept, and it is impossible to define.

Along with the impossibility of defining life precisely, it is not always an easy thing to tell the difference among living, dead, and nonliving things. Prior to the seventeenth century, many people believed that nonliving things could spontaneously turn into living things. For example, it was believed that piles of straw could turn into mice. Obviously, that is not the case. There are some very general rules to follow when trying to decide if something is living, dead, or nonliving. Scientists have identified seven basic characteristics of life. Keep in mind that, for something to be described as living, that something must display *all* seven of these characteristics (i.e., "characteristic" is plural). Although many of us have many different opinions about what "living" means, the following characteristics were designated "characteristics of living things" with the consensus of the scientific community.

- *Living things are composed of cells*: living things exhibit a high level of organization, with multicellular organisms being subdivided into cells, and cells into organelles, and organelles into molecules, and so forth.
- *Living things reproduce*: all living organisms reproduce, either by sexual or asexual means.
- *Living things respond to stimuli*: all living things respond to stimuli in their environment.
- *Living things maintain homeostasis*: all living things maintain a state of internal balance in terms of temperature, pH, water concentrations, and so on.
- *Living things require energy*: Some view life as a struggle to acquire energy (from sunlight, inorganic chemicals, or another organism) and release it in the process of forming adenosine triphosphate (ATP). The conventional view is that living organisms require energy, usually in the form ATP. They use this energy to carry out energy-requiring activities such as metabolism and locomotion.
- *Living things display heredity*: living organisms inherit traits from the parent organisms that created them.
- *Living things evolve and adapt*: All organisms have the ability to adapt or adjust to their surroundings. An example of this is adapting to environmental change resulting in an increased ability to reproduce.

Interesting Point: Again, if something follows one or just a few of the characteristics listed above, it does not necessarily mean that it is living. To be considered alive, an object must exhibit *all* of the characteristics of living things. A good example of a nonliving object that displays at least one characteristic for living is sugar crystals growing on the bottom of a syrup dispenser. On the other hand, there is a stark exception to the characteristics above. For example, mules cannot reproduce because they are sterile. Another nonliving object that exhibits many of the characteristics of life is a flame. Think about it, a flame:

- respires,
- requires nutrition,
- reproduces,
- excretes,
- grows,
- moves,
- is irritable, and
- is organized.

We all know that a flame is not alive, but how do we prove that to the skeptic? The best argument we can make is as follows:

1. nonliving materials never replicate using DNA and RNA (hereditable materials); and
2. nonliving material cannot carry out anabolic metabolism.

Did You Know?

All four divisions of Earth can be and often are present in a single location. For example, a piece of soil, of course, will have mineral material from the lithosphere. Additionally, there will be elements of the hydrosphere present as moisture within the soil, the biosphere as insects and plants, and even the atmosphere as pockets of air between soil pieces.

Structure of Earth

Earth is made up of three main compositional layers: crust, mantle, and core (see figure 2.3). The crust has variable thickness and composition: continental crust is ten to fifty kilometers thick, while the oceanic crust is eight to ten kilometers thick. The elements silicon, oxygen, aluminum, and ion make up the Earth's crust. Like the shell of an egg, the Earth's crust is brittle and can break.

Based on seismic (earthquake) waves that pass through the Earth, we know that below the crust is the mantle, a dense, hot layer of semisolid (plasticlike liquid) rock approximately 2,900 kilometers thick. The mantle, which contains silicon, oxygen, aluminum, and more iron, magnesium, and calcium than the crust, is hotter and denser because temperature and pressure inside the Earth increase with depth. According to the U.S. Geological Survey (USGS; 1999), as a comparison, the mantle might be thought of as the white of a boiled egg. The thirty-kilometer-thick transitional layer between the mantle and crust is called the Moho layer. The temperature at the top of the mantle is 870° C. The temperature at the bottom of the mantle is 2,200° C.

At the center of the Earth lies the core, which is nearly twice as dense as the mantle because its composition is metallic (Iron [Fe]–Nickel [NI] alloy) rather than

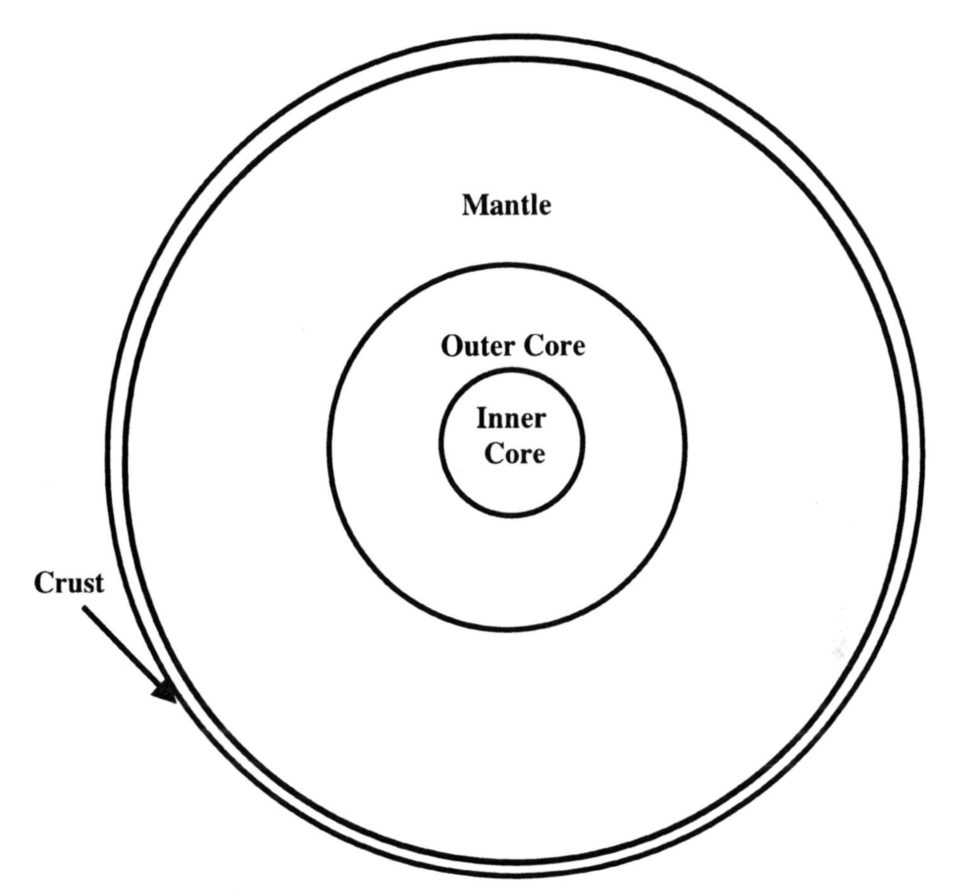

Figure 2.3. Earth's layers.

stony. Unlike the yolk of an egg, however, the Earth's core is actually made up of two distinct parts: a 2,200-kilometer-thick liquid outer core and a 1,250-kilometer-thick solid inner core. As the Earth rotates, the liquid outer core spins, creating the Earth's magnetic field.

Did You Know?

The molten magma in the mantle is so thick that it flows only about 1 inch (2.54 cm) each year.

Chapter Review Questions

2.1. How thick is Earth's continental crust? __________ kilometers.
2.2. How thick is Earth's oceanic crust? __________ kilometers.

2.3. The layer between the mantle and the crust is known as ________.
2.4. Plasticity describes magma's ability to ________.
2.5. Describe how water moves through the environment.
2.6. The solid part of the Earth is known as the ________.
2.7. Define dirt.
2.8. What is the biosphere?
2.9. What does the crust of the Earth consist of?

References and Recommended Reading

Air quality criteria. 1968. Staff report, Subcommittee on Air and Water Pollution, Committee on Public Works, U.S. Senate, 94–411.

American heritage dictionary of the English language, 4th ed. 2000. Boston: Houghton Mifflin.

Angus, S. 1997. *The Outer Hebrides: The shaping of the islands.* London: White Horse Press.

Bascom, W. 1980. *Waves and beaches.* New York: Anchor Press.

Bergman, J. 2005. Rocks and the rock cycle. www.windows.ucar.edu/tour/ling=/earth/geology/rocks_intro.htm l (accessed May 23, 2008).

Bibliography of Aeolian research (BAR). 2009. www.lbk.ars.usda.gov/ (accessed May 19, 2009).

Blue, J. 2007. Descriptor terms. USGS *Gazetteer of Planetary Nomenclature.* http://planetarynames.wr.usgs.gov/jsp/append5.jsp (accessed November 11, 2009).

Bruun, P., and A. J. Mehta. 1978. *Stability of tidal inlets: Theory and engineering.* Amsterdam: Elsevier.

California Institute of Technology. 2009. Desert varnish. http://minerals.caltech.edu/FILES/VARNISH (accessed May 19, 2009).

Campbell, N. A. 2004. *Biology: Concepts and Connections*, 4th CD-ROM ed. Menlo Park, Calif.: Benjamin-Cummings Publishing.

Easterbrook, D. J. 1999. *Surface processes and landforms.* Upper Saddle River, N.J.: Prentice Hall.

Fairbridge, R. W. 1950. Recent and Pleistocene coral reefs of Australia. *Journal of Geology* 58 (4): 330–401.

Goshorn, D. 2006. *Proceedings—DELMARVA Coastal Bays Conference III: Tri-state approaches to preserving aquatic resources.* Washington, D.C.: U.S. Environmental Protection Agency.

Halliday, W. R. 1979. Glaciospeleology cave science topics. *Caving International* 4 (July).

Hogan, C. M. 2007. Megalithic portal: Burroughston Broch, ed. Andy Burnham. www.megalithic.co.uk/article.php?sid=7891 (accessed May 20, 2009).

Huxley, T. H. 1876. *Science and education, volume III: Collected essays.* New York: D. Appleton.

Johnson, M. E., and L. K. Libbey. 1997. Global review of Upper Pleistocene Rocky Shores: Tectonic segregation, substrate variation and biological diversity. *Journal of Coastal Research* 13 (2): 297–307.

Jones, A. M. 1997. *Environmental biology.* New York: Routledge.

Jungerius, P. D., and F. van der Meulen. 1989. The development of dune blowouts, as measured with erosion pins and sequential air photos. *Catena* 16 (5): 369–76.

Keeton, W. T. 1996. *Biological science.* New York: Norton.

King, R. M. 2003. *Biology made simple.* New York: Broadway Books.

Koch, R. 1882. *Uber die Atiologie der Tuberkulose.* In *Verhandlungen des Knogresses fur Innere Medizin.* Wiesbaden, Germany: Erster Kongress.

———. 1884. *Mitt Kaiser Gesundh* 2:1–88.
———. 1893. *J. Hyg. Inf.* 14:319–33.
Larsson, K. A. 1993. Prediction of the pollen season with a cumulated activity method. *Grana* 32:111–14.
McNeil, F. S. 1954. Organic reefs and banks and associated detrital sediments. *American Journal of Science* 252 (7): 385–401.
Medicine Net. 2006. Definition of postulates: Koch's. www.medterms.com/script/main/art.asp?articlekey=7291 (accessed November 12, 2009).
Monkhouse, F. J. 1978. *Dictionary of geography*. London: Edward Arnold.
National Aeronautics and Space Administration (NASA). 2000. A mystery of Earth's wobble solved: It's the ocean. www.jpl.nasa.gov/releases/2000/chandlerwobble.html (accessed May 24, 2008).
———. 2009. Erg. http://earthobservatory.nasa.gov/ (accessed May 19, 2009).
Paterson, W. S. B. 1994. *The physics of glaciers*, 3rd ed. New York: Wiley.
Press, R., and F. Siever. 2001. *Earth*, 3rd ed. New York: W. H. Freeman.
Rieman, H. M. 1979. Deflation armor (desert pavement). *Lapidary Journal* 33 (7): 1648–50.
The scientific method: Fish health and pfiesteria. 2006. Sea Grant Maryland (SGM), University of Maryland, NOAA. http://aquaticpath.umd.edu/fishhealth/koch.html (accessed November 12, 2009).
Segerstrom, K. 1962. Deflated marine terrace as a source of dun chains: Atacama Province, Chile. In *Geological survey research 1962: Shore papers in geology and hydrology*, 91–93. Processional Paper 0450-C. Washington, D.C.: U.S. Department of the Interior, U.S. Geological Survey.
Spellman, F. R. 1996. *Stream ecology and self-purification*. Lancaster, Pa.: Technomic Publishing.
———. 1998. *Environmental science and technology: Concepts and applications*. Lancaster, Pa.: Technomic Publishing.
———. 2008. *The science of air: Concepts and applications*. Boca Raton, Fla.: CRC Press.
———. 2009. *Geology for nongeologists*. Lanham, Md.: Government Institutes.
Spellman, F. R., and N. E. Whiting. 2006. *Environmental science and technology*, 2nd ed. Lanham, Md.: Government Institutes.
Spieksma, F. T. 1991. Aerobiology in the nineties: Aerobiology and pollinosis. *International Aerobiology Newsletter* 34:1–5.
Suburban Emergency Management Project (SEMP). 2009. The secret of China's vast loess Plateau. www.semp.us/publications/biot_reader.phe?BiotID=357 (accessed May 19, 2009).
U.S. Geological Survey. 1999. Inside the Earth. http://pubs.usgs.gov/gip/dynamic/inside.html (accessed July 31, 2009).

Belly River, Glacier National Park, Montana. Photo by Frank R. Spellman.

CHAPTER 3

Weathering

> In regards to surface rock formations, weathering and erosion are both creators and executioners! As soon as rock is lifted above sea level, weather starts to break it up. Water, ice, and chemicals abrade, split, dissolve, or rot the rocky surface until it crumbles. Mixed with water and air, and microbes, and plant and animal remains, crumbled rock forms soil. Natural soil is alive—and for all [of] us this is a good thing.
>
> —F. R. Spellman, 2009

Weathering

Rocks are always breaking into smaller pieces. Thus, it logically follows that one can't obtain even the slimmest edge of understanding of geology/geography without understanding the processes that cause the breakdown of rocks, either to form new minerals that are stable on the surface of the Earth or to break the rocks down to smaller particles. Simply, weathering (which projects itself on all surface material above the water table) is the general term used for all the ways in which a rock may be broken down.

FACTORS THAT INFLUENCE WEATHERING

The factors that influence weathering include the following:

- *Rock type and structure*—each mineral contained in rocks has a different susceptibility to weathering. A rock with bedding planes, joints, and fractures provides pathways for the entry of water, leading to more rapid weathering. Differential weathering (rocks erode at differing rates) can occur when rock combinations consist of rocks that weather faster than more resistant rocks.
- *Slope*—on steep slopes weathering products may be quickly washed away by rains. Wherever the force of gravity is greater than the force of friction holding particles upon a slope, these tend to slide downhill.
- *Climate*—higher temperatures and high amounts of water generally cause chemical reactions to run faster. Rates of weathering are higher in warmer than in colder dry climates.

- *Animals*—rodents, earthworms, and ants that burrow into soil bring material to the surface where it can be exposed to the agents of weathering.
- *Time*—depends on slope, animals, and climate.

Kinds of Weathering

Although weathering processes are separated, it is important to recognize that these processes work in tandem to break down rocks and minerals to smaller fragments. Geologists recognize two kinds of weathering processes.

1. *Physical (or mechanical) weathering*—involves the disintegration of rocks and minerals (without any change in the chemical constituents of the rock) by a physical or mechanical process.
2. *Chemical weathering*—involves the decomposition of rock by chemical changes or solution by such agents as water, oxygen, carbon, or various organic acids.

PHYSICAL WEATHERING

Physical weathering involves the disintegration of a rock by physical processes. These include freezing and thawing of water (temperature changes) in rock crevices, disruption by plant roots or burrowing animals, and the changes in volume that result from chemical weathering with the rock. These and other physical weathering processes are discussed below.

- *Glacial movements*—grinding action of glaciers work to reduce huge boulders to pebbles or smaller.
- *Abrasion by streams*—gravel, pebbles, and boulders are moved along and constantly abraded by fast-flowing streams.
- *Wind-blasting effect*—with time, particles carried by wind act as sandpaper in reducing rocks to sand-sized particles.
- *Development of joints*—joints are another way that rocks yield to stress. Joints are fractures or cracks in which the rocks on either side of the fracture have not undergone relative movement. Joints form as a result of expansion due to cooling or relief of pressure as overlying rocks are removed by erosion. They form free space in a rock by which other agents of chemical or physical weathering can enter (unlike faults that show offset across the fracture). They play an important part in rock weathering as zones of weakness and water movement.
- *Crystal growth*—as water percolates through fractures and pore spaces it may contain ions that precipitate to form salt crystals. When salt crystals grow they can cause the stresses needed for mechanical rupturing of rocks and minerals.

- *Heat*—it was once thought that daily heating and cooling of rocks was a major contributor to the weathering process. This view is no longer shared by most practicing geologists. However, it should be pointed out that sudden heating of rocks from forest fires may cause expansion and eventual breakage of rock.
- *Biological activities*—plant and animal activities are important contributors to rock weathering. Plants contribute to the weathering process by extending their root systems into fractures and growing, causing expansion of the fracture. Growth of plants and their effects are evident in many places where they are planted near cement work (streets, brickwork, and sidewalks). Animal burrowing in rock cracks can break rock.
- *Frost wedging*—is often produced by alternate freezing and thawing of water in rock pores and fissures. Expansion of water during freezing causes the rock to fracture. Frost wedging is more prevalent at high altitudes where there may be many freeze-thaw cycles. One classic and striking example of weathering of Earth's surface rocks by frost wedging is illustrated by the formation of hoodoos in Bryce Canyon National Park, Utah (see figure 3.2). "Although Bryce Canyon receives a meager 18 inches of precipitation annually, it's amazing what this little bit of water can do under the right circumstances!" (NPS 2008).

Approximately two hundred freeze-thaw cycles occur annually in Bryce. During these periods, snow and ice melt in the afternoon and water seeps into the joints of the Bryce or Claron Formation. When the Sun sets, temperatures plummet and the water refreezes, expanding up to 9 percent as it becomes ice. This frost-wedging process exerts tremendous pressure or force on the adjacent rock and shatters and pries the weak rock apart. The assault from frost wedging is a powerful force, but, at the same time, rainwater (the universal solvent), which is naturally acidic, slowly dissolves away the limestone, rounding off the edges of these fractured rocks and washing away the debris. Small rivulets of water round down Bryce's rime, forming gullies. As gullies are cut deeper, narrow walls of rock known as fins begin to emerge. Fins eventually develop holes known as windows. Windows grow larger until their roofs collapse (see figure 3.1), creating hoodoos (see figure 3.2). As old hoodoos age and collapse, new ones are born (NPS 2008).

Did You Know?

Bryce Canyon National Park lies along the high eastern escarpment of the Paunsaugunt Plateau in the Colorado Plateau region of southern Utah. Its extraordinary geological character is expressed by thousands of rock chimneys (hoodoos) that occupy amphitheaterlike alcoves in the Pink Cliffs, whose bedrock host is Claron Formation of the Eocene age. (Davis and Pollock 2003)

Figure 3.1. Windows developed in rock. Upon their collapse, new hoodoos will be formed. Photo by Frank R. Spellman.

Did You Know?

Hoodoo pronunciation: 'hu-du
Noun
Etymology: West African; perhaps from voodoo
A natural column of rock in western North America often in fantastic form.
(Merriam-Webster Online, www.m-w.com)

CHEMICAL WEATHERING

Chemical weathering involves the decomposition of rock by chemical changes or solution. Rocks that are formed under conditions present deep within the Earth are exposed to conditions quite different (i.e., temperatures and pressures are lower on the surface, and copious amounts of free water and oxygen are available) when uplifted onto the surface. The chief processes are oxidation, carbonation, and hydration, and solution in water above and below the surface.

Figure 3.2. Frost-wedged-formed hoodoo after window collapse. Bryce Canyon National Park, Utah. Photo by Frank R. Spellman.

The Persistent Hand of Water

Because of its unprecedented impact on shaping and reshaping Earth, at this point in the text, it is important to point out that, given time, nothing, absolutely nothing, on Earth is safe from the heavy hand of water. The effects of water sculpting by virtue of movement and accompanying friction will be covered later in the text. For now, in regard to water exposure and chemical weathering, the main agent responsible for chemical weathering reactions is not water movement but instead weak acids formed in water.

The acids formed in water are solutions that have abundant free Hydrogen$^+$ ions. The most common weak acid that occurs in surface waters is carbonic acid. Carbonic acid (H_2CO_3) is produced when atmospheric carbon dioxide dissolves in water; it exists only in solution. Hydrogen ions are quite small and can easily enter crystal structures, releasing other ions into the water.

H_2O	+	CO_2	→	H_2CO_3	→	H^+	+	HCO_3^-
water		carbon dioxide		carbonic acid		hydrogen ion		bicarbonate ion

Types of Chemical Weathering Reactions

As mentioned, chemical weathering breaks rocks down chemically, adding or removing chemical elements, and changes them into other materials. Again, as stated, chemical weathering consists of chemical reactions, most of which involves water. Types of chemical weathering include the following:

- *Hydrolysis*—this is a water-rock reaction that occurs when an ion in the mineral is replaced by H^+ or OH^-.
- *Leaching*—ions are removed by dissolution into water.
- *Oxidation*—oxygen is plentiful near Earth's surface; thus, it may react with minerals to change the oxidation state of an ion.
- *Dehydration*—occurs when water or a hydroxide ion is removed from a mineral.
- *Complete dissolution.*

Did You Know?

Weathering occurs with no movement (in situ) and, thus, should not be confused with erosion, which involves the movement of rocks and minerals by agents such as water, wind, ice, and gravity.

The Bottom Line

Weathering is clearly a complex phenomenon. There are no cut and dried rules about how exactly different rocks will weather in different climatic environments, as much depends on minor differences in rock type and moisture availability. One thing seems certain about weathering: the processes of weathering, transport, and erosion are very much mutually interdependent and form one unified system.

Chapter Review Questions

3.1. What are the factors that influence weathering?

3.2. Disintegration of rocks and minerals by a physical or mechanical process is ________.

3.3. What involves the decomposition of rock by chemical changes or solution?

3.4. Alternate freezing and thawing of water in rock pores and fissures describes ________.

3.5. A common weak acid in surface waters is ________.

3.6. ________ cover bare rock first.

References and Recommended Reading

American Society for Testing and Materials (ASTM). 1969. *Manual on water.* Philadelphia: ASTM.

Beazley, John D. *The way nature works.* 1992. New York: Macmillan.

Brady, N. C., and R. R. Weil. 1996. *The nature and properties of soils,* 11th ed. Upper Saddle River, N.J.: Prentice Hall.

Carson, R. 1962. *Silent spring.* Boston: Houghton Mifflin.

Ciardi, J. 1997. From Stoneworks. In *The Collected Poems of John Ciardi,* ed. E. M. Cifelli. Fayetteville: University of Arkansas Press.

Davis, G. H., and G. L. Pollock. 2003. Geology of Bryce Canyon National Park, Utah. In *Geology of Utah's Parks and Monuments,* ed. D. A. Sprinkel et al. 2nd ed. Salt Lake City: Utah Geological Association.

Eswaran, H. 1993. Assessment of global resources: Current status and future needs. *Pedologie* 43:19–39.

Foth, H. D. 1978. *Fundamentals of soil science,* 6th ed. New York: Wiley.

Franck, I., and D. Brownstone. 1992. *The green encyclopedia.* New York: Prentice Hall.

Kemmer, F. N. 1979. *Water: The universal solvent.* Oak Brook, Ill.: NALCO Chemical Company.

Konigsburg, E. M. 1996. *The view from Saturday.* New York: Scholastic.

Mowet, F. 1957. *The dog who wouldn't be.* New York: Willow Books.

National Park Service (NPS). 2008. *The hoodoo* (park newsletter). Bryce, Utah: Bryce Canyon National Park.

Spellman, F. R. 1998. *The science of environmental pollution.* Boca Raton, Fla.: CRC Press.

———. 2009. *Geology for nongeologists.* Lanham, Md.: Government Institutes.

Tomera, A. N. 1989. *Understanding basic ecological concepts.* Portland, Maine: J. Weston Walch.

U.S. Department of Agriculture (USDA). 1975. *Soil classification: A comprehensive system,* soil survey staff. Washington, D.C.: Natural Resources Conservation Service.

———. 1975. *Soil taxonomy: A basic system of soil classification for making and interpreting soil surveys.* Washington, D.C.: Natural Resources Conservation Service.

———. 1994. *Keys to soil taxonomy,* soil survey staff. Washington, D.C.: Natural Resources Conservation Service.

CHAPTER 4

Running Water Systems

> Water covers most of the earth's surface. Water makes up a portion of all living things, and all living things need water in order to live. The amount of water on the earth remains relatively constant through the movement of water from ocean to air to ground to ocean. Because water constantly recycles and remains relatively constant in quantity, those of us today are drinking the same water drunk by Cleopatra, Caesar, Alexander the Great, and all of those who preceded us on earth.
>
> —F. R. Spellman (2008)

Stream Genesis and Structure

Early in the spring on a snow- and ice-covered high alpine meadow is the time and place the water cycle continues. The cycle's main component, water, has been held in reserve—literally frozen, for the long, dark winter months, but with longer, warmer spring days, the Sun is higher, more direct, and of longer duration, and the frozen masses of water respond to the increased warmth. The melt begins with a single drop, then two, then more and more. As the snow and ice melts, the drops join a chorus that continues unending; they fall from their ice-bound lip to the bare rock and soil terrain below.

The terrain the snowmelt strikes is not like glacial till—the unconsolidated, heterogeneous mixture of clay, sand, gravel, and boulders that is dug out, ground out, and exposed by the force of a huge, slow, and inexorably moving glacier. Instead, this soil and rock ground is exposed to the falling drops of snowmelt because of a combination of wind and tiny, enduring force exerted by drops of water as over season after season they collide with the thin soil cover, exposing the intimate bones of the Earth.

Gradually, the single drops increase to a small rush—they join to form a splashing, rebounding, helter-skelter cascade and many separate rivulets that trickle and then run their way down the face of the granite mountain. At an indented ledge halfway down the mountain slope, a pool forms whose beauty, clarity, and sweet iciness provides the visitor with an incomprehensible, incomparable gift—a blessing from Earth.

The mountain pool fills slowly, tranquil under the blue sky, reflecting the pines, snow, and sky around and above it, an open invitation to lie down and drink, and to peer into the glass-clear, deep waters, so clear that it seems possible to reach down over fifty feet and touch the very bowels of the mountain pool. The pool has no transition from shallow margin to depth; it is simply deep and pure. As the pool fills with more meltwater, we wish to freeze time, to hold this place and this pool in its perfect state forever; it is such a rarity to us in our modern world. But this cannot be—Mother Nature calls, prodding, urging—and for a brief instant, the water laps in the breeze against the outermost edge of the ridge, then a trickle flows over the rim. The giant hand of gravity reaches out and tips the overflowing melt onward, and it continues the downward journey, following the path of least resistance to its next destination, several thousand feet below.

When the overflow, still high in altitude but its rock-strewn bed bent downward toward the sea, meets the angled, broken rocks below, it bounces, bursts, and mists its way against steep, V-shaped walls that form a small valley, carved out over time by water and the forces of the Earth.

Within the valley confines, the meltwater has grown from drops to rivulets to a small mass of flowing water. It flows through what is at first a narrow opening, gaining strength, speed, and power as the V-shaped valley widens to form a U-shape. The journey continues as the water mass picks up speed and tumbles over massive boulders and then slows again.

At a larger but shallower pool, waters from higher elevations have joined the main body—from the hillsides, crevices, springs, rills, and mountain creeks. At the influent poolsides, all appears peaceful, quiet, and restful, but not far away, at the effluent end of the pool, gravity takes control again (actually, gravity is always in control). The overflow is flung over the jagged lip and cascades downward several hundred feet, where the waterfall again brings its load to a violent, mist-filled meeting.

The water separates and joins again and again, forming a deep, furious, wild stream that calms gradually as it continues to flow over lands that are less steep. The waters widen into pools overhung by vegetation, surrounded by tall trees. The pure, crystalline waters have become progressively discolored on their downward journey, stained brown-black with humic acid, and literally filled with suspended sediments; the once pure stream is now muddy.

The mass divides and flows in different directions, over different landscapes. Small streams divert and flow into open country. Different soils work to retain or speed the waters, and in some places the waters spread out into shallow swamps, bogs, marshes, fens, or mires. Other streams pause long enough to fill deep depressions in the land and form lakes. For a time, the water remains and pauses in its journey to the sea. But this is only a short-term pause because lakes are only a short-term resting place in the water cycle. The water will eventually move on, by evaporation or seepage into groundwater. Other portions of the water mass stay with the main flow, and the speed of flow changes to form a river, which braids its way through the landscape, heading for the sea. As it changes speed and slows, the river bottom changes from rock and stone to silt and clay. Plants begin to grow, stems thicken, and leaves broaden. The river is now full of life and the nutrients needed to sustain life. But the river courses onward, its destiny met when the flowing rich mass slows and finally spills into the sea (Spellman and Whiting 1998).

Streams

The study of running water bodies and waterways in general is known as *surface hydrology* and is a core element of environmental geography.[1] Streams are bodies of running water with a current that carry rock particles (sediment loads) and dissolved ions and flow downslope along a clearly defined and confined path and stream banks, called a *channel*. Thus, streams may vary in width from a few inches to several miles; however, in the United States they are classified as streams when they are less than sixty feet (eighteen meters) wide. Depending on their locale or certain characteristics, streams may be referred to a branches, bourns, nants, brooks, burns, creeks, becks, creeks, kills, licks, rills, river sykes, bayous, rivulets, or runs. Streams are important for several reasons:

- Streams are an important part of the water cycle; they carry most of the water that goes from the land to the sea.
- Streams are one of the main transporters of sediment load from higher to lower elevations.
- Streams carry dissolved ions, the products of chemical weathering, into the oceans and thus make the sea salty.
- Streams (along with weathering and mass wasting) are a major part of the erosional process.
- Most population centers are located next to streams because they provide a major source of water and transportation.

Key Terms

- *Spring*—the point at which a stream emerges from an underground course through unconsolidated sediments or caves.
- *Source*—the spring from which the stream originates, or other point of origin of a stream.
- *Evapotranspiration (plant water loss)*—describes the process whereby plants lose water to the atmosphere during the exchange of gases necessary for photosynthesis. Water loss by evapotranspiration constitutes a major flux back to the atmosphere.
- *Infiltration capacity*—the maximum rate soil can absorb rainfall.
- *Perennial stream*—a type of stream in which flow continues during periods of no rainfall.
- *Confluence*—the point at which two streams merge.
- *Gaining stream*—typical of humid regions, where groundwater recharges the stream.
- *Losing stream*—typical of arid regions, where streams can recharge groundwater.
- *Laminar flow*—in a stream where parallel layers of water shear over one another vertically.
- *Turbulent flow*—in a stream where complex mixing is the result.
- *Headwaters*—the part of a stream or river proximate to its source.
- *Meandering*—stream condition whereby flow follows a winding and turning course.
- *Run*—somewhat smoothly flowing segment of the stream.
- *Pool*—segment where the water is deeper and slower moving.
- *Thalweg*—line of maximum water of channel depth in a stream.
- *Riffles*—refer to shallow, high-velocity flow over mixed gravel-cobble (barlike) substrate.

- *Sinuosity*—the bending or curving shape of a stream course.
- *Wetted perimeter*—the line on which the stream's surface meets the channel walls.

Characteristics of Stream Channels

A standard rule of thumb states the following: flowing waters (rivers and streams) determine their own channels, and these channels exhibit relationships attesting to the operation of physical laws—laws that are not, as of yet, fully understood. The development of stream channels and entire drainage networks, and the existence of various regular patterns in the shape of channels, indicate that streams are in a state of dynamic equilibrium between erosion (sediment loading) and deposition (sediment deposit), and governed by common hydraulic processes. However, because channel geometry is four dimensional with a long profile, cross-section, depth, and slope profile, and because these mutually adjust over a time scale as short as years and as long as centuries or more, cause and effect relationships are difficult to establish. Other variables that are presumed to interact as the stream achieves its graded state include width and depth, velocity, size of sediment load, bed roughness, and the degree of braiding (sinuosity).

STREAM PROFILES

Mainly because of gravity, most streams exhibit a downstream decrease in gradient along their length. Beginning at the headwaters, the steep gradient becomes less as one proceeds downstream, resulting in a concave longitudinal profile. Though diverse geography provides for almost unlimited variation, a lengthy stream that originates in a mountainous area (such as the one described in the chapter opening) typically comes into existence as a series of springs and rivulets; these coalesce into a fast-flowing, turbulent mountain stream, and the addition of tributaries results in a large and smoothly flowing river that winds through the lowlands to the sea.

When studying a stream system of any length, it becomes readily apparent (almost from the start of such studies) that what we are studying is a body of flowing water that varies considerably from place to place along its length. For example, a common variable—the results of which can be readily seen—is evident whenever discharge increases, causing corresponding changes in the stream's width, depth, and velocity. In addition to physical changes that occur from location to location along a stream's course, there are a legion of biological variables that correlate with stream size and distance downstream. The most apparent and striking changes are in steepness of slope and in the transition from a shallow stream with large boulders and a stony substrate to a deep stream with a sandy substrate.

The particle size of bed material at various locations is also variable along the stream's course. The particle size usually shifts from an abundance of coarser material upstream to mainly finer material in downstream areas.

SINUOSITY

Unless forced by man in the form of heavily regulated and channelized streams, straight channels are uncommon. Streamflow creates distinctive landforms composed of straight (usually in appearance only), meandering, and braided channels, channel networks, and floodplains. Simply put, flowing water will follow a sinuous course. The most commonly used measure is the sinuosity index (SI). Sinuosity equals one in straight channels and more than one in sinuous channels.

$$SI = \frac{\text{channel distance}}{\text{down valley distance}}.$$

Meandering is the natural tendency for alluvial channels and is usually defined as an arbitrarily extreme level of sinuosity, typically an SI greater than 1.5. Many variables affect the degree of sinuosity, however, and so SI values range from near unity in simple, well-defined channels to four in highly meandering channels (Gordon, McMahon, and Finlayson 1992).

It is interesting to note that even in many natural channel sections of a stream course that appear straight, meandering occurs in the line of maximum water or channel depth (known as the thalweg). Keep in mind that streams have to meander; that is how they renew themselves. By meandering, they wash plants and soil from the land into their waters, and these serve as nutrients for the plants in the rivers. If rivers aren't allowed to meander, if they are channelized, the amount of life they can support will gradually decrease. That means less fish, ultimately—and less bald eagle, herons, and other fishing birds (Spellman 1996).

Meander flow follows a predictable pattern and causes regular regions of erosion and deposition. The streamlines of maximum velocity and the deepest part of the channel lie close to the outer side of each bend and cross over near the point of inflection between the banks. A huge elevation of water at the outside of a bend causes a helical flow of water toward the opposite bank. In addition, a separation of surface flow causes a back eddy. The result is zones of erosion and deposition, and explains why point bars develop in a downstream direction in depositional zones (Morisawa 1968).

Did You Know?

Meandering channels can be highly convoluted or merely sinuous but maintain a single thread in curves having definite geometric shape. Straight channels are sinuous but apparently random in occurrence of bends. Braided channels are those with multiple streams separated by bars and islands (Leopold 1994).

BARS, RIFFLES, AND POOLS

Implicit in the morphology and formation of meanders are bars, riffles, and pools. Bars develop by deposition in slower, less competent flow on either side of the sinuous mainstream. Onward moving water, depleted of bed load, regains competence and shears a pool in the meander—reloading the stream for the next bar. Alternating bars migrate to form riffles.

As streamflow continues along its course, a pool-riffle sequence is formed. Basically, the riffle is a mound or hillock, and the pool is a depression.

FLOODPLAIN

Stream channels influence the shape of the valley floor through which they course. This self-formed, self-adjusted flat area near to the stream is the floodplain, which loosely describes the valley floor prone to periodic inundation during over-bank discharges. What is not commonly known is that valley flooding is a regular and natural behavior of the stream. Many people learn about this natural phenomenon the hard way—that is, whenever their farms, towns, streets, and homes become inundated by a river or stream that is doing nothing more than following its "natural" periodic cycle—conforming to the Master Plan designed by the Master Planner: Mother Nature.

Did You Know?

Floodplain rivers are found where regular floods form lateral plains outside the normal channel, which seasonally become inundated, either as a consequence of greatly increased rainfall or snowmelt.

Water Flow in a Stream

Most elementary students learn early in their education process that water on Earth flows downhill (gravity)—from land to the sea. However, they may or may not be told that water flows downhill toward the sea by various routes.

For the moment, the "route" (channel, conduit, or pathway) we are concerned with is the surface water route taken by surface runoff. Surface runoff is dependent on various factors. For example, climate, vegetation, topography, geology, soil characteristics, and land use determine how much surface runoff occurs compared with other pathways.

The primary source (input) of water to total surface runoff, of course, is precipitation. This is the case even though a substantial portion of all precipitation input returns directly to the atmosphere by evapotranspiration. Evapotranspiration is a combination process, as the name suggests, whereby water in plant tissue and in the soil evaporates and transpires to water vapor in the atmosphere.

Probably the easiest way to understand precipitation's input to surface water runoff is to take a closer look at this precipitation input.

Again, a substantial portion of precipitation input returns directly to the atmosphere by evapotranspiration. It is also important to point out that, when precipitation occurs, some rainwater is intercepted, blocked, or caught by vegetation where it evaporates, never reaching the ground or being absorbed by plants. A large portion of the rainwater that reaches the surface on ground, in lakes, and in streams also evaporates directly back to the atmosphere. Although plants display a special adaptation to minimize transpiration, plants still lose water to the atmosphere during the exchange of gases necessary for photosynthesis. Notwithstanding the large percentage of precipitation that evaporates, rain- or meltwater that reaches the ground surface follows several pathways in reaching a stream channel or groundwater.

Soil can absorb rainfall to its infiltration capacity (i.e., to its maximum intake rate). During a rain event, this capacity decreases. Any rainfall in excess of infiltration capacity accumulates on the surface. When this surface water exceeds the depression storage capacity of the surface, it moves as an irregular sheet of overland flow. In arid areas, overland flow is likely because of the low permeability of the soil. Overland flow is also likely when the surface is frozen or when human activities have rendered the land surface less permeable. In humid areas, where infiltration capacities are high, overland flow is rare.

In rain events where the infiltration capacity of the soil is not exceeded, rain penetrates the soil and eventually reaches the groundwater—from which it discharges to the stream slowly and over a long period of time. This phenomenon helps to explain why streamflow through a dry weather region remains constant; the flow is continuously augmented by groundwater. This type of stream is known as a perennial stream, as opposed to an intermittent one, because the flow continues during periods of no rainfall.

Streams that course their way in channels through humid regions are fed water via the water table, which slopes toward the stream channel. Discharge from the water table into the stream accounts for flow during periods without precipitation and also explains why this flow increases, even without tributary input, as one proceeds downstream. Such streams are called gaining or effluent, opposed to losing or influent streams that lose water into the ground. It is interesting to note that the same stream can shift between gaining and losing conditions along its course because of changes in underlying strata and local climate.

Stream Water Discharge

The current velocity (speed) of water (driven by gravitational energy) in a channel varies considerably within a stream's cross-section owing to friction with the bottom and sides, with sediment, with obstructions (rocks, logs, etc.) and the atmosphere, and to sinuosity (bending or curving). Highest velocities, obviously, are found where friction is least, generally at or near the surface and near the center of the channel. In deeper streams, current velocity is greatest just below the surface due to the friction with the atmosphere; in shallower streams, current velocity is greatest at the surface due to friction with the bed. Velocity decreases as a function of depth, approaching zero at the substrate surface. A general and convenient rule of thumb is that the deepest part of the channel occurs where the stream velocity is the highest. Additionally, both width and depth of a stream increase downstream because discharge (the amount of water passing any point in a given time) increases downstream. As discharge increases, the cross-sectional shape will change, with the stream becoming deeper and wider. Velocity is important to discharge because discharge (m^3/sec) = cross-sectional area (width × average depth; m^2) × average velocity (m/sec).

$$Q = A \times V.$$

A stream is constantly seeking balance. This can be seen whenever the amount of water in a stream increases; the stream must adjust its velocity and cross-sectional area to reach balance. Discharge increases as more water is added through precipitation, tributary streams, or from groundwater seeping into the stream. As discharge increases, generally width, depth, and velocity of the stream also increase.

Transport of Material (Load)

Water flowing in a channel may exhibit laminar flow (parallel layers of water shear over one another vertically), or turbulent flow (complex mixing; see figure 4.1). In streams, laminar flow is uncommon, except at boundaries where flow is very low and in groundwater. Thus, the flow in streams generally is turbulent. Turbulence exerts a shearing force that causes particles to move along the streambed by pushing, rolling, and skipping, referred to as bed load. This same shear causes turbulent eddies that

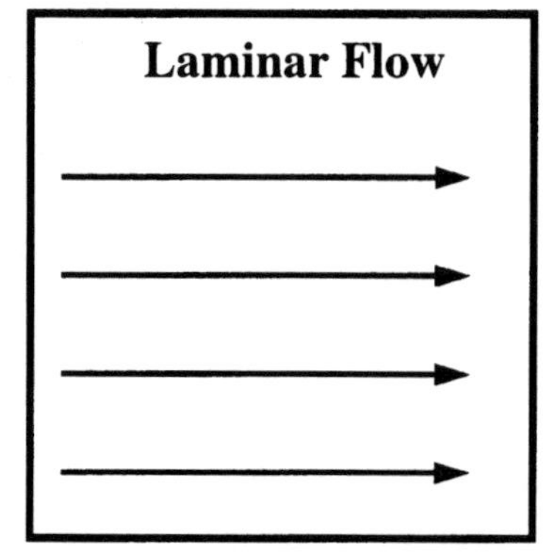

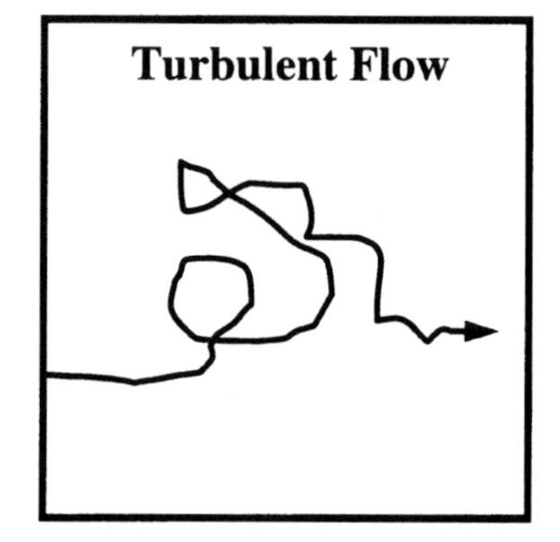

Figure 4.1. Laminar and turbulent flow.

entrain particles in suspension (called the suspended load—particles size under 0.06 millimeters). Entrainment is the incorporation of particles when stream velocity exceeds the entraining velocity for a particular particle size.

The entrained particles in suspension (suspended load) also include fine sediment, primarily clays, silts, and fine sands that require only low velocities and minor turbulence to remain in suspension. These are referred to as wash load (under 0.002 millimeters) because this load is "washed" into the stream from banks and upland areas (Gordon, McMahon, and Finlayson 1992; Spellman 1996).

Did You Know?

Entrainment is a natural extension of erosion and is vital to the movement of stationary particles in changing flow conditions. Remember, all sediments ultimately derive from erosion of basin slopes, but the immediate supply usually derives from the stream channel and banks, while the bed load comes from the streambed itself and is replaced by erosion of bank regions.

Thus, the suspended load includes the wash load and coarser materials (at lower flows). Together, the suspended load and bed load constitute the solid load. It is important to note that in bedrock streams the bed load will be a lower fraction than in alluvial streams where channels are composed of easily transported material.

A substantial amount of material is also transported as the dissolved load. Solutes (ions) are generally derived from chemical weathering of bedrock and soils, and their contribution is greatest in subsurface flows and in regions of limestone geology.

The relative amount of material transported as solute rather than solid load depends on basin characteristics, lithology (i.e., the physical character of rock), and hydrologic pathways. In areas of very high runoff, the contribution of solutes approaches or exceeds sediment load, whereas in dry regions sediments make up as much as 90 percent of the total load.

Deposition occurs when stream competence (i.e., the largest particle that can be moved as bed load and the critical erosion—competent—velocity is the lowest velocity at which a particle resting on the streambed will move) falls below a given velocity. Simply stated, the size of the particle that can be eroded and transported is a function of current velocity.

Sand particles are the most easily eroded. The greater the mass of larger particles (e.g., coarse gravel), the higher the initial current velocities must be for movement. However, smaller particles (silts and clays) require even greater initial velocities because of their cohesiveness and because they present smaller, streamlined surfaces to the flow. Once in transport, particles will continue in motion at somewhat slower velocities than initially required to initiate movement and will settle at still lower velocities.

Particle movement is determined by size, flow conditions, and mode of entrainment. Particles over 0.02 millimeters (medium-coarse sand size) tend to move by rolling or sliding along the channel bed as traction load. When sand particles fall out of the flow, they move by saltation or repeated bouncing. Particles under 0.06 millimeters (silt) move as suspended load and particles under 0.002 (clay), indefinitely, as wash load. A considerable amount of particle sorting takes place because of the different styles of particle flow in different sections of the stream (Richards 1982; Likens 1984).

Unless the supply of sediments becomes depleted the concentration and amount of transported solids increases. However, discharge is usually too low, throughout most of the year, to scrape or scour, shape channels, or move significant quantities of sediment in all but sand-bed streams, which can experience change more rapidly. During extreme events, the greatest scour occurs and the amount of material removed increases dramatically.

Sediment inflow into streams can be both increased and decreased as a result of human activities. For example, poor agricultural practices and deforestation greatly increase erosion.

Man-made structures such as dams and channel diversions can, on the other hand, greatly reduce sediment inflow.

Groundwater

Unbeknownst to most of us, the Earth possesses an unseen ocean of water. This ocean, unlike the surface oceans that cover most of the globe, is freshwater: the groundwater that lies contained in aquifers beneath Earth's crust. This gigantic freshwater source (which makes up about 1 percent of the water on Earth, about thirty-five times the amount of water in lakes and streams) forms a reservoir that feeds all the natural fountains and springs of Earth. But how does water travel into the aquifers that lie under the Earth's surface?

Groundwater sources are replenished from a percentage of the average approximately three feet of water that falls to Earth each year on every square foot of land. Water falling to Earth as precipitation follows three courses. Some runs off directly to rivers and streams (roughly six inches of that three feet), eventually working back to the sea. Evaporation and transpiration through vegetation takes up about two feet. The remaining six inches seeps into the ground, entering and filling every interstice, each hollow and cavity. Gravity pulls water toward the center of the Earth. That means that water on the surface will try to seep into the ground below it. Although groundwater comprises only 1/6 of the total (1,680,000 miles of water), if we could spread out this water over the land, it would blanket it to a depth of one thousand feet.

As mentioned, part of the precipitation that falls on land infiltrates the land surface, percolates downward through the soil under the force of gravity, and becomes groundwater. Groundwater, like surface water, is extremely important to the hydrologic cycle and to our water supplies. Almost half of the people in the United States drink public water from groundwater supplies. Overall, more water exists as groundwater than surface water in the United States, including the water in the Great Lakes. But sometimes, pumping it to the surface is not economical, and in recent years, pollution of groundwater supplies from improper disposal has become a significant problem.

We find groundwater in saturated layers called aquifers under the Earth's surface. Three types of aquifers exist: unconfined, confined, and springs.

Aquifers are made up of a combination of solid material such as rock and gravel and open spaces called pores. Regardless of the type of aquifer, the groundwater in the aquifer is in a constant state of motion. This motion is caused by gravity or by pumping.

The actual amount of water in an aquifer depends upon the amount of space available between the various grains of material that make up the aquifer. The amount of space available is called porosity. The ease of movement through an aquifer is dependent upon how well the pores are connected. For example, clay can hold a lot of water and has high porosity, but the pores are not connected, so water moves through the clay with difficulty. The ability of an aquifer to allow water to infiltrate is called permeability.

The aquifer that lies just under the Earth's surface is called the zone of saturation, an unconfined aquifer (see figure 4.2). The top of the zone of saturation is the water table. An unconfined aquifer is only contained on the bottom and is dependent on

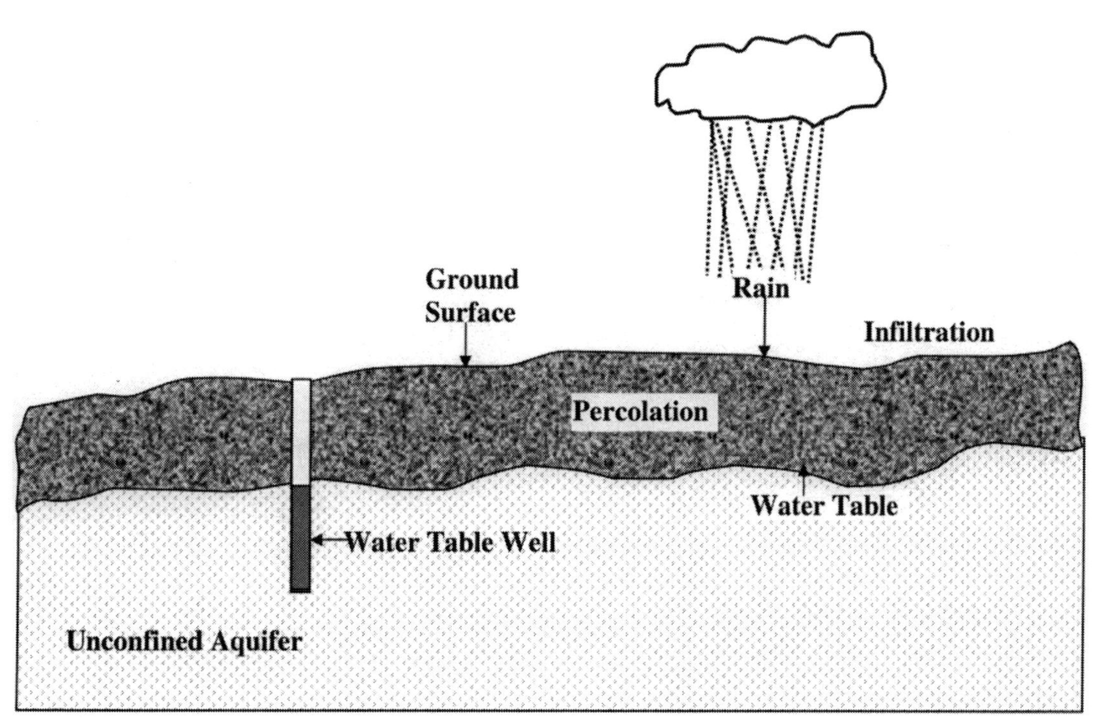

Figure 4.2. Unconfined aquifer. Source: Spellman (2009).

local precipitation for recharge. This type of aquifer is often called a water table aquifer.

Unconfined aquifers are a primary source of shallow well water (see figure 4.2). These wells are shallow (and not desirable as a public drinking water source). They are subject to local contamination from hazardous and toxic materials—fuel and oil, and septic tanks and agricultural runoff providing increased levels of nitrates and microorganisms. These wells may be classified as groundwater under the direct influence of surface water (GUDISW) and, therefore, require treatment for control of microorganisms.

A confined aquifer is sandwiched between two impermeable layers that block the flow of water. The water in a confined aquifer is under hydrostatic pressure. It does not have a free water table (see figure 4.3).

Confined aquifers are called artesian aquifers. Wells drilled into artesian aquifers are called artesian wells and commonly yield large quantities of high-quality water. An

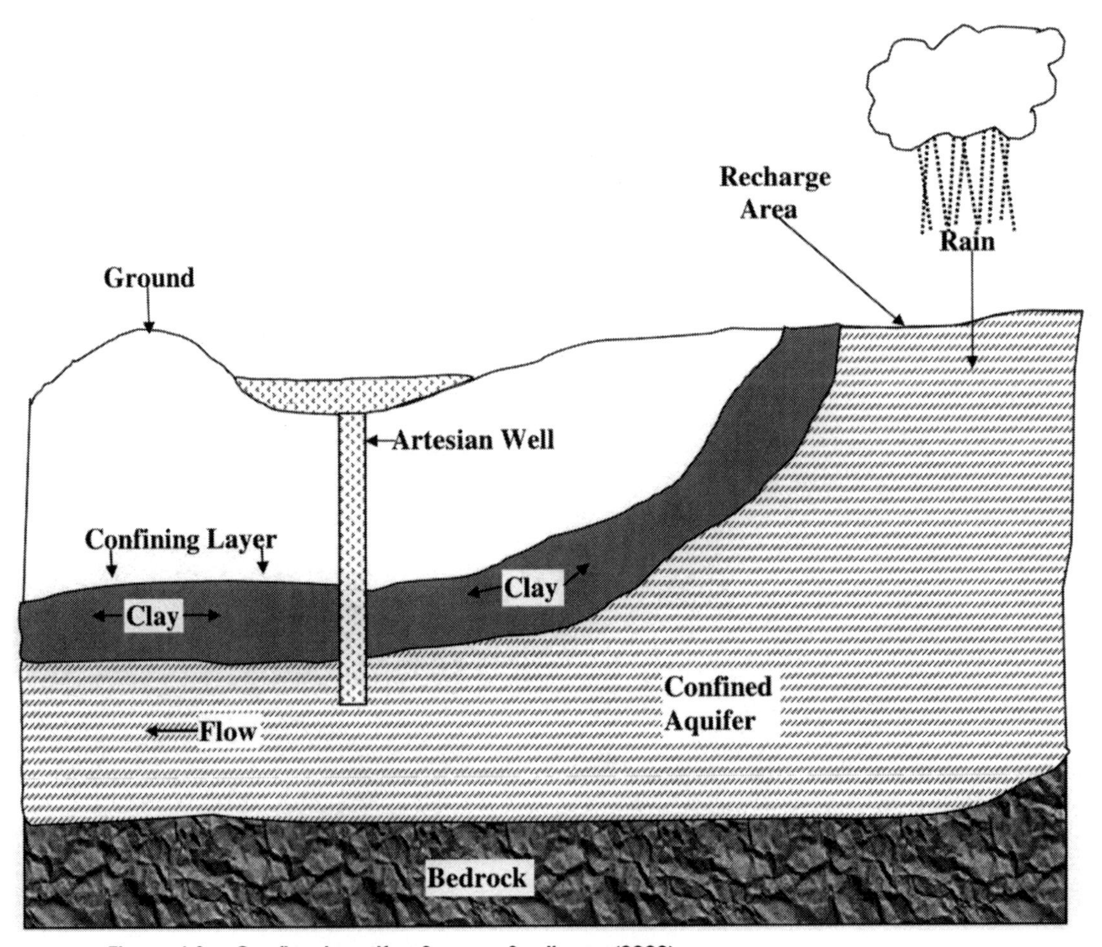

Figure 4.3. Confined aquifer. Source: Spellman (2009).

artesian well is any well where the water in the well casing would rise above the saturated strata. Wells in confined aquifers are normally referred to as deep wells and are not generally affected by local hydrological events.

A confined aquifer is recharged by rain or snow in the mountains where the aquifer lies close to the surface of the Earth. Because the recharge area is some distance from areas of possible contamination, the possibility of contamination is usually very low. However, once contaminated, confined aquifers may take centuries to recover.

Groundwater naturally exits the Earth's crust in areas called springs. The water in a spring can originate from a water table aquifer or from a confined aquifer. Only water from a confined spring is considered desirable for a public water system.

Almost all groundwater is in constant motion through the pores and crevices of the aquifer in which it occurs. The water table is rarely level; it generally follows the shape of the ground surface. Groundwater flows in the downhill direction of the sloping water table. The water table sometimes intersects low points of the ground, where it seeps out into springs, lakes, or streams.

Usual groundwater sources include wells and springs that are not influenced by surface water or local hydrologic events.

As a potable water source, groundwater has several advantages over surface water. Unlike surface water, groundwater is not easily contaminated. Groundwater sources are usually lower in bacteriological contamination than surface waters. Groundwater quality and quantity usually remains stable throughout the year. In the United States, groundwater is available in most locations.

As a potable water source, groundwater does present some disadvantages compared to surface water sources. Operating costs are usually higher because groundwater supplies must be pumped to the surface. Any contamination is often hidden from view. Removing any contaminants is very difficult. Groundwater often possesses high mineral levels and, thus, an increased level of hardness because it is in contact longer with minerals. Near coastal areas, groundwater sources may be subject to saltwater intrusion.

GEOLOGIC ACTIVITY OF GROUNDWATER

Groundwater contributes to geologic activity in a number of ways, including the following:

- *Dissolution*—because water is the main agent of chemical weathering (nothing is safe from water), groundwater is also an active weathering agent that in part or in total (limestone) leaches ions from rock.
- *Chemical cementation and replacement*—water not only carries ions away from rock but also brings chemical agents into masses of rock or rock structures. When some of the water-aided transported chemicals enter rocks or rocky masses, many act as cement and bind sedimentary rocks together. In a similar water-aided transport manner, water transports replacement molecules that work to fossilize organic substances or petrify wood.

- *Caves and caverns*—when large areas of limestone underground are dissolved by the action of groundwater, these cavities can become caves or caverns (caves with many interconnected chambers) once the water table is lowered. Once formed, a cave is open to the atmosphere and water percolating in can precipitate new material such as the common cave decorations stalactites (grow and hang from the ceiling) and stalagmites (grow from the floor upward).
- *Sinkholes*—these are common in areas underlain by limestone, carbonate rock, salt beds, or rocks that can naturally be dissolved by groundwater circulating through them (USGS 2006).
- *Karst topography*—in limestone terrains where dissolution is the main type of weathering, groundwater may work to form caves and sinkholes, and their collapse and coalescence may result in this highly irregular topography.

Chapter Review Questions

4.1. Clearly defined stream path: __________.
4.2. __________ is the process whereby plants lose water to the atmosphere during the exchange of gases necessary for photosynthesis.
4.3. __________ is the maximum rate soil can absorb rainfall.
4.4. __________ is the maximum water of channel depth in a stream.
4.5. __________ describes stream braiding.
4.6. __________ flow follows a predictable pattern and causes regular regions of erosion and deposition.
4.7. A stream is constantly seeking __________.
4.8. Complex mixing in a stream is caused by __________ flow.
4.9. Suspended stream load includes __________ and __________.
4.10. Groundwater is found in saturated layers called __________.

Note

1. Much of the material in this chapter is from F. R. Spellman, *Geology for Nongeologists* (Lanham, Md.: Government Institutes, 2009).

References and Recommended Reading

Giller, P. S., and B. Jalmqvist. 1998. *The biology of streams and rivers*. Oxford: Oxford University Press.

Gordon, N. D., T. A. McMahon, and B. L. Finlayson. 1992. *Stream hydrology: An introduction for ecologists*. Chichester, UK: Wiley.

Leopold, L. B. 1994. *A view of the river*. Cambridge, Mass.: Harvard University Press.

Likens, W. M. 1984. Beyond the shoreline: A watershed ecosystem approach. *Vert. Int. Ver. Theor. Aug. Liminol.* 22:1–22.
Morisawa, M. 1968. *Streams: Their dynamics and morphology.* New York: McGraw-Hill.
Richards, K. 1982. *Rivers: Form and processes in alluvial channels.* London: Methuen.
Spellman, F. R. 1996. *Stream ecology and self-purification.* Lancaster, Pa.: Technomic Publishing.
———. 2008. *The science of water,* 2nd ed. Boca Raton, Fla.: CRC Press.
———. 2009. *Geology for Nongeologists.* (Lanham, Md.: Government Institutes, 2009).
Spellman, F. R., and N. Whiting. 1998. *Environmental science and technology.* Lanham, Md.: Government Institutes.
U.S. Geological Survey (USGS). 2006. Sinkholes. http://ga.water.usgs.gov/edu/earthgwsinkholes.html (accessed July 6, 2008).

Matanuska Glacier, Alaska. Photos by JoAnn Garnett-Chapman.

CHAPTER 5

Glacial Landforms

Here is a problem, a wonder for all to see.
Look at this marvelous thing I hold in my hand!
This is a magic surprising, a mystery
Strange as a miracle, harder to understand.

What is it? Only a handful of dust: to your touch.
A dry, rough powder you trample beneath your feet,
Dark and lifeless; but think for a moment, how much
it hides and holds that is beautiful, bitter or sweet.

Think of the glory of color! The red of the rose,
Green of the myriad leaves and the fields of grass,
Yellow as bright as the sun where the daffodil blows,
Purple where violets nod as the breezes pass.

Strange that this lifeless thing gives vine, flower, tree.
Color and shape and character, fragrance too;
That the timber that builds the house, the ship for the sea,
Out of this powder its strength and its toughness drew!
—Celia Thaxter (1835–1894)

There was a time, geologically not that long ago—about ten to twelve thousand years ago—when many parts of the Earth were covered with massive sheets of ice.[1] Moreover, the geologic record shows that this most recent ice-sheet covering of large portions of Earth's surface was not a onetime phenomena; instead, Earth has experienced several glaciation periods as well as interglacials like the one we are presently experiencing. Although the ice that our early ancestors experienced for the majority of their lives has now retreated from most of Europe, Asia, and North America, it has left traces of its influence across the whole face of the landscape in jagged mountain peaks, gouged-out upland valleys, swamps, changed river courses, and boulder-strewn, table-flat prairies in the lowlands.

Ice covers about 10 percent of all land and ~12 percent of the oceans. Most of this ice is contained in the polar ice sea, polar sheets and ice caps, valley glaciers, and piedmont glaciers formed by valley glaciers merging on a plain. In the geology of the present time, the glaciers of today are not that significant in the grand scheme of things. However, it is the glaciation of the past with its accompanying geologic evidence left behind by ancient glaciers that is important. This geologic record indicates that the Earth's climate has undergone fluctuations in the past, and that the amount of the Earth's surface covered by glaciers has been much larger in the past than in the present. In regard to the effects of past glaciation, one need only look at the topography of the western mountain ranges in the northern part of North America to view the significant depositional processes of glaciers.

Glaciers

A *glacier* is a thick mass of slow-moving ice, consisting largely of recrystallized snow that shows evidence of downslope or outward movement due to the pull of gravity. Glaciers can only form at latitudes or elevations above the snowline (the elevation at which snow forms and remains present year round). Glaciers form in these areas if the snow becomes compacted, forcing out the air between the snowflakes. The weight of the overlying snow causes the snow to recrystallize and increase its grain size, until it increases its density and becomes a solid block of ice.

TYPES OF GLACIERS

There are various types of glaciers including the following:

- *Mountain glaciers*—these are relatively small glaciers that occur at higher elevations in mountainous regions. A good example of mountain glaciers can be seen in the remaining glaciers of Alaska and Glacier National Park, Montana (see p. 74 and figures 5.1, and 5.2). Note that the low snow/ice content of the cirque glaciers shown in figures 5.1 and 5.2 is due to Earth's recent warming trend and time of the year when photos were taken (July 2008) warm weather patterns. As cirque glaciers, like the ones shown in figures 5.1 and 5.2, grow larger, they may spread into valleys and flow down the valleys as *valley glaciers*. Valley glaciers are tongues of ice in which snow and ice accumulate and spill down a valley filling it with ice, perhaps for scores of miles. When a valley glacier extends down to sea level, it may carve a narrow valley into the coastline. These are called *fjord glaciers*, and the narrow valleys they carve and that later become filled with seawater after the ice has melted are *fjords*. When a valley glacier extends down a valley and then covers a gentle slope beyond the mountain range, it is called a *piedmont glacier*. If valley glaciers cover a mountain range, they are called *ice caps*.
- *Continental glaciers (ice sheets)*—these are the largest glaciers. They cover Greenland and Antarctica and contain about 95 percent of all glacial ice on Earth.
- *Ice shelves*—these are sheets of ice floating on water and attached to land. They may extend hundreds of miles from land and reach thicknesses of several thousand feet.
- *Polar glaciers*—these are always below the melting point at the surface and do not produce any meltwater.
- *Temperate glaciers*—these are at a temperature and pressure level near the melting

Figure 5.1. Almost empty cirque in left background. Glacier National Park, Montana. Photo by Frank R. Spellman.

point throughout the body except for a few feet of ice. This layer is subjected to annual temperature fluctuations.

Glacier Characteristics

The primary characteristics displayed by glaciers are changes in size and movement. A glacier changes in size by the addition of snowfall, compaction, and recrystallization. This process is known as *accumulation*. Glaciers also shrink in size (due to temperature increases). This process is known as *ablation*.

Earth's gravity, pushing, pulling, and tugging almost everything toward Earth's surface, is involved with the movement of glaciers. Gravity moves glaciers to lower elevations by two different processes:

- *Basal sliding*—this type of glacier movement occurs when a film of water at the base of the glacier reduces friction by lubricating the surface and allowing the whole glacier to slide across its underlying bed.
- *Internal flow*—called creep, forms fold structures and results from deformation of the ice crystal structure; the crystals slide over each other like a deck of cards. This type of flow is conducive to the formation of crevasses in the upper portions of the

Figure 5.2. Remnants of cirque glacier, left background, near Elizabeth Lake, Belly River region, Glacier National Park, Montana. Photo by Frank R. Spellman.

glacier. Generally, crevasses form when the lower portion of a glacier flows over sudden changes in topography.

Did You Know?

Within a glacier the velocity constantly changes. The velocity is low next to the base of the glacier and where it is in contact with valley walls. The velocity increases toward the center and upper parts of the glacier.

Glaciation

Glaciation is a geological process that modifies land surface by the action of glaciers. For those who study glaciation and glaciers, the fact that glaciations have occurred so recently in North America and Europe accounts for extant evidence, allowing the opportunity to study the undeniable results of glacial erosion and deposition. This is the case, of course, because the forces involved with erosion—weathering, mass wasting, and stream erosion—have not had enough time to remove the traces of glaciation from Earth's surface. Glaciated landscapes are the result of both glacial erosion (glaciers transport rocks and erode surfaces) and glacial deposition (glaciers transport material that melts and deposits material).

GLACIAL EROSION

Glacial erosion has a powerful effect on land that has been buried by ice and has done much to shape our present world. Both valley and continental glaciers acquire tens of thousands of boulders and rock fragments, which, frozen into the sole of the glacier, act like thousands of files, gouging and rasping the rocks (and everything else) over which the glaciers pass. The rock surfaces display fluting, striation, and polishing effects of glacial erosion. The form and direction of these grooves can be used to show the direction in which the glaciers move.

Glacial erosion manifests itself in small-scale erosional features, landform production by mountain glaciers and by ice caps and ice sheets. These are described in the following.

- *Small-scale erosional features*—these include glacial striations and polish. *Glacial striations* are long, parallel scratches and glacial grooves that are produced at the bottom of temperate glaciers by rocks embedded in the ice scraping against the rock underlying the glacier. *Glacial polish* is characteristic of rock that has a smooth surface produced as a result of fine-grained material embedded in the glacier acting like sandpaper on the underlying surface.
- *Landforms produced by mountain glaciers*—these erosion-produced features include the following:
  - *Cirques*—these are bowl-shaped valleys formed at the heads of glaciers and below arêtes and horned mountains; often contain a small lake called a *tarn.*
  - *Glacial valleys*—these are valleys that once contained glacial ice and become eroded into a U-shape in cross-section. V-shaped valleys are the result of stream erosion.
  - *Arêtes*—these are sharp ridges formed by headward glacial erosion.
  - *Horns*—these are sharp, pyramidal mountain peaks formed when headward erosion of several glaciers intersect.
  - *Hanging valleys*—Yosemite's Bridalveil Falls is a waterfall that plunges over a hanging valley. Generally, hanging valleys result in tributary streams that are not able to erode to the base level of the main stream; therefore, the tributary stream is left at a higher elevation than the main stream, creating a hanging valley and sometimes spectacular waterfalls.
  - *Fjords*—these are submerged, glacially deepened, narrow inlets with sheer, high sides, a U-shaped cross-profile, and a submerged seaward sill largely formed of end moraine.
- *Landforms produced by ice caps and ice sheets*
  - *Abrasional features*—these are small-scale abrasional features in the form of glacial

polish and striations that occur in temperate environments beneath ice caps and ice sheets.
- *Streamlined forms*—sometimes called "basket of eggs" topography, the land beneath a moving continental ice sheet is molded into smooth, elongated forms called *drumlins*. Drumlins are aligned in the direction of ice flow, their steeper, blunter ends point toward the direction from which the ice came.

GLACIAL DEPOSITS

All sediment deposited as a result of glacial erosion is called *glacial drift*. The sediment-deposited, glacial drift consists of rock fragments that are carried by the glacier on its surface, within the ice, and at its base.

- *Ice land deposits*
  - *Till (or rock flour)*—this is nonsorted glacial drift deposited directly from ice. Consisting of a random mixture of different-sized fragments of angular rocks in a matrix of fine-grained, sand- to clay-sized fragments, till was produced by abrasion within the glacier. After undergoing diagenesis and turning to rock, till is called *tillite*.
  - *Erratics*—this is a glacially deposited rock, fragment, or boulder that rests on a surface made of different rock. Erratics are often found miles from their source, and by mapping the distribution pattern of erratics, geologists can often determine the flow directions of the ice that carried them to their present locations. No one has described a glacial erratic better than William Wordsworth (1807) in his classic poem "The Leech-Gatherer" (or "Resolution and Independence").

> As a huge stone is sometimes seen to lie
> Couched on the bald top of an eminence;
> Wonder to all who do the same espy,
> By what means it could thither come, and whence;
> So that it seems a thing endued with sense:
> Like a Sea-beast crawled forth, that on a shelf
> Of rock or sand reposeth, there to sun itself.

- *Moraines*—these are mounds, ridges, or ground coverings of unsorted debris, deposited by the melting away of a glacier. Depending on where it formed in relation to the glacier, moraines can be as follows:
  - *Ground moraines*—these are till-covered areas deposited beneath the glacier and result in a hummocky topography with lots of small, enclosed basins.
  - *End moraines and terminal moraines*—these are ridges of unconsolidated debris deposited at the low-elevation end of a glacier as the ice retreats due to ablation (melting). They usually reflect the shape of the glacier's terminus.
  - *Lateral moraines*—these are till deposits that were deposited along the sides of mountain glaciers.

    - *Medial Moraines*—when two valleys' glaciers meet to form a larger glacier, the rock debris along the sides of both glaciers merge to form a medial moraine (runs down the center of a valley floor).
    - *Glacial marine drift (icebergs)*—these are glaciers that reach lakeshores or oceans and calve off into large icebergs that then float on the water surface until they melt. The rock debris that the icebergs contain is deposited on the lakebed or ocean floor when the iceberg melts.
- *Stratified drift*—this is glacial drift that can be picked up and moved by meltwater streams that can then deposit that material as stratified drift.
    - *Outwash plains*—melt runoff at the end of a glacier is usually choked with sediment and form braided streams, which deposit poorly sorted stratified sediment in an outwash plain—they usually are flat, interlocking alluvial fans.
    - *Outwash terraces*—these form if the outwash streams cut down into their outwash deposits, forming river terraces called outwash terraces.
    - *Kettle holes*—these are depressions (sometimes filled by lakes, e.g., Minnesota, the land of a thousand lakes) due to melting of large blocks of stagnant ice, found in any typical glacial deposit.
    - *Kames*—these are isolated hills of stratified material formed from debris that fell into openings in retreating or stagnant ice.
    - *Eskers*—these are long, narrow, and often branching, sinuous ridges of poorly sorted gravel and sand formed by deposition from former glacier streams.

Glacial Ages

The general term *glacial age* or, more commonly, *ice age* denotes a geological period of long-term reduction in the temperature of the Earth's surface and atmosphere, resulting in an expansion of continental ice sheets, polar ice sheets, and alpine glaciers. The point is that when we speak of *ice ages* we are speaking of past occurrences, though by some definitions we are still in an ice age (because the Antarctic and Greenland ice sheets still exist). This text uses the term *ice age* to refer to the past and not to the present.

Before beginning a discussion of the past, it is important to define the era referred to as "the past." Tables 5.1 and 5.2 are provided to assist us in making this definition. Table 5.1 gives the entire expanse of time from Earth's beginning to present. Table 5.2 provides the sequence of geological epochs over the past sixty-five million years, as dated by modern methods. The Paleocene through Pliocene together make up the Tertiary Period; the Pleistocene and the Holocene compose the Quaternary Period.

When we think about the prehistoric past, two things generally come to mind—ice ages and dinosaurs. Of course, in the immense span of time prehistory covers, those two eras represent only a brief moment in time, so let's look at what we know about the past and about Earth's climate and conditions. One thing to consider—geological history shows us that the normal climate of the Earth was so warm that subtropical weather reached to 60° N and S, and polar ice was entirely absent.

Only during less than about 1 percent of Earth's history did glaciers advance and

Table 5.1. Geologic Eras and Periods

Era	*Period*	*Millions of Years before Present*
Cenozoic	Quaternary	2.5–present
	Tertiary	65–2.5
Mesozoic	Cretaceous	135–65
	Jurassic	190–135
	Triassic	225–190
Paleozoic	Permian	280–225
	Pennsylvanian	320–280
	Mississippian	345–320
	Devonian	400–345
	Silurian	440–400
	Ordovician	500–440
	Cambrian	570–500
Precambrian		4,600–570

reach as far south as what is now the temperate zone of the northern hemisphere. The latest such advance, which began about one million years ago, was marked by geological upheaval and (perhaps) the advent of human life on Earth. During this time, vast ice sheets advanced and retreated, grinding their way over the continents.

A TIME OF ICE

Nearly two billion years ago, the oldest known glacial epoch occurred. A series of deposits of glacial origin in southern Canada, extending east to west about one thousand miles, shows us that, within the last billion years or so, apparently at least six major phases of massive, significant climatic cooling and consequent glaciation occurred at intervals of about 150 million years. Each lasted perhaps as long as fifty million years.

Examination of land and oceanic sediment core samples clearly indicate that, in more recent times (the Pleistocene epoch to the present), many alternating episodes of warmer and colder conditions occurred over the last two million years (during the

Table 5.2. Sequence of Geological Epochs

Epoch	*Million Years Ago*
Holocene	.01–0
Pleistocene	1.6–.01
Pliocene	5–1.6
Miocene	24–5
Oligocene	35–24
Eocene	58–35
Paleocene	65–58

middle and early Pleistocene epochs). In the last million years, at least eight such cycles have occurred, with the warm part of the cycle lasting a relatively short interval.

During the Great Ice Age (the Pleistocene epoch), ice advances began, a series of them that at times covering over one-quarter of the Earth's land surface. Great sheets of ice thousands of feet thick, glaciers moved across North America over and over, reaching as far south as the Great Lakes. An ice sheet thousands of feet thick spread over northern Europe, sculpting the land and leaving behind lakes, swamps, and terminal moraines as far south as Switzerland. Each succeeding glacial advance was apparently more severe than the previous one. Evidence indicates that the most severe began about fifty thousand years ago and ended about ten thousand years ago. Several interglacial stages separated the glacial advances, melting the ice. Average temperatures were higher than ours today.

"Temperatures were higher than today?" Yes. Think about that as we proceed.

Because one-tenth of the globe's surface is still covered by glacial ice, scientists consider the Earth still to be in a glacial stage. The ice sheet has been retreating since the climax of the last glacial advance, and world climates, although fluctuating, are slowly warming.

From observations and from well-kept records, we know that the ice sheet is in a retreating stage. The records clearly show that a marked worldwide retreat of ice has occurred over the last hundred years. World famous for its fifty glaciers and two hundred lakes, Glacier National Park in Montana does not present the same visual experiences it did a hundred years ago. In 1937, a ten-foot pole put into place at the terminal edge of one of the main glaciers holds a "1939" sign. The sign is still in place—but the terminal end of the glacier has retreated several hundred feet back up the slope of the mountain. Swiss resorts built during the early 1900s to offer scenic glacial views now have no ice in sight. Theoretically, if glacial retreat continues, melting all of the world's ice supply, sea levels would rise more than two hundred feet, flooding many of the world's major cities. New York and Boston would become aquariums.

The question of what causes ice ages is one scientists still grapple with. Theories range from changing ocean currents to sunspot cycles. On one fact we are absolutely certain, however. An ice age event occurs because of a change in the Earth's climate.

But what could cause such a drastic change?

Climate results from uneven heat distribution over the Earth's surface. It's caused by the Earth's tilt—the angle between the Earth's orbital plane around the Sun and its rotational axis. This angle is currently 23.5 degrees.

The angle has changed. It has not always been 23.5 degrees. The angle, of course, affects the amount of solar energy that reaches the Earth and where it falls.

The heat balance of the Earth, which is driven mostly by the concentration of carbon dioxide (CO_2) in the atmosphere, also affects long-term climate.

If the pattern of solar radiation changes or if the amount of CO_2 changes, climate change can result. Abundant evidence that the Earth does undergo climatic change exists, and we know that climatic change can be a limiting factor for the evolution of many species.

Evidence (primarily from soil core samples and topographical formations) tells us that change in climate includes events such as periodic ice ages characterized by glacial and interglacial periods. Long, glacial periods lasted up to one hundred thousand years, where temperatures decreased about 9° F and ice covered most of the planet. Short periods lasted up to twelve thousand years, with temperatures decreasing by 5° F, and ice covering 40° latitude and above. Smaller periods (e.g., the "Little Ice Age," which occurred about 1000–1850 AD) had about a 3° F drop in temperature. Note that, despite its name, "Little Ice Age" was a time of severe winters and violent storms, not a true glacial period.

These ages may or may not be significant—but consider that we are presently in an interglacial period and that we may be reaching its apogee. What does that mean? No one knows with any certainty—the "we do not know what we do not know" syndrome. Let's look at the effects of ice ages—the effects we think we know about.

Changes in sea levels could occur. Sea level could drop by about one hundred meters in a full-blown ice age, exposing the continental shelves. Increased deposition during melt would change the composition of the exposed continental shelves. Less evaporation would change the hydrological cycle. Significant landscape changes could occur—on the scale of the Great Lakes formation. Drainage patterns throughout most of the world and topsoil characteristics would change. Flooding on a massive scale could occur.

How would these changes affect us? That depends—on whether you live in northern Europe, Canada, Seattle, Washington, around the Great Lakes, or near a seashore.

We are not sure what causes ice ages, but we have some theories (don't people always have theories?). To generate a full-blown (massive ice sheet covering most of the globe) ice age, scientists point out that certain periodic or cyclic events or happenings must occur. Periodic fluctuations would have to affect the solar cycle, for instance. However, we have no definitive proof that this has ever occurred.

Another theory speculates that periods of volcanic activity could generate masses of volcanic dust that would block or filter heat from the Sun. This would cool down the Earth. Some speculate that the carbon dioxide cycle would have to be periodic/cyclic to bring about periods of climate change. There is reference to a so-called factor 2 reduction, causing a 7° F temperature drop worldwide. Others speculate that another global ice age could be brought about by increased precipitation at the poles due to changing orientation of continental landmasses. Others theorize that a global ice age would result if changes in the mean temperatures of ocean currents occurred. But the question is how? By what mechanism?

Are these plausible theories? No one is sure—this is speculation only.

Speculation aside, what are the most probable causes of ice ages on Earth? According to the Milankovitch hypothesis, ice age occurrences are governed by a combination of factors: (1) the Earth's change of altitude in relation to the Sun (the way it tilts in a forty-one-thousand-year cycle and at the same time wobbles on its axis in a twenty-two-thousand-year cycle) making the time of its closest approach to the Sun come at different seasons; and (2) the ninety-two-thousand-year cycle of eccentricity in its orbit round the Sun, changing it from an elliptical to a near circular orbit, the severest period of an ice age coinciding with the approach to circularity.

So what does all this mean? We have a lot of speculation about ice ages and their causes and effects. This is the bottom line: we know that ice ages occurred—we know that they caused certain things to occur (formation of the Great Lakes, etc.), and while there is a lot we do not know, we recognize the possibility of recurrent ice ages.

Lots of possibilities exist. Right now, no single theory is sound, and doubtless many factors are involved. Keep in mind that the possibility does exist that we are still in the Pleistocene Ice Age. It may reach another maximum in another sixty thousand plus years or so.

WARM WINTER

Earlier, when we discussed possible causes of glaciation and subsequent climatic cooling, we were left hanging without adequate explanation. This is what I call the real inconvenient truth: again, we simply don't know what we don't know. In this section, we discuss how we know what we think we do know about climatic change.

The headlines we see in the paper sound authoritative: "1997 Was the Warmest Year on Record"; "Scientists Discover Ozone Hole Is Larger than Ever"; and "Record Quantities of Carbon Dioxide Detected in Atmosphere." Or maybe you saw the one that read "January 1998 Was the Third Warmest January on Record." Other reports indicate we are undergoing a warming trend. But conflicting reports abound. What do we know about climate change?

Two environmentally significant events took place late in 1997: El Niño's return and the Kyoto Conference: Summit on Global Warming and Climate Change. News reports blamed El Niño—the global coupled ocean-atmosphere phenomenon—for just about anything and everything that had to do with weather conditions throughout the world. Some occurrences were indeed El Niño related or generated: the out-of-control fires, droughts, floods, and stretches of dead coral; no sign of fish in the water; and few birds around certain Pacific atolls. The devastating storms that struck the west coasts of South America, Mexico, and California were also probably El Niño related. El Niño's effect on the 1997 hurricane season, one of the mildest on record, is not in question, either.

But does a connection exist between El Niño and global warming or global climate change?

On December 7, 1997, the Associated Press reported that, while delegates at the global climate conference in Kyoto haggled over greenhouse gases and emission limits, a compelling question had emerged: "Is global warming fueling El Niño?" The National Oceanic and Atmospheric Administration (NOAA 2008) stated emphatically that El Niños are not caused by global warming. Clear evidence exists that El Niños have been present for thousands, and some indications suggest maybe millions, of years. One thing seems certain; nobody knows for sure.

Why aren't we sure? We can't be sure because we need more information than we have today. Our amount of recorded data is paltry, lacking at best; our information suggests, however, that El Niño is getting stronger and more frequent.

Some scientists fear that El Niño's increasing frequency and intensity (records show that two of this century's three worst El Niños came in 1982 and 1997) may be linked to global warming. At the Kyoto Conference, experts said the hotter atmosphere is heating up the world's oceans, setting the stage for more frequent and extreme El Niños.

Weather-related phenomena seem to be intensifying throughout the globe. Can we be sure that this is related to global warming yet? No. Without more data—more time—more science, we cannot be sure.

Should we be concerned? Yes. According to the Associated Press (1997) coverage of the Kyoto Conference, scientist Richard Fairbanks reported that he found startling evidence of our need for concern. During two months of scientific experiments on Christmas Island (the world's largest atoll in the Pacific Ocean) conducted in autumn 1997, he discovered a frightening scene. The water surrounding the atoll was 7° F higher than average for the time of year, which upset the balance of the environmental system. According to Fairbanks, 40 percent of the coral was dead, the warmer water had killed off or driven away fish, and the atoll's plentiful bird population was almost completely gone.

El Niños have an acute impact on the globe; that is not in doubt. However, we do not know if it is caused by or intensified because of global warming. What do we know about global warming and climate change?

USA Today (1997) discussed the results of a report issued by the Intergovernmental Panel on Climate Change. They interviewed Jerry Mahlman of NOAA and Princeton University, and presented the following information about what most scientists agree on:

- There is a natural "greenhouse effect" and scientists know how it works . . . and without it, Earth would freeze.
- The Earth undergoes normal cycles or warming and cooling on grand scales. Ice ages occur every twenty to one hundred thousand years.
- Globally, average temperatures have raised 1° F in the past one hundred years, within the range that might occur normally.
- The level of man-made carbon dioxide in the atmosphere has risen 30 percent since the beginning of the Industrial Revolution in the nineteenth century and is still rising.
- Levels of man-made carbon dioxide will double in the atmosphere over the next one hundred years, generating a rise in global average temperatures of about 3.5° F (larger than the natural swings in temperature that have occurred over the past ten thousand years).
- By 2050, temperatures will rise much higher in northern latitudes than the increase in global average temperatures. Substantial amounts of northern sea ice will melt, and snow and rain in the northern hemisphere will increase.
- As the climate warms, the rate of evaporation will rise, further increasing warming. Water vapor also reflects heat back to Earth.

WHAT WE THINK WE KNOW ABOUT GLOBAL WARMING

What is global warming? To answer this question we need to discuss "greenhouse effect." As mentioned, water vapor, carbon dioxide, and other atmospheric gases (greenhouse gases) help warm the Earth. Earth's average temperature would be closer to zero than its actual 60° F without the greenhouse effect. But the average temperature could increase, changing orbital climate, as gases are added to the atmosphere.

How does the greenhouse effect actually work? Let's take a closer look at this phenomenon.

Greenhouse Effect

Earth's greenhouse effect, of course, took its name because of similarity of effect. Because greenhouse glass walls and ceilings are largely transparent to shortwave radiation from the Sun, surfaces and objects inside the greenhouse absorb the radiation. The radiation, once absorbed, transforms into long-wave (infrared) radiation (heat) and radiates back from the greenhouse interior. But the glass prevents the long-wave radiation from escaping again, absorbing the warm rays. The interior of the greenhouse becomes much warmer than the air outside because of the heat trapped inside.

Earth and its atmosphere undergoes a process very similar to this—the greenhouse effect. Shortwave and visible radiation reaching Earth is absorbed by the surface as heat. The long heat waves radiate back out toward space, but the atmosphere absorbs many of them, trapping them. This natural and balanced process is essential to supporting our life-systems on Earth. Changes in the atmosphere can radically change the amount of absorption (and, therefore, the amount of heat) the Earth's atmosphere retains. In recent decades, scientists speculate that various air pollutants have caused the atmosphere to absorb more heat. At the local level, with air pollution, the greenhouse effect causes heat islands in and around urban centers, a widely recognized phenomenon.

The main contributors to this effect are the greenhouse gases: water vapor, carbon dioxide, carbon monoxide, methane, volatile organic compounds (VOCs), nitrogen oxides, chlorofluorocarbons (CFCs), and surface ozone. These gases cause a general climatic warming by delaying the escape of infrared radiation from the Earth into space. Scientists stress this is a natural process—indeed, if the "normal" greenhouse effect did not exist, the Earth would be 33° C cooler than it presently is (Hansen et al. 1986).

Human activities are now rapidly intensifying the natural phenomenon of Earth's greenhouse effect, which may lead to problems of warming on a global scale. Much debate, confusion, and speculation about this potential consequence is underway because scientists cannot yet agree about whether the recently perceived worldwide warming trend is because of greenhouse gases, due to some other cause, or whether it is simply a wider variation in the normal heating and cooling trends they have been studying. Unchecked, the greenhouse effect may lead to significant global warming, with profound effects upon our lives and our environment. Human impact on greenhouse effect is real; it has been measured and detected. The rate at which the greenhouse effect is intensifying is now more than five times what it was during the last century (Hansen et al. 1989).

Supporters of the global warming theory base their assumptions on humans' altering of the Earth's normal and necessary greenhouse effect. The human activities they blame for increases of greenhouse gases include burning of fossil fuels, deforestation, and use of certain aerosols and refrigerants. These gases have increased how much heat remains trapped in the Earth's atmosphere, gradually increasing the temperature of the whole globe.

From information based on recent or short-term observation, many scientists note that the last decade has been the warmest since temperature recordings began in the late nineteenth century. They see that the general rise in temperature over the last century coincides with the Industrial Revolution and its accompanying increase in fossil fuel use. Other evidence supports the global warming theory. In places that are synonymous with ice and snow—the Arctic and Antarctica, for example, we see evidence of receding ice and snow cover.

Trying to pin down definitively whether or not changing our anthropogenic activities could have any significant effect on lessening global warming, though, is difficult. Scientists look at temperature variations over thousands and even millions of years, taking a long-term view at Earth's climate. The variations in Earth's climate are wide enough that they cannot definitively show that global warming is anything more than another short-term variation. Historical records have shown the Earth's temperature does vary widely, growing colder with ice ages and then warming again, and because we cannot be certain of the causes of those climate changes, we cannot be certain of what appears to be the current warming trend.

Still, debate abounds for the argument that our climate is warming and our activities are part of the equation. The 1980s saw nine of the twelve warmest temperatures ever recorded, and the Earth's average surface temperature has risen approximately 0.6° C (1° F) in the last century (Spellman 2009). *Time* magazine (1998) reports that scientists are increasingly convinced that, because of the buildup in the atmosphere of carbon dioxide and other gases produced in large part by the burning of fossil fuels, the Earth is getting hotter. Each month from January through July 1998, for example, set a new average global temperature record, and if that trend continued, the surface temperature of the Earth could rise by about 1.8 to 6.3° F by 2100. At the same time, others offer as evidence that the 1980s also saw three of the coldest years: 1984, 1985, and 1986.

What is really going on? We cannot be certain. Assuming that we are indeed seeing long-term global warming, we must determine what causes it. But again, we face the problem that scientists cannot be sure of the greenhouse effect's precise causes. Our current, possible trend in global warming may simply be part of a much longer trend of warming since the last ice age. We have learned much in the past two centuries of science, but little is actually known about the causes of the worldwide global cooling and warming that sent the Earth through major and smaller ice ages. The data we need reaches back over millennia. We simply do not possess enough long-term data to support our theories.

Currently, scientists can point to six factors they think could be involved in long-term global warming and cooling.

1. Long-term global warming and cooling could result if changes in the Earth's position relative to the Sun occur (i.e., the Earth's orbit around the Sun), with higher temperatures when the two are closer together and lower when further apart.

2. Long-term global warming and cooling could result if major catastrophes (meteor impacts or massive volcanic eruptions), which throw pollutants into the atmosphere that can block out solar radiation, occur.
3. Long-term global warming and cooling could result if changes in albedo (reflectivity of Earth's surface) occur. If the Earth's surface were more reflective, for example, the amount of solar radiation radiated back toward space instead of absorbed would increase, lowering temperatures on Earth.
4. Long-term global warming and cooling could result if the amount of radiation emitted by the Sun changes.
5. Long-term global warming and cooling could result if the shape and relationship of the land and oceans change.
6. Long-term global warming and cooling could result if the composition of the atmosphere changes.

"If the composition of the atmosphere changes"—this final factor, of course, defines our present concern: have human activities had a cumulative impact large enough to affect the total temperature and climate of Earth? Right now, we cannot be sure. The problem concerns us, and we are alert to it, but we are not certain.

The Bottom Line

When news media personnel, self-proclaimed and self-described inventors of the Internet, would-be presidential candidates, and doomsayers in general make their dire warnings about global climate change (specifically that the Earth is getting warmer), I remind my students that the dire predictions and actual consequences could be even worse. But before I explain my point, let's consider two points that should be clear (to those of us who have studied the geologic past and common-sense-based facts): (1) the transition from ice age to interglacial (a warm period) is well documented with considerable actual evidence to back it up; and (2) keep in mind that, when we left the last glacial age some ten thousand years ago, we certainly cannot point the finger of blame for the warming trend that began then on humans and their actions. Simply put, the fact is, at that time, humans made little contribution of carbon dioxide, CFCs, or any other chemical substance to Earth's atmosphere—there simply were no industrial smoke stacks pouring their poison into our atmosphere. Yet, the Earth warmed up anyway.

Now, let's get back to the argument I make to my students that there is a situation worse than global warming—global warming, when less fuel is needed to keep us all warm and increased levels of carbon dioxide create junglelike masses that envelope the Earth and thus liberate megaliters of fresh, clean air to breathe. No sir, that is not the worst, I tell my students (who are all ears, of course, because there is nothing that gets their attention more than doom and gloom, death and destruction, and rock 'n' roll, of course): in my opinion, the more dire consequence of humanity's actions are not global warming but, instead, global cooling—another full-blown ice age. If we assume we are in an interglacial that is on a warming trend—human assisted, of course—the question is, what do we do if we plunge into global cooling, another ice age?

From many points of view, including geographical, plunging into an ice age right now or in the near future could and would be measured in catastrophic proportions. Consider that if the ice sheets start to grow again to the point of historical growth in both the northern and southern hemispheres, there will be many questions to be answered by many experts in many fields including geography. For example, the expert in population geography (a division of human geography that is often grouped with or blurred by demography) would be concerned with the potential effects of expanding ice sheets. With the expansion of ice sheets to the lower latitudes, the population expert would be concerned with demographic phenomena effects on natality, mortality, growth rates, and so forth, through both space and time. Moreover, would global cooling increase or decrease population numbers? What would be the effect on population movement and mobility; on occupational structure; on grouping of people in settlements; on the way geographical character of places would be impacted; and on the way in which places in turn would react to immigration growth? These are important questions, and many probably do not have answers before the fact (proactive); rather, these answers can only be ascertained after the fact (reactive).

Let's look at one possible hypothetical worst-case scenario that could result because of massive global cooling, that is, the advent or advance of another global ice age. At the present time, there are several million people (approximately thirty-four million in 2008) who live north of the U.S.-Canadian border (including Alaska) on the North American continent. If another ice age develops, where are these people to live? They will have to move south—away from the bitter cold and massive sheets of ice, to a warmer and more livable climate, hopefully.

Assuming the mass horde of humans can successfully move away from the advancing ice sheets and bitter cold without enduring or initiating mass hysteria, utter chaos, murder, and destruction, are the new immigrants going to be able to sustain themselves? Will they find employment? Will they find housing? Will there be enough food to share? For those residing in marginal areas, near the advancing ice sheets but also in relatively safe land areas, a more important question might be, how are these people to keep warm? Remember, at the present time, we are suffering from a dwindling supply of energy—high-cost energy; this situation will only worsen.

It is interesting to note that, after a discussion such as the one presented above, and given time to mull over the unemotional facts concerning global warming versus global cooling and then deciding if global warming is better than global cooling, or vice versa, not one of the hundreds of students I have had over the past twenty years opted for global cooling.

Chapter Review Questions

5.1. ________ glaciers are relatively small glaciers that occur at higher elevations in mountainous regions.

5.2. ________ glaciers extend down to sea level where they may carve a narrow valley into the coastline.

5.3. ________ type of glacier movement occurs when a film of water at the base

of the glacier reduces friction by lubricating the surface and allowing the whole glacier to slide across its underlying bed.
5.4. ________ is a bowl-shaped valley formed at the heads of glaciers.
5.5. ________ is a small cirque lake.
5.6. ________ are sharp ridges formed by headward glacial erosion.
5.7. ________ are streamlined, elongated forms.
5.8. Rock flour: ________.
5.9. Glacially deposited rock, fragment, or boulder that rests on a surface made of different rock: ________.
5.10. ________ are long, narrow, and often branching, sinuous ridges of poorly sorted gravel and sand formed by deposition from former glacier streams.

Note

1. Much of the information in this chapter is condensed from F. R. Spellman, *Geology for Nongeologists* (Lanham, Md.: Government Institutes, 2009).

References and Recommended Reading

Associated Press, *Lancaster New Era* (Lancaster, Pa.). 1998. Ozone hole over Antarctica at record size. September 28.

———. 1998. Tougher air pollution standards too costly, Midwestern states say. September 25.

Associated Press, *Virginian-Pilot* (Norfolk). 1997. Does warming feed El Niño? December 7, A-15.

Chernicoff, S. 1999. *Geology*. Boston: Houghton Mifflin.

Dolan, E. F. 1991. *Our poisoned sky*. New York: Cobblehill Book.

Hansen, J. E., et al. 1986. Climate sensitivity to increasing greenhouse gases. In *Greenhouse effect and sea level rise: A challenge for this generation*, ed. M. C. Barth and J. G. Titus. New York: Van Nostrand Reinhold.

———. 1989. Greenhouse effect of chlorofluorocarbons and other trace gases. *Journal of Geophysical Research* 94 (November): 16,417–21.

National Oceanic and Atmospheric Administration (NOAA). 2008. Global warming: Frequently asked questions. http://lwf.ncdc.noaa.gov/oa/climate/globalwarming.html (accessed November 21, 2008).

Spellman, F. R. 2009. *Geology for nongeologists*. Lanham, Md.: Government Institutes.

Tarbuck, E. J., and F. K. Lutgens. 2000. *Earth science*. Upper Saddle River, N.J.: Prentice Hall.

Time. 1998. Global warming: It's here . . . and almost certain to get worse. August 24.

USA TODAY. 1997. Global warming: Politics and economics further complicate the issue. December 1, A-1, 2.

Lava rocks. Craters of the Moon National Park, Idaho. Photo by Frank R. Spellman.

Lava flow. Craters of the Moon National Park, Idaho. Photo by Frank R. Spellman.

Lava flow. Craters of the Moon National Park, Idaho. Photo by Frank R. Spellman.

Cinder dome. Craters of the Moon National Park, Idaho. Photo by Frank R. Spellman.

CHAPTER 6

Volcanic Landforms and Plate Tectonics

In regards to formation of the Craters of the Moon landforms, Idaho, Shoshone legend speaks of a serpent on a mountain who, angered by lightning, coiled around and squeezed the mountain until liquid rock flowed, fire shot from cracks, and the mountain exploded.

Robert Limbert (1924) had the following to say about light playing on cobalt blue lavas of the Blue Dragon Flows (i.e., flows in Craters of the Moon National Monument): "It is the play of light at sunset across this lava that charms the spectator. It becomes a twisted, wavy sheen. With changing conditions of light and air, it varies also, even while one stands and watches. It is a place of color and silence."

Craters of the Moon volcanic area now lies dormant, but its eight eruptive periods formed 60 lava flows which traveled as far as 45 miles from their vents.

—U.S. Geological Survey (USGS 2009)

Readers might want to keep the above Shoshone Legend and Limbert's words in mind while viewing the photographs of Craters of the Moon National Monument shown on pages 92 and 93.

Igneous Rocks and Magma Eruption

Igneous (Latin: *fire*) rocks are those rocks that have solidified from an original molten silicate state.[1] The occurrence and distribution of igneous rocks and igneous rock types can be related to the operation of plate tectonics. The molten rock material from which igneous rocks form is called magma. Magma, characterized by a wide range of chemical compositions and with high temperature, is a mixture of liquid rock, crystals, and gas. Magmas are large bodies of molten rock deeply buried within the Earth. These magmas are less dense than surrounding rocks and will therefore move upward. In the upward movement, sometimes magmatic materials are poured out upon the surface of the Earth as, for example, when lava flows from a volcano. These igneous rocks are volcanic or *extrusive rocks*; they form when the magma cools and crystallizes on the surface of the Earth. Under certain other conditions, magma does not make it to the Earth's surface and cools and crystallizes within the Earth's crust. These intruding rock materials harden and form *intrusive* or *plutonic rocks.*

Magma

Magma is molten silicate material and may include already formed crystals and dissolved gases. The term magma applies to silicate melts within the Earth's crust. When magma reaches the surface it is referred to as lava. The chemical composition of magma is controlled by the abundance of elements in the Earth. These include oxygen, silicon, aluminum, hydrogen, sodium, calcium, iron, potassium, and manganese, which make up 99 percent. Because oxygen is so abundant, chemical analyses are usually given in terms of oxides. Silicon dioxide (SiO_2, also known as silica) is the most abundant oxide. Because magma gas expands as pressure is reduced, magmas have an explosive character. The flow (or viscosity) of magma depends on temperature, composition, and gas content. Higher silicon dioxide content and lower temperature magmas have higher viscosity.

Magma consists of three types: basaltic, andesitic, and rhyolitic. Table 6.1 summarizes the characteristics of each type.

Intrusive Rocks

Intrusive (or plutonic rocks) are rocks that have solidified from molten mineral mixtures beneath the surface of the Earth. Intrusive rocks that are deeply buried tend to cool slowly and develop a coarse texture. On the other hand, those intrusive rocks near the surface that cool more quickly are finer textured. The shape, size, and arrangement of the grains comprising it determine the texture of igneous rocks. Because of crowded conditions under which mineral particles are formed, they are usually angular and irregular in outline. Typical intrusive rocks include

- *gabbro*—a heavy, dark-colored igneous rock consisting of coarse grains of feldspar and augite;
- *peridotite*—a rock in which the dark minerals are predominant;
- *granite*—the most common and best-known of the coarse-textured intrusive rocks; and

Table 6.1. Characteristics of Magma Types

Magma Type	*Solidified Volcanic*	*Solidified Plutonic*	*Chemical Composition*	*Temperature*
Basaltic	Basalt	Gabbro	45–55% silicon dioxide	1000–1200° C
Andesitic	Andesite	Diorite	55–65% silicon dioxide	800–1000° C
Rhyolitic	Rhyolite	Granite	65–75% silicon dioxide	650– 800° C

- *syenite*—resembles granite but is less common in its occurrence and contains little or no quartz.

Extrusive Rocks

Extrusive (or volcanic) rocks are those that pour out of craters of volcanoes or from great fissures or cracks in the Earth's crust and make it to the surface of the Earth in a molten state (liquid lava). Extrusive rocks tend to cool quickly and typically have small crystals (because fast cooling does not allow large crystals to grow). Some cool so rapidly that no crystallization occurs and it produces volcanic glass.

Some of the more common extrusive rocks are felsite, pumice, basalt, and obsidian.

- *Felsite*—very fine-textured igneous rocks.
- *Pumice*—frothy lava that solidifies while steam and other gases bubble out of it.
- *Basalt*—world's most abundant fine-grained extrusive rock.
- *Obsidian*—volcanic glass; cools so fast that there is no formation of separate mineral crystals.

Bowen's Reaction Series

The geologist Norman L. Bowen back in the early 1900s was able to explain why certain types of minerals tend to be found together while others are almost never associated with one another. Bowen found that minerals tend to form in specific sequences in igneous rocks, and these sequences could be assembled into a composite sequence. The idealized progression that he determined is still accepted as the general model (see figure 6.1) for the evolution of magmas during the cooling process.

In order to better understand Bowen's Reaction Series, it is important to define key terms:

- *Magma*—molten igneous rock.
- *Felsic*—white pumice.
- *Pumice*—textured form of volcanic rock; a solidified frothy lava.
- *Extrusion*—magma intruded or emplaced beneath the surface of the Earth.
- *Feldspar*—the family of minerals including microcline, orthoclase, and plagioclase.
- *Mafic*—a mineral containing iron and magnesium.
- *Aphanitic*—mineral grains too small to be seen without a magnifying glass.
- *Phaneritic*—mineral grains large enough to be seen without a magnifying glass.
- *Reaction series*—a series of minerals in which a mineral reacts to change to another mineral.
- *Rock-forming mineral*—the minerals commonly found in rocks. Bowen's Reaction Series lists all of the common ones in igneous rocks.

- *Specific gravity*—the relative mass or weight of a material compared to the mass or weight of an equal volume of water.

Some igneous rocks are named according to textural criteria:

- Scoria—porous
- Pumice
- Obsidian—glass
- Tuff—cemented ash
- Breccia—cemented fragments
- Permatite—extremely large crystals
- Aplite—sugary texture, quartz and feldspar
- Porphyry—fine matrix, large crystals

THE DISCONTINUOUS REACTION SERIES

The left side of figure 6.1 shows a group of mafic or iron-magnesium-bearing minerals—olivine, pyroxene, amphibole, and biotite. If the chemistry of the melt is just right, these minerals react discontinuously to form the next mineral in the series. If there is enough silica in the igneous magma melt, each mineral will change to the next mineral lower in the series as the temperature drops. Descending down Bowen's Reaction Series, the minerals increase in the proportions of silica in their composition. In basaltic melt, as shown in figure 6.1, olivine will be the first mafic mineral (silicate

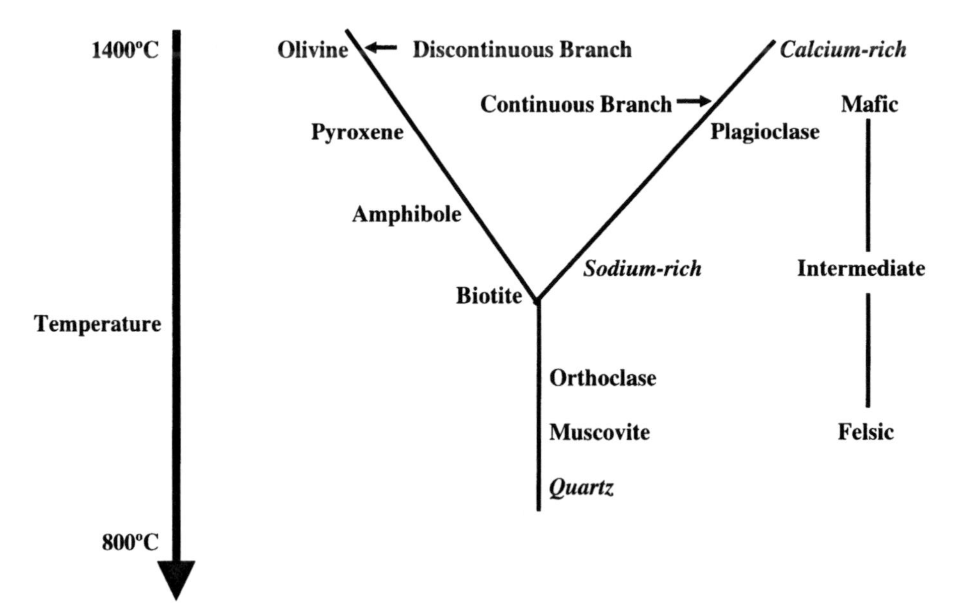

Figure 6.1. Bowen's reaction series.

mineral rich in magnesium and iron) to form. When the temperature is low enough to form pyroxene, all of the olivine will react with the melt to form pyroxene, and pyroxene will crystallize out of the melt. At the crystallization temperature of amphibole, all the pyroxene will react with the melt to form amphibole, and amphibole will crystallize. At the crystallization temperature of biotite, all of the amphibole will react to form biotite, and biotite will crystallize. Thus, all igneous rocks should only have biotite; however, this is not the case. In crystallizing olivine, if there is not enough silica to form pyroxene, then the reaction will not occur and olivine will remain. Additionally, in crystallizing olivine the temperature drops too fast for the reaction to take place (volcanic magma eruption); then the reaction will not have time to occur, the rock will solidify quickly, and the mineral will remain olivine.

THE CONTINUOUS REACTION SERIES

The right side of figure 6.1 shows the plagioclases. Plagioclase minerals have the formula $(Ca, Na)(Al, Si)_3O_8$. The highest temperature plagioclase has only calcium (Ca). The lowest temperature plagioclase has only sodium (Na). In between, these ions mix in a continuous series from 100 percent Ca and 0 percent Na at the highest temperature to 50 percent Ca and 50 percent Na at the middle temperature to 0 percent Ca and 100 percent Na at the lowest temperature. In a basaltic melt, for example, the first plagioclase to form could be 100 percent Ca and 0 percent Na plagioclase. As the temperature drops the crystal reacts with the melt to form 99 percent Ca and 1 percent Na plagioclase, and 99 percent Ca and 1 percent Na plagioclase crystallizes. Then those react to form 98 percent Ca and 2 percent Na and the same composition would crystallize, and so forth. All of this happens continuously provided there is enough time for the reactions to take place and enough sodium, aluminum, and silica in the melt to form each new mineral. The end result will be a rock with plagioclases with the same ratio of Ca to Na as the starting magma.

Key Point!

In regard to the Bowen Reaction Series, on both sides of the reaction series shown in figure 6.1, the silica content of the minerals increases as the crystallization trend heads downward. Biotite has more silica than olivine. Sodium plagioclase has more silica than calcium plagioclase.

Eruption of Magma

The volcanic processes that lead to the deposition of extrusive igneous rocks can be studied in action today and help us to explain the textures of ancient rocks with respect to depositional processes. Some of the major features of volcanic processes and landforms are discussed in this section. The information below is from USGS's (2008) *Principal Types of Volcanoes.*

TYPES OF VOLCANOES

Geologists generally group volcanoes into four main kinds—cinder cones, composite volcanoes, shield volcanoes, and lava domes.

- *Cinder cones*—are the simplest type of volcano. They are built from particles and blobs of congealed lava ejected from a single bent. As the gas-charged lava is blown violently into the air, it breaks into small fragments that solidify and fall as cinders around the vent to form a circular or oval cone. Most cinder cones have a bowl-shaped crater at the summit and rarely rise more than a thousand feet or so above their surroundings. Cinder cones are numerous in western North American as well as throughout other volcanic terrains of the world.
- *Composite volcanoes*—some of Earth's grandest mountains are composite volcanoes—sometimes called stratovolcanoes. They are typically steep-sided, symmetrical cones of large dimension built of alternating layers of lava flows, volcanic ash, cinders, blocks, and bombs and may rise as much as eight thousand feet above their bases. Most composite volcanoes have a crater at the summit that contains a central vent or a clustered group of vents. Lavas either flow through breaks in the crater wall or issue from fissures on the flanks of the cone. Lava, solidified within the fissures, forms dikes that act as ribs, which greatly strengthen the cone. The essential feature of a composite volcano is a conduit system through which magma from a reservoir deep in the Earth's crust rises to the surface. The volcano is built up by the accumulation of material erupted through the conduit and increases in size as lava, cinders, ash, and so on, are added to its slopes.
- *Shield volcanoes*—are built almost entirely of fluid lava flows. Flow after flow pours out in all directions to form a central summit vent, or group of vents, building a broad, gently sloping cone of flat, domical shape, with a profile much like that of a warrior's shield. They are built up slowly by the accretion of thousands of highly fluid lava flows called basalt lava that spread widely over great distances and then cool as thin, gently dipping sheets. Lava also commonly erupts from vents along fractures (rift zones) that develop on the flanks of the cone. Some of the largest volcanoes in the world are shield volcanoes (see figure 6.2).
- *Lava domes*—are formed by relatively small bulbous masses of lava too viscous to flow any great distance; consequently, on extrusion, the lava piles over and around its vent. A dome grows largely by expansion from within. As it grows its outer surface cools and hardens, then shatters, spilling loose fragments down its sides. Some domes form craggy knobs or spines over the volcanic vent, whereas others form short, steep-sided lava flows known as "coulees" (from French, "to flow"). Volcanic domes commonly occur within the craters or on the flanks of large composite volcanoes.

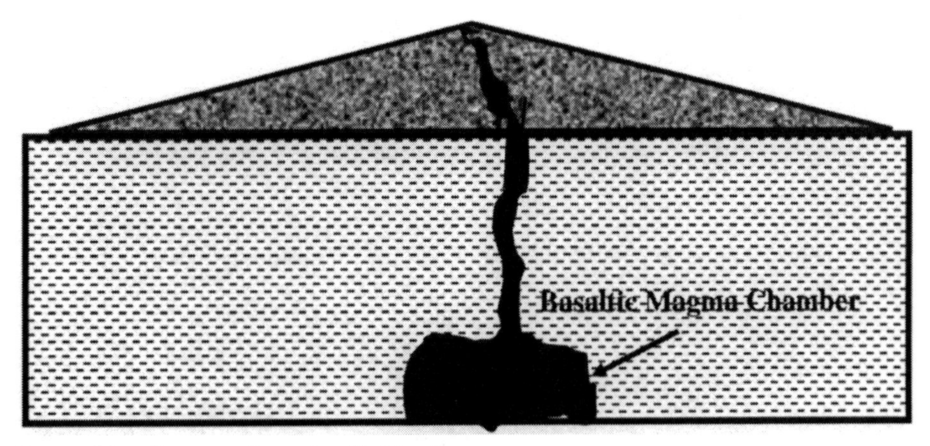

Figure 6.2. Cross section of a shield volcano.

TYPES OF VOLCANIC ERUPTIONS

The type of volcanic eruption is often labeled with the name of a well-known volcano where characteristic behavior is similar—hence the use of such terms as "Strombolian," "Vulcanian," "Vesuvian," "Pelean," "Hawaiian," and others.

- *Strombolian-type eruption*—this type of eruption is in constant action with huge clots of molten lava bursting from the summit crater to form luminous arcs through the sky. Collecting on the flanks of the cone, lava clots combined to steam down the slopes in fiery rivulets.
- *Vulcanian-type eruption*—this type of eruption is characterized by very viscous lavas; a dense cloud of ash-laden gas explodes from the crater and rises high above the peak. Steaming ash forms a whitish cloud near the upper level of the cone.
- *Pelean-type eruption* (or "Nuee Ardente"—glowing cloud)—this type of eruption is characterized by its explosiveness. It erupts from a central crater with violent explosions that eject great quantities of gas, volcanic ash, dust, incandescent lava fragments, and large rock fragments.
- *Hawaiian-type eruption* (or quiet)—is characterized by less viscous lavas that permit the escape of gas with a minimum of explosive violence. In fissure-type eruptions, a fountain of fiery lava erupts to a height of several hundred feet or more. Such lava may collect in old pit craters to form lava lakes, form cones, or feed radiating flows.
- *Vesuvian eruption*—is characterized by great quantities of ash-laden gas that is violently discharged to form a cauliflower-shaped cloud high above the volcano.
- *Phreatic (or steam-blast) eruption*—is driven by explosive, expanding steam resulting from cold ground- or surface water coming into contact with hot rock or magma. The distinguishing feature of phreatic explosions is that they only blast out fragments of preexisting solid rock from the volcanic conduit; no new magma is erupted.
- *Plinian eruption*—is a large explosive event that forms enormous dark columns of

tephra (solid material ejected) and gas high into the stratosphere. Such eruptions are named for Pliny the Younger, who carefully described the disastrous eruption of Vesuvius in 79 AD. This eruption generated a huge column of tephra into the fall. Many thousands of people evacuated areas around the volcano, but about two thousand were killed, including Pliny the Elder.

Types of Lava

Two Hawaiian words, pahoehoe and aa, are used to describe how lava flows. *Pahoehoe* (Pa-hoy-hoy) is the name for smooth or ropy lava. Cooler lava hardens on the surface; hotter, more fluid lava flows under it, often leaving caves or tubes behind. *Aa* (ah-ah) is the name of rough, jagged lava. This molten lava is much less fluid and usually moves slower. A crust never hardens on the surface, but chunks of cooler rock tumble along the top and sides instead. Aa can be impassable.

Lava Flow Terminology

- *Lava flow*—lava flows are associated with volcanoes, and others are the result of fissure flow. These masses of molten rock pour onto the Earth's surface during an effusive eruption. Both moving lava and the resulting solidified deposit are referred to as lava flows. Because of the wide range in (1) viscosity of the different lava types (basalt, andesite, dacite, and rhyolite); (2) lava discharge during eruptions; and (3) characteristics of the erupting vent and topography over which lava travels, lava flows come in a great variety of shapes and sizes.
- *Lava cascade*—a cascade of water is a small waterfall formed as water descends over rocks. In similar fashion, a lava cascade refers to the rush of descent of lava over a cliff. In Hawaii, lava cascades typically occur when lava spills over the edge of a crater, a fault scarp, or a sea cliff into the ocean.
- *Lava drapery*—is the cooled, congealed rock on the face of a cliff, crater, or fissure formed by lava pouring or cascading over their edges.
- *Lava channels*—are narrow, curved, or straight open pathways through which lava moves on the surface of a volcano. The volume of lava moving down a channel fluctuates so that the channel may be full or overflowing at times and nearly empty at other times. During overflow, some of the lava congeals and cools along the banks to form natural levees that may eventually enable the lava channel to build a few meters above the surround ground.
- *Standing waves*—in a fast-moving lava flow, these appear to be stationary relative to the lava that moves over the land through them, similar to the standing waves in a water stream. In Hawaii, standing waves as high as three meters have been observed.

- *Lava spillways*—are confined lava channels on the sides of a volcanic cone or shield that form when lava overflows the rim of the vent.
- *Lava surge*—intermittent surges or accelerations in the forward advance of lava can occur when the supply of lava to a flow suddenly increases or a flow front gives way. The supply of lava may increase as a consequence of a higher discharge of lava from the vent, a sudden change in the vent geometry so that a great volume of lava escapes (e.g., the collapse of a vent wall), or by the escape of ponded lava from along a channel. Lava surges may be accompanied by thin, short-lived breakouts of fluid lava from the main channel and flow front.
- *Methane explosion*—sudden explosions of methane gas occur frequently near the edges of active lava flows. Methane gas is generated when vegetation is covered and heated by molten lava. The explosive gas travels beneath the ground through cracks and fills abandoned lava tubes for long distances around the margins of the flow. Methane gas explosions have occurred at least one hundred meters from the leading edge of a flow, blasting rocks and debris in all directions.
- *Volcanic domes*—are rounded, steep-sided mounds built by very viscous magma, usually either dacite or rhyolite. Such magmas are typically too viscous (resistant to flow) to move far from the vent before cooling and crystallizing. Domes may consist of one of more individual lava flows. Volcanic domes are also referred to as lava domes. (USGS 2000)

Did You Know?

The longest historical, dome-building eruption is still occurring at Santiaguito Dome, which is erupting on the southeast flank of Santa Maria volcano in Guatemala; the dome began erupting in 1922.

Intrusions

Intrusive (or plutonic igneous) rocks have been intruded or injected into the surrounding rocks. Some of these intrusions are invisible because they are imbedded at great depth; consequently, igneous intrusive bodies may be seen only after the underlying rocks have been removed by erosion.

Intrusions are of two types: concordant intrusions, which are parallel to layers of rocks, and discordant intrusions, which cut across layers. Some of the more common intrusive bodies (plutons) are discussed below.

Concordant Intrusions

- *Sills*—are tabular bodies of igneous rocks that spread out as essentially thin, horizontal sheets between beds or layers of rocks.

- *Laccoliths*—are lenslike, mushroom-shaped, or blisterlike intrusive bodies, usually near the surface, that have relatively flat under surfaces and arched or domed upper surfaces. They differ from sills in that they are thicker in the center and become thinner near their margins.
- *Lopoliths*—are megasills, usually of gabbro or diorite, which may cover hundreds of square kilometers and be kilometers thick. They often have a concave structure and are differentiated, that is, they take so long to harden that heavy minerals have a chance to sink and light minerals can rise.

Discordant Intrusions

- *Dikes*—are thin, wall-like sheets of magma intruded into fractures in the crust.
- *Stocks* or *plutons*—are small irregular intrusions.
- *Batholiths*—are the largest of igneous intrusions and are usually granitic and cover hundreds or thousands of square kilometers.

Did You Know?

Most obsidian is black, but red, green, and brown obsidian is known. Obsidian forms when magma is cooled so quickly that individual minerals cannot crystallize.

Volcanic Landforms

Volcanic landforms (or volcanic edifices) are controlled by the geological processes that form them and act on them after they have formed. Four principal types of volcanic landforms are formed: plateau basalts or lava plains, volcanic mountains, craters, and calderas.

PLATEAU BASALTS AND LAVA PLAINS

These are formed when great floods of lava are released by fissure eruptions instead of central vents and spread in sheetlike layers over the Earth's surface, forming broad plateaus. Some of these plateaus are quite extensive. For example, the Columbia River Plateau of Oregon, Washington, Nevada, and Idaho is covered by two hundred thousand square miles of basaltic lava.

VOLCANIC MOUNTAINS

These are mountains that are composed of the volcanic products of central eruptions and are classified as cinder cones (conical hills), composite cones (stratovolcanoes), and lava domes (shield volcanoes).

VOLCANIC CRATERS

These are circular, funnel-shaped depressions, usually less than one kilometer in diameter, that form as a result of explosions that emit gases and tephra.

CALDERAS

These are much larger depressions, circular to elliptical in shape, with diameters ranging from one kilometer to fifty kilometers. Calderas form as a result of a collapse of a volcanic structure. The collapse results from evacuation of the underlying magma chamber.

Thermal Areas

Thermal areas are locations where volcanic or other igneous activity takes place, as is evidenced by the presence or action of volcanic gases, steam, or hot water escaping from the Earth.

FUMAROLES

These are vents where gases, either from a magma body at depth, or steam from heated groundwater, emerge at the surface of the Earth.

HOT SPRINGS

Hot springs or thermal springs are areas where hot water comes to the surface of the Earth. Cool groundwater moves downward and is heated by a body of magma or hot rock. A hot spring results if this hot water can find its way back to the surface, usually along fault zones.

GEYSERS

A geyser results if the hot spring has a plumbing system that allows for the accumulation of steam from the boiling water. When the steam pressure builds so that it is higher than the pressure of the overlying water in the system, the steam will move rapidly toward the surface, causing the eruption of the overlying water. Some geysers, like Old Faithful in Yellowstone Park, erupt at regular intervals, but most geysers are quite erratic in their performance. The time between eruptions is controlled by the time it takes for the steam pressure to build in the underlying plumbing system.

Plate Tectonics

Within the past forty-five or fifty years, geologists have developed the theory of plate tectonics (tectonics: Greek, "builder"). The theory of plate tectonics deals with the formation, destruction, and large-scale motions of great segments of the Earth's surface (crust), called *plates*. This theory relies heavily on the older concepts of continental drift (developed during the first half of the twentieth century) and seafloor spreading (understood during the 1960s), which help to explain the cause of earthquakes and volcanic eruptions, and the origin of fold mountain systems.

Crustal Plates

Earth's crustal plates are composed of great slabs of rock (lithosphere), about one hundred kilometers thick, that cover many thousands of square miles (they are thin in comparison to their length and width); they float on the ductile asthenosphere, carrying both continents and oceans. Many geologists recognize at least eight main plates and numerous smaller ones. These *main* plates include the following:

- African Plate covering Africa—continental plate
- Antarctic Plate covering Antarctica—continental plate
- Australian Plate covering Australia—continental plate
- Eurasian Plate covering Asia and Europe—continental plate
- Indian Plate covering the Indian subcontinent and part of the Indian Ocean—continental plate
- Pacific Plate covering the Pacific Ocean—oceanic plate
- North American Plate covering North America and northeast Siberia—continental plate
- South American Plate covering South America—continental plate

The *minor* plates include the following:

- Arabian Plate
- Caribbean Plate
- Juan de Fuca Plate
- Cocos Plate
- Nazea Plate
- Philippine Plate
- Scotia Plate

Plate Boundaries

As mentioned, the asthenosphere is the ductile, soft, plasticlike zone in the upper mantle on which the crustal plates ride. Crustal plates move in relation to one another at one of three types of plate boundaries: convergent (collision boundaries), divergent (spreading boundaries), and transform boundaries. These boundaries between plates are typically associated with deep-sea trenches, large faults, fold mountain ranges, and mid-oceanic ridges.

CONVERGENT BOUNDARIES

Convergent boundaries (or active margins) develop where two plates slide toward each other, commonly forming either a subduction zone (if one plate subducts or moves underneath the other) or a continental collision (if the two plates contain continental crust). To relieve the stress created by the colliding plates, one plate is deformed and slips below the other.

DIVERGENT BOUNDARIES

Divergent boundaries occur where two plates slide apart from each other. Oceanic ridges, which are examples of these divergent boundaries, are where new oceanic, melted lithosphere materials well up, resulting in basaltic magmas that intrude and erupt at the oceanic ridge, in turn creating new oceanic lithosphere and crust (new ocean floor). Along with volcanic activity, the mid-oceanic ridges are also areas of seismic activity.

TRANSFORM PLATE BOUNDARIES

Transform, or shear/constructive boundaries, do not separate or collide; rather, they slide past each other in a horizontal manner with a shearing motion. Most transform boundaries occur where oceanic ridges are offset on the sea floor. The San Andreas Fault in California is an example of a transform boundary.

Chapter Review Questions

6.1. ________ are large bodies of molten rock deeply buried within the Earth.
6.2. Magma consists of three types: ________, ________, and ________.
6.3. ________ is a heavy, dark-colored igneous rock consisting of coarse grains of feldspar and augite.
6.4. ________ is a finely textured igneous rock.
6.5. The ________ is the simplest type of volcano.
6.6. ________ are confined lava channels on the sides of a volcanic cone or shield that form when lava overflows the rim of the vent.
6.7. ________ have been intruded or injected into the surrounding rocks.

6.8. ________ are tabular bodies of igneous rocks that spread out as essentially thin, horizontal sheets between beds or layers of rocks.
6.9. ________ are small, irregular intrusions.
6.10. Magma body gas vents: ________.
6.11. Earth's ________ are composed of great slabs of rock.
6.12. Plate that covers Pacific Ocean: ________.
6.13. ________ boundaries develop where two plates slide toward each other.
6.14. ________ boundaries develop where two plates slide apart from each other.
6.15. ________ boundaries slide past each other.

Note

1. Much of the information in this chapter is adapted from F. R. Spellman, *Geology for Nongeologists* (Lanham, Md.: Government Institutes, 2009).

References and Recommended Reading

Abbott, P. L. 1996. *Natural disasters.* New York: William C. Brown Publishing.
Anderson, J. G., et al. 1986. Strong ground motion from the Michoacan, Mexico, earthquake. *Science* 233:1043–49.
Atkinson, L., and C. Sancetta. 1993. Hail and farewell. *Oceanography* 6 (34).
Browning, J. M. 1973. Catastrophic rock slides. Mount Huascaran, north-central Peru, May 32, 1970. *Bulletin American Association of Petroleum Geologists* 57:1335–41.
Coch, N. K. 1995. *Geohazards, natural and human.* New York: Prentice Hall.
Eagleman, J. 1983. *Severe and unusual weather.* New York: Van Nostrand Reinhold.
Francis, P. 1993. *Volcanoes: A planetary perspective.* New York: Oxford University Press.
GeoMan. 2008. Bowen's Reaction Series. http://jersey.uoregon.edu/~mstrick/AskGeoMan/geoQuerry32.html (accessed November 13, 2009).
Holmes, A. 1978. *Principles of physical geology,* 3rd ed. New York: Wiley.
Keller, E. A. 1985. *Environmental geology,* 4th ed. New York: Merrill Publishing.
Kiersh, G. A. 1964. Vaiont reservoir disaster. *Civil Engineering* 34: 32–39.
Limbert, R. 1924. Among the Craters of the Moon. *National Geographic* (March).
Lyman, J., and R. H. Fleming. 1940. Composition of Seawater. *J. Mar. Res.* 3:134–46.
McKnight, T. 2004. *Geographica: The complete illustrated atlas of the world.* New York: Barnes and Noble Books.
Murck, B. W., B. J. Skinner, and S. C. Porter. 1997. *Dangerous Earth: An introduction to geologic hazards.* New York: Wiley.
Oreskes, N., ed. 2003. *Plate tectonics: An insider's history of the modern theory of the Earth.* New York: Westview.
Skinner, B. J., and S. C. Porter. 1995. *The dynamic Earth: An introduction to physical geology,* 3rd ed. New York: Wiley.
Spellman, F. R., and N. E. Whiting. 2006. *Environmental science and technology,* 2nd ed. Lanham, Md.: Government Institutes.
Stanley, S. M. 1999. *Earth system history.* New York: W. H. Freeman (see 211–22).

Stephens, J. C., et al. 1984. Organic soils subsidence. *Geological Society of American Reviews in Engineering Geology* 6:3.

Sverdrup, H. U., M. W. Johnson, and R. H. Fleming. 1942. *The oceans: Their physics, chemistry and general biology*. New York: Prentice Hall.

Swanson, D. A., T. H. Wright, and R. T. Helz. 1975. Linear vent systems and estimated rates of magma production and eruption of the Yakima basalt on the Columbia Plateau. *American Journal of Science* 275:877–905.

Tilling, R. I. 1984. *Eruptions of Mount St. Helens: Past, present, and future*. Washington, D.C.: Department of the Interior, U.S. Geological Survey.

Turcotte, D. L., and G. Schubert. 2002. *Geodynamics*, 2nd ed. New York: Wiley.

U.S. Geological Survey (USGS). 1989. Lessons learned from the Loma Prieta, California, earthquake of October 17, 1989. Circular 1045.

———. 2000. Photo glossary of volcanic terms. http://volcanoes.usgs.gov/images/pglossary/index.php (accessed June 1, 2008).

———. 2008. Principal types of volcanoes. http://pubs.usgs.gov/gip/volc/types.html (accessed May 31, 2008).

———. 2009. Craters of the Moon: History and culture. www.NPS.gov/crmo/historyculture (accessed May 28, 2009).

Williams, H., and A. R. McKinney. 1979. *Volcanology*. New York: Freeman & Copper.

Bryce Canyon, Utah, is a perfect example of the result of wind, water, ice, freezing, and thawing working in combination to sculpt Earth's landforms. Photo by Frank R. Spellman.

CHAPTER 7

Wind-Eroded Landforms

Wind-eroded landforms, such as those commonly found in deserts, have fascinated humans throughout the ages.[1] In this chapter we not only discuss wind-eroded landforms but also the processes of mass wasting and desertification. In a text about geography that deals with land and the people who occupy or do not occupy those various land regions, it is only fitting that a discussion of wind-eroded landforms, mass wasting, and desertification would be covered in one chapter. Moreover, it is even more fitting that the deserts of the world would be the primary focus of such a discussion. This makes sense when you consider that deserts cover approximately one-third of the Earth's land surface.

Wind Erosion

During a recent research outing to several national parks in the western United States, I stopped at several locations and photographed various natural wonders. One of the focal points of study was on the weathering processes that are discussed in this chapter. The natural bridges, such as the one shown in figure 7.1 and the natural arches shown in figures 7.2–7.5, all are a result of some form of weathering; thus, they are highlighted in this chapter.

There was a time not that long ago when many believed the main difference between natural bridges (see figure 7.1) and the natural arches shown in figures 7.2–7.5 was that the natural bridges were formed by water erosion and natural arches were formed by wind erosion. Contrary to popular belief or myth, however, wind is not a significant factor in the formation of natural arches or other natural formations. Substantial studies have shown that natural arches and natural bridges are formed by many different processes of erosion (working solely or in combination) that contribute to the natural, selective removal of rock. Every process relevant to natural arch formation involves the action of water, gravity, temperature variation, or tectonic pressure on rock.

Again, wind is not a significant agent in natural arch formation. Wind does act to disperse the loose grains that result from microscopic erosion. Moreover, sandstorms can scour and polish already existing arches. The bottom line (point to remember) is that wind alone never creates them (Barnes 1987; Vreeland 1994).

As prefaced above, wind action or erosion is very limited in extent and effect. It is largely confined to desert regions, but even there it is limited to a height of about eighteen inches above ground level. Wind does have the power, however, to transport,

Figure 7.1. Natural Bridge. Natural Bridge, Virginia. Photo by Frank R. Spellman.

deposit, and erode sediment. In this chapter, we will discuss each of the aspects of the wind because they are important in any study of geology.

Did You Know?

Wind is common in deserts because the air near the surface is heated and rises, cooler air comes in to replace hot, rising air, and this movement of air results in winds. Also, arid desert regions have little or no soil moisture to hold rock and mineral fragments.

Wind Sediment Transport

Sediment near the ground surface is transported by wind in a process called *saltation* (Latin, *saltus*, "leap"). Similar to what occurs in the bed load of streams, wind saltation refers to short jumps (leaps) of grains dislodged from the surface and leaping a short distance. As the grains fall back to the surface, they dislodge other grains that then get carried by wind until they collide with the ground to dislodge other particles. Above

Figure 7.2. Natural Window Arch. Monument Valley, Utah. Photo by Frank R. Spellman.

ground level, wind can swoop down to the surface and lift smaller particles, suspending them (making airborne) in the wind, and they may travel long distances (see figure 7.6).

Did You Know?

Sand ripples occur as a result of large grains accumulating as smaller grains are transported away. Ripples form in lines perpendicular to wind direction. Windblown dust is sand-sized particles that generally do not travel very far in the wind, but smaller sized fragments can be suspended in the wind for much larger distances.

Wind-Driven Erosion

As mentioned, wind by itself has little if any effect on solid rock. But, in arid and semiarid regions, wind can be an effective geologic agent anywhere that it is strong enough (possesses high enough velocity) to pick up a load of rock fragments, which

Figure 7.3. Natural Window Arch. Arches National Park, Moab, Utah. Photo by Frank R. Spellman.

may become effective tools of erosion in the land-forming process. Wind can erode by *deflation* and *abrasion*.

DEFLATION

The process of deflation (or blowing away) is the lowering of the land surface due to removal of fine-grained particles by the wind. Deflation concentrates the coarser grained particles at the surface, eventually resulting in a relatively smooth surface composed only of the coarser grained fragments that cannot be transported by the wind. Such a coarse-grained surface is called *desert pavement* (see figure 7.7). Some of these coarser grained fragments may exhibit a dark, enamel-like coat of iron or manganese called *desert varnish*.

Deflation may create several types of distinctive features. For example, *lag gravels* are formed when the wind blows away finer rock particles, leaving behind a residue of coarse gravel and stones. *Blowouts* may be developed where wind has scooped out soft, unconsolidated rocks and soil.

ABRASION

The wind abrades (sandblasts) by picking up sand and dust particles that are transported as part of its load. Abrasion is restricted to a distance of about a meter or two

Figure 7.4. Natural Window Arch. Arches National Park, Moab, Utah. Photo by Frank R. Spellman.

above the surface because sand grains are lifted a short distance. The destructive action of these windblown abrasives may wear away wooden telephone poles and fence posts, and abrade, scour, or groove solid rock surfaces.

Wind abrasion also plays a part in the development of such landforms as isolated rocks (pedestals and table rocks) that have had their bases undercut by windblown sand and grit (see figure 7.8). *Ventifacts* are another interesting and relatively common product of wind erosion. These are any bedrock surface or stone or pebble that has been abraded or shaped by windblown sediment in a process similar to sandblasting. Ventifacts are formed when the wind blows sand against the side of the stone, shaping it into a flat, polished surface. At a much larger scale, elongate ridges called *yardangs* form by the abrasion and streamlining of rock structures oriented parallel to the prevailing wind direction.

Wind Deposition

The velocity of the wind and the size, shape, and weight of the rock particles determine the manner in which wind carries its load. Wind-transported materials are most commonly derived from floodplains, beach sand glacial deposits, volcanic explosions, and dried lake bottoms—places containing light ash and loose, weathered rock fragments.

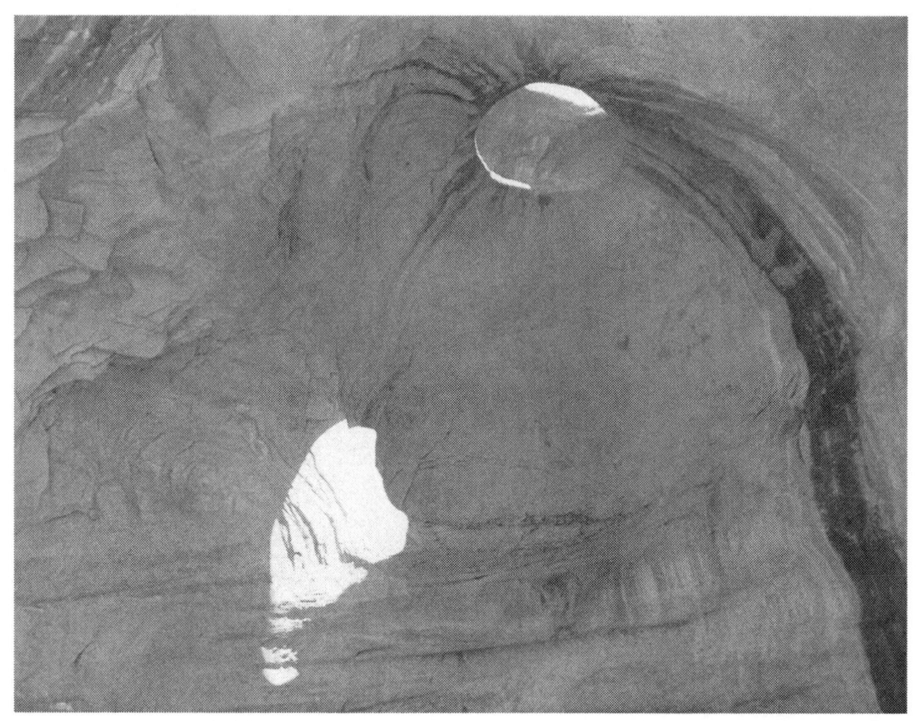

Figure 7.5. Eye to the Sky, Monument Valley, Utah. Photo by Frank R. Spellman.

The wind is capable of transporting large quantities of material for very great distances. The wind deposits sediment when its velocity decreases to the point where the particles can no longer be transported. Initially (in a strong wind), part of the sediment load rolls or slides along the ground (bed load). Some sand particles move by a series of leaping or bounding movements (saltation). And lighter dust may be transported upward (suspension) into higher, faster moving wind currents, traveling many thousands of miles.

As mentioned, the wind will begin to deposit its load when its velocity is decreased or when the air is washed clean by falling rain or snow. A decrease in wind velocity may also be brought about when the wind strikes some barrier-type obstacle (fences, trees, rocks, or human-made structures) in its path. As the air moves over the top of the obstacle, streamlines converge and the velocity increases. After passing over the obstacle, the streamlines diverge and the velocity decreases. As the velocity decreases, some of the load in suspension can no longer be held in suspension and, thus, drops out to form a deposit. The major types of windblown or eolian deposits are dunes and loess.

DUNES

Sand dunes are asymmetrical mounds with a gentle slope in the upwind direction and steep slope on the downwind side (see figure 7.9). Dunes vary greatly in size and shape

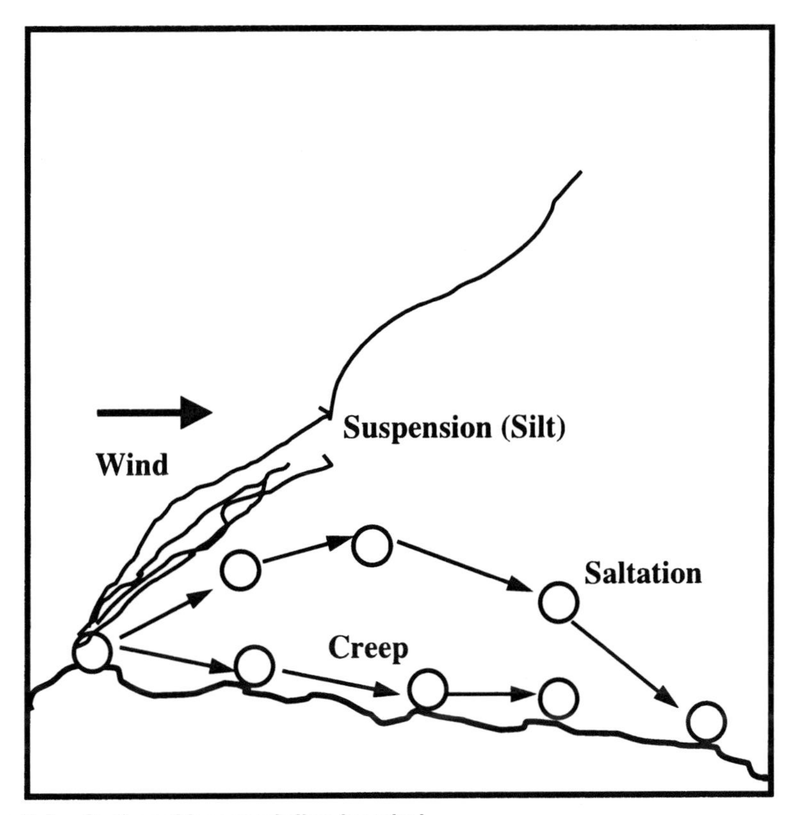

Figure 7.6. Sediment transportation by wind.

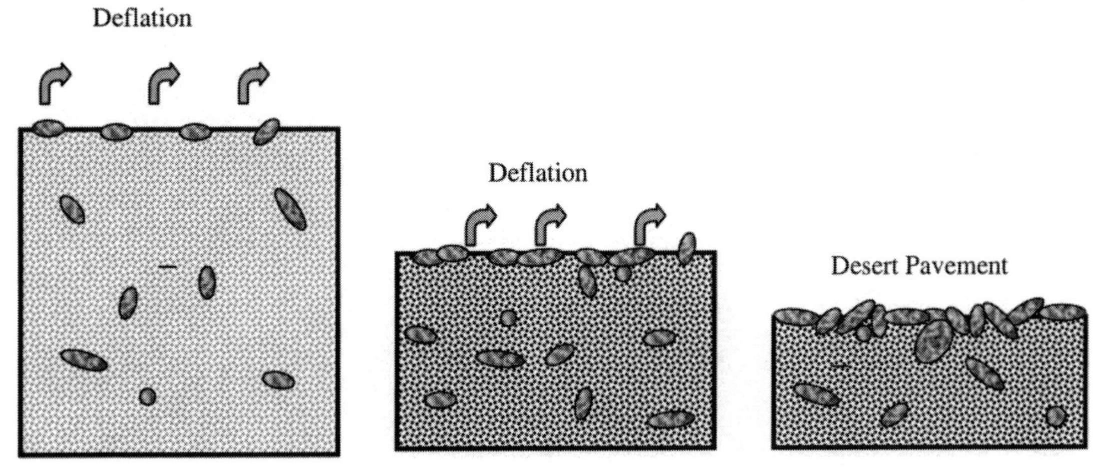

Figure 7.7. Wind-driven deflation processes.

Figure 7.8. Seeming to defy gravity, Balanced Rock (Arches National Park, Moab, Utah) has a harder cap rock that somewhat protects the more easily eroded base; eventually the double hammering of water and wind erosion will cause it to disintegrate, leaving a pile of rocky debris as a reminder of the power of erosion. Photo by Frank R. Spellman.

and form when there is a ready supply of sand, a steady wind, and some kind of obstacle or barrier such as rocks, fences, or vegetation to trap some of the sand. Sand dunes form when moving air slows down on the downwind side of an obstacle. Dunes may reach heights up to five hundred meters and cover large areas. Types of sand dunes include barchan, transverse, longitudinal, and parabolic.

- *Barchan dunes*—these are crescent-shaped dunes characterized by two long, curved extensions pointing in the direction of the wind and a curved slip face on the

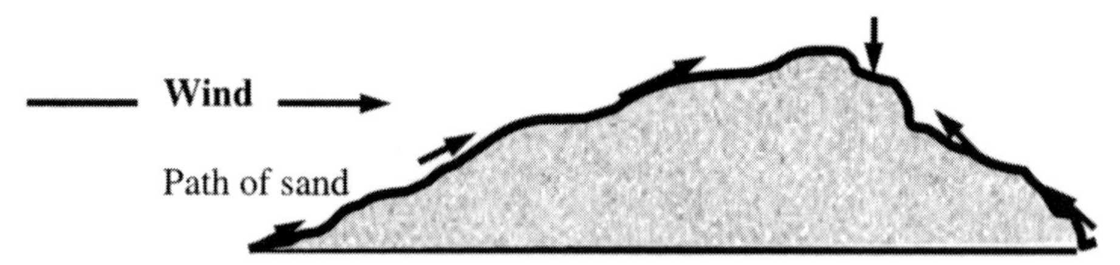

Figure 7.9. Profile of typical sand dune. Arrows denote paths of wind currents.

downwind side of the dune (see figure 7.10a). These dunes are formed in areas where winds blow steadily and from a single direction.

- *Transverse dunes*—these form along seacoasts and lakeshores and may be fifteen feet high and half a mile in length. Transverse dunes develop with their long axis at right angles to the wind (see figure 7.10b).
- *Longitudinal dunes*—these are long, ridgelike dunes that develop parallel to the wind (see figure 7.10c).
- *Parabolic dunes*—these are U-shaped dunes open and facing upwind. They are usually stabilized by vegetation and occur where there is abundant vegetation, a constant wind direction, and an abundant sand supply (see figure 7.10d).

LOESS

Loess is a yellowish, fine-grained, nonstratified material carried by the wind and accumulated in deposits of dust. The materials forming loess are derived from surface dust originating primarily in deserts, river floodplains, deltas, and glacial outwash deposits. Loess is cohesive and possesses the property of forming steep bluffs with vertical faces such as the deposits found in the pampas of Argentina and the lower Mississippi River Valley.

Mass Wasting

Mass wasting, or mass movement, takes place as earth materials (loose, uncemented mixture of soil and rock particles known as *regolith*) move downslope in response to gravity without the aid of a transporting medium such as water, ice, or wind—though these factors play a role in regolith movement. This type of erosion is apt to occur in any area with slopes steep enough to allow downward movement of rock debris. Some of the factors that help gravity overcome this resistance are discussed below.

GRAVITY

The heavy hand of gravity constantly pulls everything, everywhere, toward Earth's surface. On a flat surface, parallel to Earth's surface, the constant force of gravity acts downward. This downward force prevents gravitational movement of any material that remains on or parallel to a flat surface.

On a slope, the force of gravity can be resolved into two components: a component acting perpendicular to the slope and a component acting tangential to the slope. Thus, material on a slope is pulled inward in a direction that is perpendicular (the glue) to the slope (see figure 7.11a). This helps prevent material from sliding downward. However, as stated previously, on a slope, another component of gravity exerts a force (a constant tug) that acts to pull material down a slope, parallel to the surface

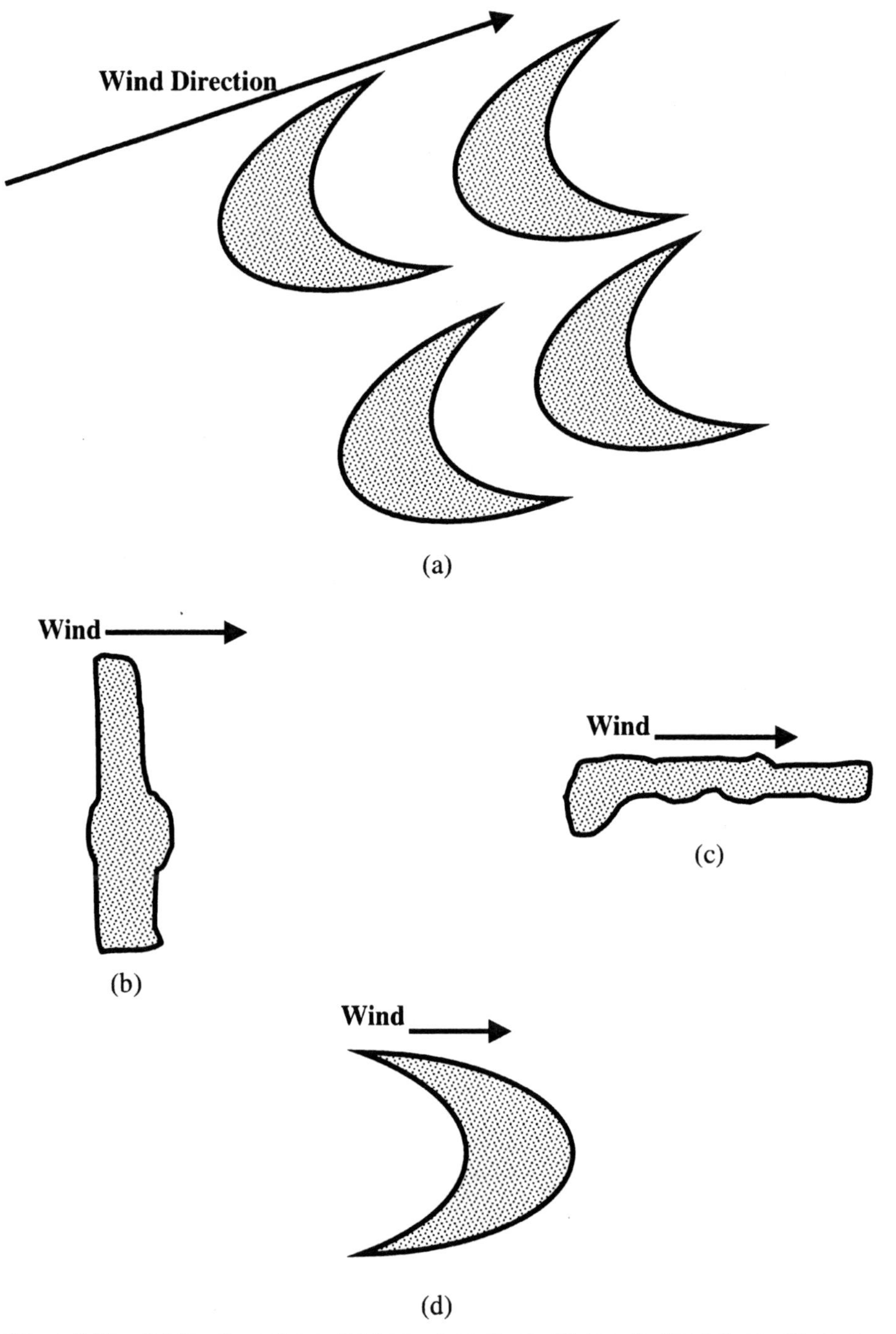

Figure 7.10. (a) Barchan dune; (b) transverse dune; (3) longitudinal dune; (4) parabolic dune.

of the slope. Known as *shear stress*, this force of gravity exerts stress in direct relationship to the steepness of the slope, that is, shear stress increases as the slope steepens. In response to increased shear stress, the perpendicular force (the glue) of gravity decreases (see figure 7.11b).

Did You Know?

When shear on a slope decreases, material may still be stuck to the slope and prevented from moving downward by the force of friction. It may be held in place by the frictional contact between the particles making up that material. Contact between the surfaces of the particles creates a certain amount of tension that holds the particles in place at an angle. The steepest angle at which loose material on a slope remains motionless is called the angle of repose (generally about 35 degrees). Particles with angled edges that catch on each other also tend to have a higher angle of repose than those that have become rounded through weathering and that simply roll over each other.

WATER

Even though mass wasting may occur in either wet or dry materials, water greatly facilitates downslope movements; it is an important agent in the process of mass wasting. Water will either help hold material together (act as glue—demonstrated in building beach sandcastles with slightly dampened sand), increasing its angle of repose, or cause it to slide downward like a liquid (acting like a lubricant). Water may soften clays, making them slippery, add weight to the rock mass, and, in large amounts, actually force rock particles apart, thus reducing soil cohesion.

FREEZING AND THAWING

Earlier the erosive power of frost wedging (water contained in rock and soil expands when frozen) was pointed out. Mass wasting in cold climates is governed by the fact that water is frozen as ice during long periods of the year, especially in high-altitude regions. Ice, although it is solid, does have the ability to flow (glacial-movement effect), and alternate periods of freezing and thawing can also contribute to movement, and in some instances ice expansion may be great enough to force rocks downhill.

UNDERCUTTING

Undercutting occurs when streams erode their banks or surf action along a coast undercuts a slope, making it unstable. Undercutting can also occur when human-made excavations remove support and allow overlying material to fall.

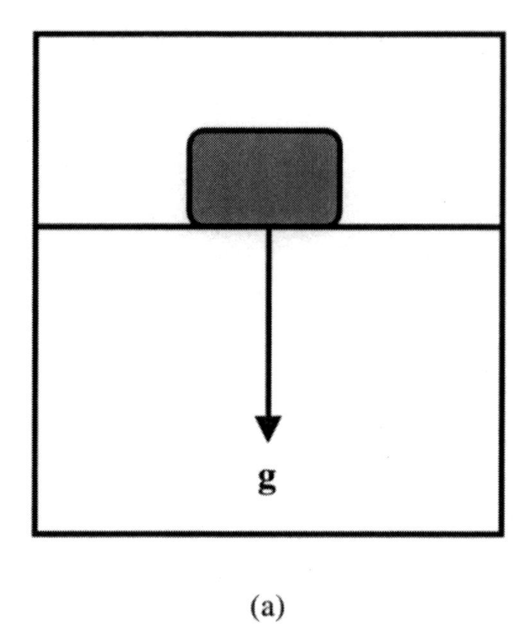

(a)

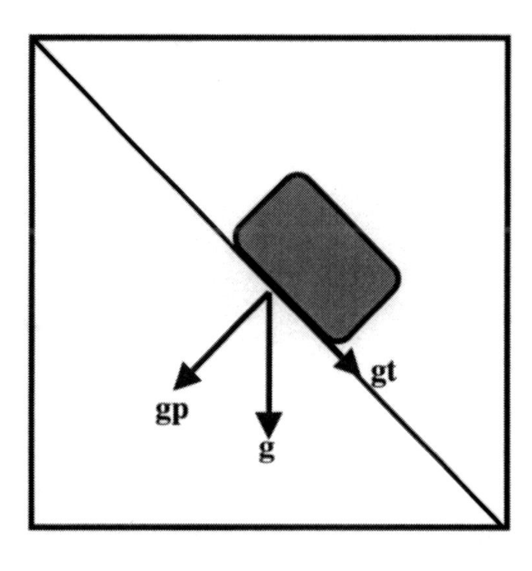

(b)

Figure 7.11. (a) Gravity acting perpendicular to the surface. (b) The perpendicular component (the glue) of gravity, gp, helps to hold the material in place on the slope.

ORGANIC ACTIVITIES

Whenever animals burrow into the ground, they disturb soil materials, casting rocks out of their holes as they dig; these are commonly piled up downslope. Eventually, weather conditions and the constant force applied by gravity can put these piles into motion. Animals also contribute to mass wasting whenever they walk on soil surfaces; their motions can knock materials downhill.

SHOCK WAVES OR VIBRATIONS

A sudden strong shock or vibration, such as an earthquake, faulting, blasting, and heavy traffic, can trigger slope instability. Minor shocks like heavy vehicles rambling down the road, trees blowing in the wind, or human-made explosions can also trigger mass-wasting events such as landslides.

Kinds of Mass Movements

A landslide is a mass movement that occurs suddenly and violently. In contrast, soil creep is mass movement that is almost imperceptible. These processes can be divided into two broad categories: rapid and slow movements. *Rapid movements* include landslides, slumps, mudflows, and earthflows. *Slow movements* include soil creep and solifluction.

RAPID MOVEMENTS

- *Landslides*—these are by far the most spectacular and violent of all mass movements. Landslides are characterized by the sudden movement of great quantities of rock and soil downslope. Such movements typically occur on steep slopes that have large accumulations of weathered material. Precipitation in the form of rain or snow may seep into the mass of steeply sloping rock debris, adding sufficient weight to start the entire mass sliding.
- *Slumps*—these special landslides occur along a curved surfaces. The upper surface of each slump block remains relatively undisturbed, as do the individual blocks. Slumps leave arcuate (Latin, curved like a bow) scars or depressions on the hillslope. Heavy rains or earthquakes usually trigger slumps. Slump is a common occurrence along the banks of streams or the walls of steep valleys.
- *Mudflows*—these are highly fluid, high-velocity mixtures of sediment and water that have a consistency of wet concrete. Mass wasting of this type typically occurs as certain arid or semiarid mountainous regions are subjected to unusually heavy rains.
- *Earthflows*—these are usually associated with heavy rains and move at velocities between several centimeters and hundreds of meters per year. They usually remain

active for long periods of time. They generally tend to be narrow, tonguelike features that begin at a scarp or small cliff.

SLOW MOVEMENTS

- *Soil creep*—this continuous movement, usually so slow as to be imperceptible, normally occurs on almost all slopes that are moist but not steep enough for landslides. Soil creep is usually accelerated by frost wedging, alternate thawing and freezing, and certain plant and animal activities. Evidence for creep is often seen in bent trees, offsets in roads and fences, and inclined utility poles.
- *Solifluction*—this downslope movement is typical of areas where the ground is normally frozen to considerable depth—arctic, subarctic, and high mountain regions. The actual soil flowage occurs when the upper portion of the mantle rock thaws and becomes water saturated. The underlying, still frozen subsoil acts as a slide for the sodden mantle rock that will move down even the gentlest slope.

Did You Know?

Landslides constitute a major geologic hazard because they are widespread, occur in all fifty states and U.S. territories, and cause $1–2 billion in damages and more than twenty-five fatalities on average each year. Expansion of urban and recreational developments into hillside areas leads to more people that are threatened by landslides each year. Landslides commonly occur in connection with other major natural disasters such as earthquakes, volcanoes, wildfires, and floods. (USGS 2008)

Desertification

Deserts are areas where the amount of precipitation received is less than the potential evaporation (<10 inches/year); they cover roughly 30 percent of the Earth's land surface—areas we think of as arid. *Desertification* occurs in hot areas far from sources of moisture, areas isolated from moisture by high mountains, in coastal areas along which there are onshore winds and cold-water currents, and high-pressure areas where descending air masses produce warm, dry air.

According to the U.S. Geological Survey (USGS 1997), the world's great deserts were formed by natural processes interacting over long intervals of time. During most of these times, deserts have grown and shrunk independent of human activities. Desertification does not occur in linear, easily mappable patterns. Deserts advance erratically, forming patches on their borders. Scientists question whether desertification, as a process of global change, is permanent or how and when it can be halted or reversed.

Chapter Review Questions

7.1. Wind is more of a ________ of sediment than a creator of sediment.
7.2. Sediment near the ground surface is transported by wind in a process called ________.
7.3. The process of ________ is the lowering of the land surface due to removal of fine-grained particles by the wind.
7.4. A ________ is any bedrock surface or stone or pebble that has been abraded or shaped by windblown sediment in a process similar to sandblasting.
7.5. A dune is a major type of windblown ________ deposit.
7.6. Crescent-shaped dunes: ________.
7.7. Parabolic dunes are ________-shaped.
7.8. Yellowish, fine-grained, nonstratified, wind-carried material: ________.
7.9. ________ takes place when earth materials move downslope in response to gravity.
7.10. ________ are special landslides that occur along a curved surface.

Note

1. Much of the information in this chapter is from F. R. Spellman, *Geology for Nongeologists* (Lanham, Md.: Government Institutes, 2009).

References and Recommended Reading

Barnes, F. A. 1987. *Canyon country arches and bridges.* Self-published.
Goodwin, P. H. 1998. *Landslides, slumps, and creep.* New York: Franklin Watts.
Jennings, T. 1999. *Landslides and avalanches.* North Mankato, Minn.: Thame-side Press.
U.S. Geological Survey (USGS). 1997. Desertification. http://pubs.usgs.gov/gip/deserts/desertification (accessed July 8, 2008).
———. 2008. Landslide hazards program. http://landslides.usgs.gov/ (accessed July 22, 2008).
Vreeland, R. H. 1994. *Nature's bridges and arches, volume 1: General information*, 2nd ed. Self-published.
Walker, J. 1992. *Avalanches and landslides.* New York: Gloucester Press.

Gulf of Thailand, South China Sea. Photo by Frank R. Spellman.

CHAPTER 8

Coastal Features

Coastal features are produced by the interaction of a variety of geological forces. Tectonic forces move bedrock up or down along a particular coastline; rivers deposit sediment into the ocean; and waves erode, transport, and deposit material depending on the shape of the shoreline.

Oceans

Oceans are the storehouse of Earth's water. Oceans cover about 71 percent of Earth's surface. The average depth of Earth's oceans is about 3,800 meters, with the greatest ocean depth recorded at 11,036 meters in the Mariana Trench. At the present time, the oceans contain a volume of about 1.35 billion cubic kilometers (96.5 percent of Earth's total water supply), but the volume fluctuates with the growth and melting of glacial ice.

Composition of ocean water has remained constant in composition throughout geologic time. The major constituents dissolving in ocean water (from rivers and precipitation and the result of weathering and degassing of the mantle by volcanic activity) is composed of about 3.5 percent, by weight, of dissolved salts including chloride (55.07 percent), sodium (30.62 percent), sulfate (7.72 percent), magnesium (3.68 percent), calcium (1.17 percent), potassium (1.10 percent), bicarbonate (0.40 percent), bromine (0.19 percent), and strontium (0.02 percent).

The most significant factor related to ocean water that everyone is familiar with is the salinity of the water—how salty it is. *Salinity*, a measure of the amount of dissolved ions in the oceans, ranges between thirty-three and thirty-seven parts per thousand. Often the concentration is the amount (by weight) of salt in water, as expressed in "parts per million" (ppm). Water is saline if it has a concentration of more than one thousand ppm of dissolved salts; ocean water contains about thirty-five thousand ppm of salt (USGS 2007). Chemical precipitation, absorption onto clay minerals, and plants and animals prevent seawater from containing even higher salinity concentrations. However, salinity does vary in the oceans because surface water evaporates, rain and stream water is added, and ice forms or thaws.

Did You Know?

Salinity is higher in midlatitude oceans because evaporation exceeds precipitation. Salinity is also higher in restricted areas of the oceans like the Red Sea (up to forty parts per thousand). Salinity is lower near the equator because precipitation is higher, and salinity is lower near the mouths of major rivers because of input of fresh water.

Along with salinity, another important property of seawater is temperature. The temperature of surface seawater varies with latitude, from near 0° C near the poles to 29° C near the equator. Some isolated areas can have temperatures up to 37° C. Temperature decreases with ocean depth.

Ocean Floor

The bottoms of the oceans' basins (ocean floors) are marked by mountain ranges, plateaus, and other relief features similar to (although not as rugged as) those on the land.

As shown in figure 8.1, the floor of the ocean has been divided into four divisions: the continental shelf, continental slope, continental rise, and deep-sea floor or abyssal plain.

- *Continental shelf*—this is the flooded, nearly flat true margin of the continents. Varying in width to about 40 miles and a depth of approximately 650 feet, continental shelves slope gently outward from the shores of the continents (see figure 8.1). Continental shelves occupy approximately 7.5 percent of the ocean floor.

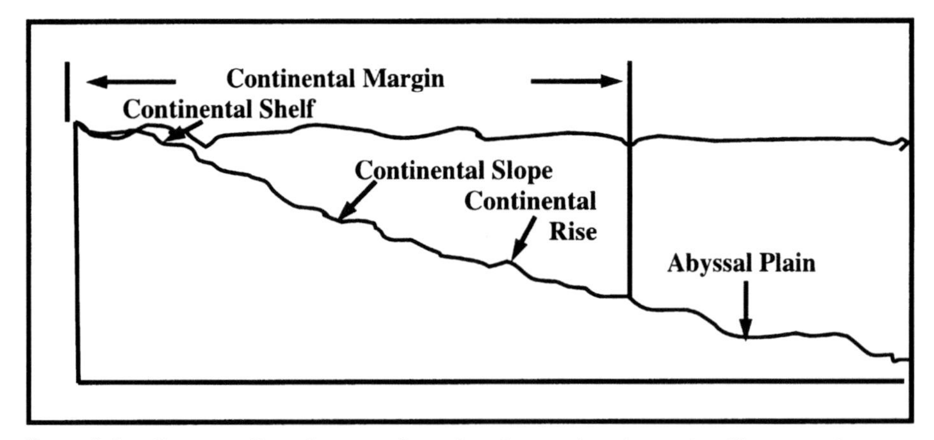

Figure 8.1. Cross section of ocean floor showing major elements of topography.

- *Continental slope*—this is a relatively steep slope descending from the continental shelf (see figure 8.1) rather abruptly to the deeper parts of the ocean. These slopes occupy about 8.5 percent of the ocean floor.
- *Continental rise*—this is a broad, gentle slope below the continental slope containing sediment that has accumulated along parts of the continental slope.
- *Abyssal plain*—this is a sediment-covered, deep-sea plain about twelve to eighteen thousand feet below sea level. This plain makes up about 42 percent of the ocean floor.

The deep ocean floor does not consist exclusively of the abyssal plain. In places, there are areas of considerable relief. Among the more important such features are:

- *Seamounts*—these are isolated, mountain-shaped elevations more than three thousand feet high.
- *Mid-oceanic ridge*—these are submarine mountains, extending more than thirty-seven thousand miles through the oceans, generally ten thousand feet above the abyssal plain.
- *Trench*—this is a deep, steep-sided trough in an abyssal plain.
- *Guyots*—this is a seamount that is flat topped and was once a volcano. They rise from the ocean bottom and usually are covered by three to six thousand feet of water.

Ocean Tides, Currents, and Waves

Water is the master sculptor of Earth's surfaces. The ceaseless, restless motion of the sea is an extremely effective geologic agent. Besides shaping inland surfaces, water sculpts the coast. Coasts include sea cliffs, shores, and beaches. Seawater set in motion erodes cliffs, transports eroded debris along shores, and dumps it on beaches. Therefore, most coasts retreat or advance. In addition to the unceasing causes of motion—wind, density of seawater, and rotation of the Earth—the chief agents in this process are tides, currents, and waves.

TIDES

The periodic rise and fall of the sea (once every twelve hours and twenty-six minutes) produces the tides. Tides are due to the gravitational attraction of the Moon and, to a lesser extent, the Sun and the Earth. The Moon has a larger effect on tides and causes the Earth to bulge toward the Moon. It is interesting to note that at the same time the Moon causes a bulge on Earth, a bulge occurs on the opposite side of the Earth due to inertial forces (further explanation is beyond the scope of this text). The effect of the tides is not too noticeable in the open sea, the difference between high and low tide amounting to about two feet. The tidal range may be considerably greater near shore, however. It may range from less than two feet to as much as fifty feet. The tidal range will vary according to the phase of the Moon and the distance of the Moon from the Earth. The type of shoreline and the physical configuration of the ocean floor will also affect the tidal range.

CURRENTS

The oceans have localized movements of masses of seawater called ocean currents. These are the result of drift of the upper fifty to one hundred meters of the ocean due to drag by wind. Thus, surface ocean currents generally follow the same patterns as atmospheric circulation, with the exception that atmospheric currents continue over the land surface while ocean currents are deflected by the land. Along with wind action, current may also be caused by tides, variation in salinity of the water, rotation of the Earth, and concentrations of turbid or muddy water. Temperature changes in water affect water density, which, in turn, causes currents—these currents cause seawater to circulate vertically.

WAVES

Waves, varying greatly in size, are produced by the friction of wind on open water. Wave height and power depend upon wind strength and fetch—the amount of unobstructed ocean over which the wind has blown. In a wave, water travels in loops. Essentially an up-and-down movement of the water, the diameter of the loops decreases with depth. The diameter of loops at the surface is equal to wave height (h). Breakers are formed when the wave comes into shallow water near the shore. The lower part of the wave is retarded by the ocean bottom, and the top, having greater momentum, is hurled forward causing the wave to break. These breaking waves may do great damage to coastal property as they race across coastal lowlands driven by winds, gale, or hurricane velocities.

Coastal Erosion, Transportation, and Deposition

The geologic work of the sea, like previously discussed geologic agents, consist of erosion, transportation, and deposition. The sea accomplishes its work of coastal landform sculpting largely by means of waves and wave-produced currents; their effect on the seacoast may be quite pronounced. The coast and accompanying coastal deposits and landform development represent a balance between wave energy and sediment supply.

WAVE EROSION

Waves attack shorelines and erode by a combination of several processes. The resistance of the rocks composing the shoreline and the intensity of wave action to which it is subjected are the factors that determine how rapidly the shore will be eroded. Wave erosion works chiefly by hydraulic action, corrosion, and attrition. As waves strike a sea cliff, *hydraulic action* crams air into rock crevices, putting tremendous pressure on the surrounding rock; as waves retreat, the explosively expanding air enlarges cracks and breaks off chunks of rock (scree). Chunks hurled by waves against the cliff break off more scree (via a sandpapering action)—a process called *corrasion/abrasion*. When the sea rubs and grinds rocks together, the scree that is formed is thrown into the cliffs, reducing broken rocks to pebbles and sand grains—a process called *attrition* (Lambert 2007).

Several features are formed by marine erosion—different combinations of wave action, rock type, and rock beds produce these features. Some of the more typical erosion-formed features of shorelines are discussed below.

- *Sea cliffs or wave-cut cliffs*—these are formed by wave erosion of underlying rock followed by the caving in of the overhanging rocks. As waves eat farther back inland, they leave a wave-cut beach or platform. Such cliffs are essentially vertical and are common at certain localities along the New England and Pacific coasts of North America.
- *Wave-cut bench*—these are the result of wave action not having enough time to lower the coastline to sea level. Because of the resistance to erosion, a relatively flat wave-cut bench develops. If subsequent uplift of the wave-cut bench occurs, it may be preserved above sea level as a wave-cut bench.
- *Headlands*—these are fingerlike projections of resistant rock extending out into the water. Indentations between headlands are termed *coves*.
- *Sea caves, sea arches, and stacks*—these are formed by continued wave action on a sea cliff. Wave action hollows out cavities or caves in the sea cliffs. Eventually, waves may cut completely through a headland to form a sea arch; if the roof of the arch collapses, the rock left separated from the headland is called a stack.

MARINE TRANSPORTATION

Waves and currents are important transporting agents. Rip currents and undertow carry rock particles back to the sea, and long-shore currents will pick up sediments (some of it in solution), moving them out from the shore into deeper water. Materials carried in solution or suspension may drift seaward for great distances and eventually be deposited far from shore. During the transportation process, sediments undergo additional erosion, becoming reduced in size.

MARINE DEPOSITION

Marine deposition takes place whenever currents and waves suffer reduced velocity. Some rocks are thrown up on the shore by wave action. Most of the sediments thus deposited consist of rock fragments derived from the mechanical weathering of the continents, and they differ considerably from terrestrial or continental deposits. Due to input of sediments from rivers, deltas may form; due to beach drift, such features as spits and hooks, bay barriers, and tombolos may form. Depositional features along coasts are discussed below.

- *Beaches*—these are transitory coastal deposits of debris that lie above the low-tide limit in the shore zone.
- *Barrier islands*—these are long, narrow accumulations of sand lying parallel to the shore and separated from the shore by a shallow lagoon.
- *Spits and hooks*—these are elongated, narrow embankments of sand and pebble extending out into the water but attached by one end to the land.
- *Tombolos*—these are bars of sand or gravel connecting an island with the mainland or another island.
- *Wave-built terraces*—these are structures built up from sediments deposited in deep water beyond a wave-cut terrace.
- *Deltas*—these form where sediment supply is greater than the ability of waves to remove sediment.

Chapter Review Questions

8.1. __________ is a measure of the amount of dissolved ions in the oceans.
8.2. __________ is a deep, steep-sided trough in an abyssal plain.
8.3. __________ is a type of seamount that is flat topped and once was a volcano.
8.4. __________ are due to the gravitational attraction of the Moon.
8.5. __________ are fingerlike projections of resistant rock extending out into the water.

References and Recommended Reading

Gross, G. M. *Oceanography: A view of the Earth*. Englewood Cliffs, N.J.: Prentice Hall.
Lambert, D. 2007. *The field guide to geology*. New York: Checkmark Books.
Pinet, P. R. 1996. *Invitation to oceanography*. St. Paul, Minn.: West Publishing.
U.S. Geological Survey (USGS). 2007. The water cycle: Water storage in oceans. http://ga.water.usgs.gov/edu/watercycleoceans.html (accessed July 11, 2008).

III

CLIMATE AND WEATHER ASPECTS

Clouds at 36,000 feet. Photo by Frank R. Spellman.

CHAPTER 9

Climate and Weather

An understanding of the impact of climate (and its associated weather patterns) is essential to any basic study of the geography of Earth. *Climate* is defined as the average weather conditions of a place, usually measured annually. This includes temperature and rainfall. *Weather*, often confused as climate, is a day to day snapshot of local conditions, including temperature, rainfall, wind, and condition of the atmosphere. Thus, in order to understand Earth's climate, we must begin with a discussion of the atmosphere. In regard to Earth's atmosphere, probably no one has described it better than Shakespeare in *Hamlet*: "This most excellent canopy, the air, look you, this brave o'erhanging firmament, / this majestical roof fretted with golden fire."

Formation of Earth's Atmosphere

Several theories of cosmogony attempt to explain the origin of the universe. Without speculating on the validity of any one theory, the following is simply the author's view.

The time: 4,500 billion years ago.

Before the universe there was time—only time; otherwise, the vast void held only darkness—everywhere.

Overwhelming darkness—everywhere.

Not dim . . . not murky . . . not shadowy, or not unlit. Simple nothingness—nothing but darkness, a shade of black so intense we cannot fathom or imagine it today. Light had no existence—this was black of blindness, of burial in the bowels of the Earth, the blackness of no other choice.

With time—eons of time—darkness came to a sudden, smashing, shattering, annihilating, scintillating, cataclysmic end—and there was light . . . light everywhere. This new force replaced darkness and lit up the expanse without end, creating a brightness fed by billions of glowing round masses so powerful as to renounce and overcome the darkness that had come before.

With the light was heat energy, which shone and warmed and transformed into mega-mega-mega trillions of super-excited ions, molecules, and atoms—heat of unimaginable proportions, forming gases—gases we don't even know how to describe, how to quantify, let alone how to name. But gases they were—and they were everywhere.

With light, energy, heat, and gases present, the stage was set for the greatest show of all time, anywhere—ever: the formation of the universe.

Over time—time in stretches we cannot imagine, so vast we cannot contemplate them meaningfully—the heat, light, energy, and gases all came together and grew, like an expanding balloon, into one solid, glowing, growing mass. But it continued to grow, with the pangs, sweating, and moans accompanying any birthing, until it had reached the point of no return—explosion level. And it did; it exploded with the biggest bang of all time (with the biggest bang hopefully of all time).

The Big Bang sent masses of hot gases in all directions—to the farthest reaches of anything, everything—into the vast, wide, measureless void. Clinging together as they rocketed, soared, and swirled, forming galaxies that gradually settled into their arcs through the void, constantly propelled away from the force of their origin, these masses began their eternal evolution.

Two masses concern us . . . the Sun and Earth.

> Busy old fool, unruly Sun,
> Why dost thou thus,
> Through windows and through
> curtains, call on us?
> —John Donne, "The Sun Rising"

Thankfully, the Sun calls on us.

Forces well beyond the power of the Sun (beyond anything imaginable) stationed this massive gaseous orb approximately ninety-three million miles from the dense, molten core enveloped in cosmic gases and the dust of time that eventually became the insignificant mass we now call Earth.

Distant from the Sun, Earth's mass began to cool, slowly; the progress was slower than we can imagine, but cool it did. While the dust and gases cooled, Earth's inner core, mantle, and crust began to form—no more a quiet or calm evolution than the revolution that cast it into the void had been.

Downright violent was this transformation—the cooling surface only a facade for the internal machinations going on inside, outgassing from huge, deep destructive vents (we would call them volcanoes today) and erupting continuously—never stopping—blasting away, delivering two main ingredients: magma and gas.

The magma worked to form the primitive features of Earth's early crust. The gases worked to form Earth's initial atmosphere—our point of interest: the atmosphere. Without atmosphere, what is there?

About four billion years before the present, Earth's early atmosphere was chemically reducing, consisting primarily of methane, ammonia, water vapor, and hydrogen—for life as we know it today, an inhospitable brew.

Earth's initial atmosphere was not a calm, quiet, quiescent environment. To the contrary, it was an environment best characterized as dynamic—ever changing—where bombardment after bombardment of intense, bond-breaking, ultraviolet light, along with intense lightning and radiation from radionuclides, provided energy to bring about chemical reactions that resulted in the production of relatively complicated molecules, including amino acids and sugars (building blocks of life).

About 3.5 billion years before the present, primitive life formed in two radically different theaters: on Earth and below the primordial seas near hydrothermal vents that spotted the wavering, water-covered floor.

Initially, on Earth's unstable surface, these very primitive life-forms derived their energy from fermentation of organic matter formed by chemical and photochemical processes, then gained the ability to produce organic matter (CH_2O) by photosynthesis.

Thus the stage was set for the massive biochemical transformation that resulted in the production of almost all the atmosphere's oxygen (O_2).

The O_2 initially produced was quite toxic to primitive life-forms. However, much of this oxygen was converted to iron oxides by reaction with soluble iron. This process formed enormous deposits of iron oxides—the existence of which provides convincing evidence for the liberation of O_2 in the primitive atmosphere.

Eventually, enzyme systems developed that enabled organisms to mediate the reaction of waste-product oxygen with oxidizable organic matter in the sea. Later, the mode of waste-gradient disposal was utilized by organisms to produce energy by respiration, which is now the mechanism by which nonphotosynthetic organisms obtain energy. In time, O_2 accumulated in the atmosphere. In addition to providing an abundant source of oxygen for respiration, the accumulated atmospheric oxygen formed an ozone (O_3) shield—the O_3 shield absorbs bond-rupturing ultraviolet radiation.

With the O_3 shield protecting tissue from destruction by high-energy ultraviolet radiation, the Earth, although still hostile to life-forms we are familiar with, became a much more hospitable environment for life (self-replacing molecules), and life-forms were enabled to move from the sea (where they flourished next to the hydrothermal gas vents) to the land. And from that point to the present, Earth's atmosphere became more life-form friendly—everywhere.

Earth's Thin Skin

Shakespeare likened it to a majestic overhanging roof (constituting the transition between its surface and the vacuum of space); others have likened it to the skin of an apple. Both these descriptions of our atmosphere are fitting, as is its being described as the Earth's envelope, veil, or gaseous shroud. The atmosphere is more like the apple skin, however. This thin skin, or layer, contains the life-sustaining oxygen (21 percent) required by all humans and many other life-forms; the carbon dioxide (0.03 percent) so essential for plant growth; the nitrogen (78 percent) needed for chemical conversion to plant nutrients; the trace gases such as methane, argon, helium, krypton, neon, xenon, ozone, and hydrogen; and varying amounts of water vapor and airborne particulate matter. Life on Earth is supported by this atmosphere, solar energy, and other planets' magnetic fields.

Gravity holds about half the weight of a fairly uniform mixture of these gases in the lower eighteen thousand feet of the atmosphere; approximately 98 percent of the material in the atmosphere is below one hundred thousand feet.

Atmospheric pressure varies from one thousand millibars (mb) at sea level to ten mb at one hundred thousand feet. From one to two hundred thousand feet, the pressure drops from 9.9 mb to 0.1 mb and so on.

The atmosphere is considered to have a thickness of forty to fifty miles; however, here we are primarily concerned with the troposphere, the part of the Earth's atmosphere that extends from the surface to a height of about twenty-seven thousand feet above the poles, about thirty-six thousand feet in midlatitudes, and about fifty-three thousand feet over the equator. Above the troposphere is the stratosphere, a region that increases in temperature with altitude (the warming is caused by absorption of the Sun's radiation by ozone) until it reaches its upper limit of 260,000 feet.

The Troposphere

Extending above Earth approximately twenty-seven thousand feet, the troposphere is the focus of this text because people, plants, animals, and insects live here and depend on this thin layer of gases. Moreover, all of the Earth's weather takes place within the troposphere. The troposphere begins at ground level and extends 7.5 miles up into the sky where it meets with the second layer called the stratosphere.

Did You Know?

It was pointed out earlier that the gases that are so important to life on Earth are primarily contained in the troposphere. Also note that another important substance is contained in the troposphere: water vapor. Along with being the most remarkable of the trace gases contained in the troposphere, water vapor is also the most variable. Considerable attention is paid to water since it is in most of the air surrounding Earth. Unlike the other trace gases in the atmosphere, water vapor alone exists in gas, solid, and liquid forms. It also functions to add and remove heat from the air when it changes from one form to another.

Water vapor (in conjunction with airborne particles, obviously) is essential for the stability of Earth's ecosystem. This water vapor–particle combination interacts with the global circulation of the atmosphere and produces the world's weather, including clouds and precipitation.

The Stratosphere

The stratosphere begins at the 7.5 mile point and reaches 21.1 miles into the sky. In the rarified air of the stratosphere, the significant gas is ozone (life-protecting ozone—not to be confused with pollutant ozone), which is produced by the intense, ultraviolet radiation from the Sun. In quantity, the total amount of ozone in the atmosphere is so small that, if it were compressed to a liquid layer over the globe at sea level, it would have a thickness of less than 3/16 inch.

Ozone contained in the stratosphere can also impact (add to) ozone in the troposphere. Normally, the troposphere contains about twenty parts per billion of ozone. On occasion, however, via the jet stream, this concentration can increase to five to ten times higher than average.

In this book, the focus is on the troposphere and stratosphere because these two layers directly impact life as we know it and are or can be heavily influenced by pollution and its effects.

Did You Know?

The troposphere, stratosphere, mesosphere, and thermosphere act together as a giant safety blanket. They keep the temperature on the Earth's surface from dipping to extreme, icy cold that would freeze everything solid or from soaring to blazing heat that would burn up all life.

Moisture in the Atmosphere

Clouds! Clouds everywhere! Where is that Sun?

On a hot day when clouds build up signifying that a storm is imminent, we do not always appreciate what is happening.

What is happening?

This cloud buildup actually signals that one of the most vital processes in the atmosphere is occurring: the condensation of water is raised to higher levels and cooled within strong updrafts of air created either by convection currents, turbulence, or physical obstacles like mountains. The water originated from the surface—evaporated from the seas, from the soil, or transpired by vegetation. Once within the atmosphere, however, a variety of events combine to convert the water vapor (produced by evaporation) to water droplets. The air must rise and cool to its dew point, of course. At dew point, water condenses around minute, airborne particulate matter to make tiny cloud droplets forming clouds—clouds from which precipitation occurs.

Whether created by the Sun heating up a hillside, by jet aircraft exhausts, or factory chimneys, there are actually only ten major cloud types. The deliverers of countless millions of tons of moisture from the Earth's atmosphere, they form even from the driest desert air containing as little as 0.1 percent water vapor. They not only provide a visible sign of motion but also indicate change in the atmosphere portending weather conditions that may be expected up to forty-eight hours ahead. In this chapter, we take a brief look at the nature and consequences of these cloud-forming processes.

Cloud Formation

The atmosphere is a highly complex system, and the effects of changes in any single property tend to be transmitted to many other properties. The most profound effect on the atmosphere is the result of alternate heating and cooling of the air, which causes adjustments in relative humidity and buoyancy; this causes condensation, evaporation, and cloud formation.

The temperature structure of the atmosphere (along with other forces that propel the moist air upward) is the main force behind the form and size of clouds. Exactly how does temperature affect atmospheric conditions? For one thing, temperature (that is, heating and cooling of the surface atmosphere) causes vertical air movements. Let's take a look at what happens when air is heated.

Let's start with a simple parcel of air in contact with the ground. As the ground is heated, the air in contact with it will warm also. This warm air increases in temperature and expands. Remember, gases expand on heating much more than liquids or solids, so this expansion is quite marked. In addition, as the air expands, its density falls (meaning that the same mass of air now occupies a larger volume). You've heard that warm air rises? Because of its lessened density, this parcel of air is now lighter than the surrounding air and tends to rise. Conversely, if the air cools, the opposite occurs—it contracts, its density increases, and it sinks. Actually, alternate heating and cooling are intimately linked with the process of evaporation, condensation, and precipitation.

But how does a cloud actually form? Let's look at another example.

On a sunny day, some patches of ground warm up more quickly than others because of differences in topography (soil and vegetation, etc.). As the surface temperature increases, heat passes to the overlying air. Later, by midmorning, a bulbous mass of warm, moisture-laden air rises from the ground. This mass of air cools as it meets lower atmospheric pressure at higher altitudes. If cooled to its dew point temperature, condensation follows and a small cloud forms. This cloud breaks free from the heated patch of ground and drifts with the wind. If it passes over other rising air masses, it may grow in height. The cloud may encounter a mountain and be forced higher still into the air. Condensation continues as the cloud cools, and if the droplets it holds become too heavy, they fall as rain.

Major Cloud Types

Earlier, I mentioned there are ten major cloud types. These include the following:

Stratiform genera
- Species
- Cirrus
- Cirrostratus
- Cirrocumulus

Altostratus
Altocumulus
Stratus
Stratocumulus
Nimbostratus
Cumuliform genera
Species
Cumulus
Cumulonimbus

From the list above it is apparent that the cloud groups are classified into a system that uses Latin words to describe the appearance of clouds as seen by an observer on the ground. Table 9.1 summarizes the four principal components of this classification system (Ahrens 1994).

Did You Know?

Clouds play an important role in boundary-layer meteorology and air quality. Convective clouds transport pollutants vertically, allowing an exchange of air between the boundary layer and the free troposphere. Cloud droplets formed by heterogeneous nucleation on aerosols grow into rain droplets through condensation, collision, and coalescence. Clouds and precipitation scavenge pollutants from the air. Once inside the cloud or rainwater, some compounds dissociate into ions and/or react with one another through aqueous chemistry (i.e., cloud chemistry is an important process in the oxidation of sulfur dioxide to sulfate). Another important role for clouds is the removal of pollutants trapped in rainwater and its deposition onto the ground. Clouds can also affect gas-phase chemistry by attenuating solar radiation below the cloud base that has a significant impact on photolysis reactions.

—EPA/600/R-99/030

Further classification identifies clouds by height of cloud base. For example, cloud names containing the prefix *cir-*, as in cirrus clouds, are located at high levels, while

Table 9.1. Summary of Components of Cloud Classification System

Latin Root	*Translation*	*Example*
Cumulus	heaped/puffy	fair-weather cumulus
Stratus	layered	altostratus
Cirrus	curl of hair/wispy	cirrus
Nimbus	ran	cumulonimbus

cloud names with the prefix *alto-*, as in altostratus, are found at middle levels. This module introduces several cloud groups. The first three groups are identified based upon their height above the ground. The fourth group consists of vertically developed clouds, while the final group consists of a collection of miscellaneous cloud types.

A *stratus* cloud is a featureless, gray, low-level cloud. Its base may obscure hilltops or occasionally extend right down to the ground, and because of its low altitude, it appears to move very rapidly on breezy days. Stratus can produce drizzle or snow, particularly over hills, and may occur in huge sheets covering several thousand miles.

Cumulus clouds also seem to scurry across the sky, reflecting their low altitude. These small, dense, white, fluffy, flat-based clouds are typically short lived, lasting no more than ten to fifteen minutes before dispersing. They are typically formed on sunny days, when localized convection currents are set up: these currents can form over factories or even brush fires, which may produce their own clouds.

Cumulus may expand into low-lying, horizontally layered, massive *stratocumulus* or into extremely dense, vertically developed, giant *cumulonimbus* with a relatively hazy outline and a glaciated top that can be up to seven miles in diameter. These clouds typically form on summer afternoons; their high, flattened tops contain ice, which may fall to the ground in the form of heavy showers of rain or hail.

Rising to middle altitudes, the bluish-gray, layered *altostratus* and rounded, fleecy, whitish-gray *altocumulus* appear to move slowly because of their greater distance from the observer.

Cirrus (meaning tuft of hair) clouds are made up of white, narrow bands of thin, fleecy parts, are relatively common over northern Europe, and generally ride the jet stream rapidly across the sky.

Cirrocumulus are high-altitude clouds composed of a series of small, regularly arranged cloudlets in the form of ripples or grains; they are often present with cirrus clouds in small amounts.

Cirrostratus are high-altitude, thin, hazy clouds, usually covering the sky and giving a halo effect surrounding the Sun or Moon.

Did You Know?

Clouds whose names incorporate the word *nimbus* or the prefix *nimbo-* are clouds from which precipitation is falling.

Let's summarize the information related to how moisture accumulates in and precipitates from the atmosphere. The process of evaporation (converting moisture into vapor) supplies moisture into the lower atmosphere. The prevailing winds then circulate the moisture and mix it with drier air elsewhere.

Water vapor is only the first stage of the precipitation cycle; the vapor must be converted into liquid form. This is usually achieved by cooling, either rapidly, as in convection, or slowly, as in cyclonic storms. Mountains also cause uplift, but the rate will depend upon their height and shape and the direction of the wind.

To actually produce precipitation, the cloud droplets must become large enough to reach the ground without evaporating. The cloud must possess the right physical properties to enable the droplets to grow.

If the cloud lasts long enough for growth to take place, then precipitation will usually occur. Precipitation results from a delicate balance of counteracting forces, some leading to droplet growth and others to droplet destruction.

Precipitation and Evapotranspiration

Because it determines the intensity and distribution of many of the processes operating within the system, precipitation is one of the most important regulators of the hydrological cycle. The rate of evapotranspiration is closely related to precipitation and, thus, is also an integral part of the hydrological cycle.

The principal actions brought on by weather systems that affect land and sea and the humans, animals, and vegetation thereon are winds and precipitation. The latter comes in a variety of forms as discussed below. Most weather of consequence to people occurs in storms. These may be local in origin but more commonly are carried to locations in wide areas along pathways followed by active air masses consisting of highs and lows. The key ingredient in storms is water, either as a liquid or as a vapor. The vapor acts like a gas and thus contributes to the total pressure of the atmosphere, making up a small but vital fraction of the total (NASA 2008a).

Precipitation is found in a variety of forms. Which form actually reaches the ground depends upon many factors: for example, atmospheric moisture content, surface temperature, intensity of updrafts, and method and rate of cooling.

Water vapor in the air will vary in amount depending on sources, quantities, processes involved, and air temperature. Heat, mainly as solar irradiation but with some contributed by the Earth and human activity and some from change of state processes, will cause some water molecules either in water bodies (oceans lakes, rivers, etc.) or in soils to be excited thermally and escape from their sources. This is called evaporation; if water is released from trees and other vegetation the process is known as evapotranspiration. The evaporated water, or moisture, that enters the air is responsible for a state called humidity. Absolute humidity is the weight of water vapor contained in a given volume of air. The *mixing ratio* refers to the mass of the water vapor within a given mass of dry air. At any particular temperature, the maximum amount of water vapor that can be contained is limited to some amount; when that amount is reached, the air is said to be saturated for that temperature. If less than the maximum amount is present, then the property of air that indicates this is its relative humidity (RH), defined as the actual water vapor amount compared to the saturation amount at the given temperature; this is usually expressed as a percentage. RH also indicates how much moisture the air can hold above its stated level, which, after attaining, could lead to rain.

When a parcel of air attains or exceeds RH = 100 percent, condensation will occur, and water in some state will begin to organize as some type of precipitation. One familiar form is dew, which occurs when the saturation temperature or some quantity of moisture reaches a temperature at the surface at which condensations sets in, leaving the moisture to coat the ground (especially obvious on lawns).

The term *dew point* has a more general use, being that temperature at which an air parcel must be cooled to become saturated. Dew frequently forms when the current air mass contains excessive moisture after a period of rain but the air is now clear; the dew precipitates out to coat the surface (noticeable on vegetation). Ground fog is a variant in which lowered temperatures bring on condensation within the near-surface air as well as the ground.

The other types of precipitation are listed in table 9.2 along with descriptive characteristics related to each type.

Evaporation and transpiration are complex processes that return moisture to the atmosphere. The rate of evapotranspiration depends largely on two factors: (1) how saturated (moist) the ground is and (2) the capacity of the atmosphere to absorb the moisture.

As mentioned, precipitation occurs when the dew point is reached. It was also mentioned that it is quite possible to have an air mass or cloud containing water vapor cooled below the dew point without precipitation occurring. In this state, the air mass is said to be *supercooled.*

How, then, are droplets of water formed? Water droplets form around microscopic foreign particles already present in the air. These particles on which the droplets form are called *hygroscopic nuclei.* They are present in the air primarily in the form of dust, in the form of salt from seawater evaporation, and from combustion residue. These foreign particles initiate the formation of droplets that eventually fall as precipitation. To have precipitation, larger droplets or drops must form. This may be brought about by two processes: (1) coalescence (collision) or (2) the Bergeron process.

COALESCENCE

Simply put, *coalescence* is the fusing together of smaller droplets into larger ones. The variation in the size of the droplets has a direct bearing on the efficiency of this process. Raindrops come in different sizes and can reach diameters up to seven millimeters. Having larger droplets greatly enhances the coalescence process.

But what actually goes on inside a cloud to cause rain to fall? To answer this question, we must take a look inside a cloud to see exactly what processes occur to make rain—rain that actually falls as rain. Rainmaking is based on the essentials of the Bergeron process.

Table 9.2. Types of Precipitation

Type	*Approximate Size*	*State of Water*	*Description*
Mist	0.005 to 0.05 mm	Liquid	Droplets large enough to be felt on face when air is moving one meter/second. Associated with stratus clouds.
Drizzle	Less than 0.5 mm	Liquid	Small, uniform drops that fall from stratus clouds, generally for several hours.
Rain	0.5 to 5 mm	Liquid	Generally produced by nimbostratus or cumulonimbus clouds. When heavy, size can be highly variable from one place to another.
Sleet	0.5 to 5 mm	Solid	Small, spherical to lumpy ice particles that form when raindrops freeze while falling through a layer of sub-freezing air. Because the ice particles are small, any damage is generally minor. Sleet can make travel hazardous.
Glaze	Layers 1 mm to 2 cm thick	Solid	Produced when supercooled raindrops freeze on contact with solid objects. Glaze can form a thick covering of ice having sufficient weight to seriously damage trees and power lines.
Rime	Variable accumulation	Solid	Deposits usually consisting of ice feathers that point into the wind. These delicate, frostlike accumulations form as supercooled cloud or fog droplets encounter objects and freeze on contact.
Snow	1 mm to 2 cm	Solid	The crystalline nature of snow allows it to assume many shapes, including six-sided crystals, plates, and needles. Produced in supercooled clouds where water vapor is deposited as crystals that remain frozen during their descent.
Hail	5 mm or larger	Solid	Precipitation in the form of hard, rounded pellets or irregular lumps of ice. Produced in large, convective, cumulonimbus clouds, where frozen ice particles and supercooled water coexist.
Graupel	2 mm to 5 mm	Solid	Sometimes called "soft hail," graupel forms as rime collects on snow crystals to produce irregular masses of "soft" ice. Because these particles are softer than hailstones, they normally flatten out upon impact.

Source: NASA (2008b).

BERGERON PROCESS

Named after the Swedish meteorologist who suggested it, the *Bergeron process* is probably the more important process for the initiation of precipitation. To gain understanding of how the Bergeron process works, let's look at what actually goes on inside a cloud to cause rain.

Within a cloud made up entirely of water droplets there will be a variety of droplet sizes. The air will be rising within the cloud anywhere from ten to twenty centimeters per second (depending on the type of cloud). As the air rises, the drops become larger through collision and coalescence; many will reach drizzle size. Then the updraft intensifies up to fifty centimeters per second (and more), which reduces the downward movement of the drops, allowing them more time to become even larger. When the cloud becomes approximately one kilometer deep, small raindrops of seven hundred micrometers diameter are formed.

The droplets, because of their small size, do not freeze immediately, even when the temperatures fall below 0° C. Instead, the droplets remain unfrozen in a supercooled state. However, when the temperature drops as low as −10° C, ice crystals may start to develop among the water droplets. This mixture of water and ice would not be particularly important but for a peculiar characteristic or property of water. Therefore, at −10° C, air saturated with respect to liquid water is super-saturated relative to ice by 10 percent and at −20° C by 21 percent. Thus, ice crystals in the cloud tend to grow and become heavier at the expense of the water droplets.

Eventually, the ice crystals sink to the lower levels of the cloud where temperatures are only just below freezing. When this occurs, they tend to combine (the supercooled droplets of water act as an adhesive) and form snowflakes. When the snowflakes melt, the resulting water drops may grow further by collision with cloud droplets before they reach the ground as rain. The actual rate at which water vapor is converted to raindrops depends on three main factors: (1) the rate of ice crystal growth, (2) supercooled vapor, and (3) the strength of the updrafts (mixing) in the cloud.

We stated that, in order for precipitation to occur, water vapor must condense, which occurs when water vapor ascends and cools. Three mechanisms by which air rises, allowing for precipitation to occur, are convectional, orographic, and frontal.

CONVECTIONAL PRECIPITATION

Convectional precipitation is the spontaneous rising of moist air due to instability (NASA 2008b). This type of precipitation is usually associated with thunderstorms and occurs in the summer because localized heating is required to initiate the convection cycle. We have discussed that upward-growing clouds are associated with convection. Since the updrafts (commonly called a "thermal") are usually strong, cooling of the air is rapid and lots of water can be condensed quickly, usually confined to a local area, and a sudden summer downpour may occur as a result.

Convectional thunderstorm clouds are also described as supercells (NASA 2008b). Convective thunderstorms are the most common type of atmospheric instability that produces lightning followed by thunder. Lightning is one of the most spectacular phenomena witnessed in storms.

Did You Know?

A lightning bolt can attain an electric potential up to thirty million volts and current as much as ten thousand amps. It can cause air temperatures to reach 10,000 C. But a bolt's duration is extremely short (fractions of a second). Although a bolt can kill people it hits, most can survive.

OROGRAPHIC PRECIPITATION

Orographic precipitation is a straightforward process, characteristic of mountainous regions; almost all mountain areas are wetter than the surrounding lowlands. This type of precipitation arises when air is forced to rise over a mountain or mountain range. The wind, blowing along the surface of the Earth, ascends along topographic variations. Where air meets this extensive barrier, it is forced to rise. This ascending wind usually gives rise to cooling and encourages condensation and, thus, orographic precipitation on the windward side of the mountain range.

FRONTAL PRECIPITATION

Frontal precipitation results when two different fronts (or the boundary between two air masses characterized by varying degrees of precipitation), at different temperatures, meet. The warm-air mass (since it is lighter) moves up and over the colder air mass. The cooling is usually less rapid than in the vertical convection process because the warm-air mass moves up at an angle, more of a horizontal motion.

Another important part or process of the hydrological cycle (though it is often neglected because it can rarely be seen) is *evapotranspiration*. More complex than precipitation, evaporation and transpiration is a land-atmosphere interface process whereby a major flow of moisture is transferred from ground level to the atmosphere. It returns moisture to the air, replenishing that lost by precipitation, and it also takes part in the global transfer of energy. The rate of evapotranspiration depends largely on two factors: (1) how moist the ground is and (2) the capacity of the atmosphere to absorb the moisture. Therefore, the greatest rates are over the tropical oceans, where moisture is always available and the long hours of sunshine and steady trade winds evaporate vast quantities of water.

Just how much moisture is returned to the atmosphere via transpiration? In answering this question, table 9.3 makes clear, for example, that, in the United States

Table 9.3. Water Balance in the United States (in bgd*)

Precipitation	*4,200*
Evaporation and transpiration	3,000
Runoff	1,250
Withdrawal	310
Irrigation	142
Industry (utility cooling water)	142
Municipal	26
Consumed (irrigation loss)	90
Returned to streams	220

*bgd = billion gallons per day
Source: National Academy of Sciences, *Water Balance in the U.S.* (Washington, D.C.: National Research Council Publication 100-B. 1962).

alone, about two-thirds of the average rainfall over the U.S. mainland is returned via evaporation and transpiration.

EVAPORATION

Evaporation is the process by which a liquid is converted into a gaseous state. Evaporation takes place (except when air reaches saturation at 100 percent humidity) almost on a continuous basis. It involves the movement of individual water molecules from the surface of Earth into the atmosphere, a process occurring whenever a vapor pressure gradient exists from the surface to the air (i.e., whenever the humidity of the atmosphere is less than that of the ground). Evaporation also requires energy (derived from the Sun or from sensible heat from the atmosphere or ground): 2.48×10^6 joules to evaporate each kilogram of water at 10° C.

TRANSPIRATION

A related process, *transpiration* is the loss of water from a plant by evaporation. Most water is lost from the leaves through pores known as stomata, whose primary function is to allow gas exchange between the plant's internal tissues and the atmosphere. Transpiration from the leaf surfaces causes a continuous upward flow of water from the roots via the xylem, which is known as the transpiration stream.

Transpiration occurs mainly by day, when the stomata open up under the influence of sunlight. Acting as evaporators, they expose the pure moisture (the plant's equivalent of perspiration) in the leaves to the atmosphere. If the vapor pressure of the air is less than that in the leaf cells, the water is transpired.

As you might guess, because of transpiration, far more water passes through a plant than is needed for growth. In fact, only about 1 percent or so is actually used in plant growth. Nevertheless, the excess movement of moisture through the plant is important to the plant because the water acts as a solvent, transporting vital nutrients from the soil into the roots and carrying them through cells of the plant. Obviously, without this vital process plants would die.

EVAPOTRANSPIRATION: THE PROCESS

Although evapotranspiration plays a vital role in cycling water over Earth's landmasses, it is seldom appreciated. In the first place, distinguishing between evaporation and transpiration is often difficult. Both processes tend to be operating together, so the two are normally combined to give the composite term *evapotranspiration*.

Governed primarily by atmospheric conditions, energy is needed to power the process. Wind also plays an important role, by acting to mix the water molecules with the air and transport them away from the surface. The primary limiting factor in the process is lack of moisture at the surface (soil is dry). Evaporation can continue only so long as there is a vapor pressure gradient between the ground and the air.

The Atmosphere in Motion

There are scientists and engineers out there in the real world who will tell us that perpetual motion in or for any machine is a pipe dream—it's wishful thinking and it is impossible. Have you ever pondered the most dynamic perpetual motion machine of them all—Earth's atmosphere?

Have you ever wondered why the Earth's atmosphere is in perpetual motion? Probably not—but it is. It must be in a state of perpetual motion because it constantly strives to eliminate the constant differences in temperature and pressure between different parts of the globe. How are these differences eliminated or compensated for? The answer is, by its motion, which produces winds and storms. In this chapter, the horizontal movements that transfer air around the globe are considered.

Global Air Movement

Basically, winds are the movement of the Earth's atmosphere, which by its weight exerts a pressure on the Earth that we can measure using a barometer. Winds are often confused with air currents, but they are different. Wind is the horizontal movement of air or motion along the Earth's surface. Air currents, on the other hand, are vertical air motions collectively referred to as updrafts and downdrafts.

Throughout history, man has been both fascinated by and frustrated by winds. Man has written about winds almost from the time of the first written word. For example, Herodotus (like Homer and many others) wrote about winds in his *Histories*. Wind has had such an impact upon human existence that Man has given winds names that describe a particular wind, specific to a particular geographical area. Table 9.4 lists some of these winds, their colorful names, and the region where they occur. Some

Table 9.4. Assorted Winds of the World

Wind Name	Location
aajej	Morocco
alm	Yugoslavia
biz roz	Afghanistan
haboob	Sudan
imbat	North Africa
datoo	Gibraltar
nafhat	Arabia
besharbar	Caucasus
Samiel	Turkey
tsumuji	Japan
brickfielder	Australia
chinook	America
williwaw	Alaska

of these names are more than just colorful—the winds are actually colored. For example, the *harmattan* blows across the Sahara filled with red dust; mariners called this red wind the "sea of darkness."

To state that Earth's atmosphere is constantly in motion is to state the obvious. Anyone observing the constant weather changes around them is well aware of this phenomenon. Although obvious, the importance of the dynamic state of our atmosphere is much less obvious.

As mentioned, the constant motion of Earth's atmosphere (air movement) consists of both horizontal (wind) and vertical (air currents) dimensions. The atmosphere's motion is the result of thermal energy produced from the heating of the Earth's surface and the air molecules above. Because of differential heating of the Earth's surface, energy flows from the equator poleward.

Although air movement plays the critical role in transporting the energy of the lower atmosphere, bringing the warming influences of spring and summer and the cold chill of winter, the effects of air movements on our environment are often overlooked, even though wind and air currents are fundamental to how nature functions. All life on Earth has evolved with mechanisms dependent on air movement: pollen is carried by winds for plant reproduction; animals sniff the wind for essential information; and wind power was the motive force that began the earliest stages of the industrial revolution. Now we see the effects of winds in other ways, too: wind causes weathering (erosion) of the Earth's surface; wind influences ocean currents; and air pollutants and contaminants such as radioactive particles transported by the wind impact our environment.

CAUSES OF AIR MOTION

In all dynamic situations, forces are necessary to produce motion and changes in motion—winds and air currents. The air (made up of various gases) of the atmosphere is subject to two primary forces: (1) gravity and (2) pressure differences from temperature variations.

Gravity (gravitational forces) holds the atmosphere close to the Earth's surface. Isaac Newton's law of universal gravitation states that every body in the universe attracts another body with a force equal to

$$F = G\frac{m_1 m_2}{r2}$$

where

F = Force,
m1 and m2 = the masses of the two bodies
G = universal constant of 6.67 × 10 − 11N × m2/kg2
R = distance between the two bodies

Important point: The force of gravity decreases as an inverse square of the distance between the two bodies.

Thermal conditions affect density, which in turn causes gravity to affect vertical air motion and planetary air circulation. This affects how air pollution is naturally removed from the atmosphere.

Although forces in other directions often overrule gravitational force, the ever-present force of gravity is vertically downward and acts on each gas molecule, accounting for the greater density of air near the Earth.

Atmospheric air is a mixture of gases, so the gas laws and other physical principles govern its behavior. The pressure of a gas is directly proportional to its temperature. Pressure is force per unit area (P = F/A), so a temperature variation in air generally gives rise to a difference in pressure or force. This difference in pressure resulting from temperature differences in the atmosphere creates air movement—on both large and local scales. This pressure difference corresponds to an unbalanced force, and when a pressure difference occurs, the air moves from a high- to a low-pressure region.

In other words, horizontal air movements (called advective winds) result from temperature gradients, which give rise to density gradients and, subsequently, pressure gradients. The force associated with these pressure variations (pressure gradient force) is directed at right angles to (perpendicular to) lines of equal pressure (called isobars) and is directed from high to low pressure.

Look at figure 9.1. The pressures over a region are mapped by taking barometric readings at different locations. Lines drawn through the points (locations) of equal pressure are called isobars. All points on an isobar are of equal pressure, which means no air movement along the isobar. The wind direction is at right angles to the isobar in the direction of the lower pressure. In figure 9.1, notice that air moves down a pressure gradient toward a lower isobar like a ball rolls down a hill. If the isobars are close together, the pressure gradient force is large, and such areas are characterized by high wind speeds. If isobars are widely spaced (again, see figure 9.1), the winds are light because the pressure gradient is small.

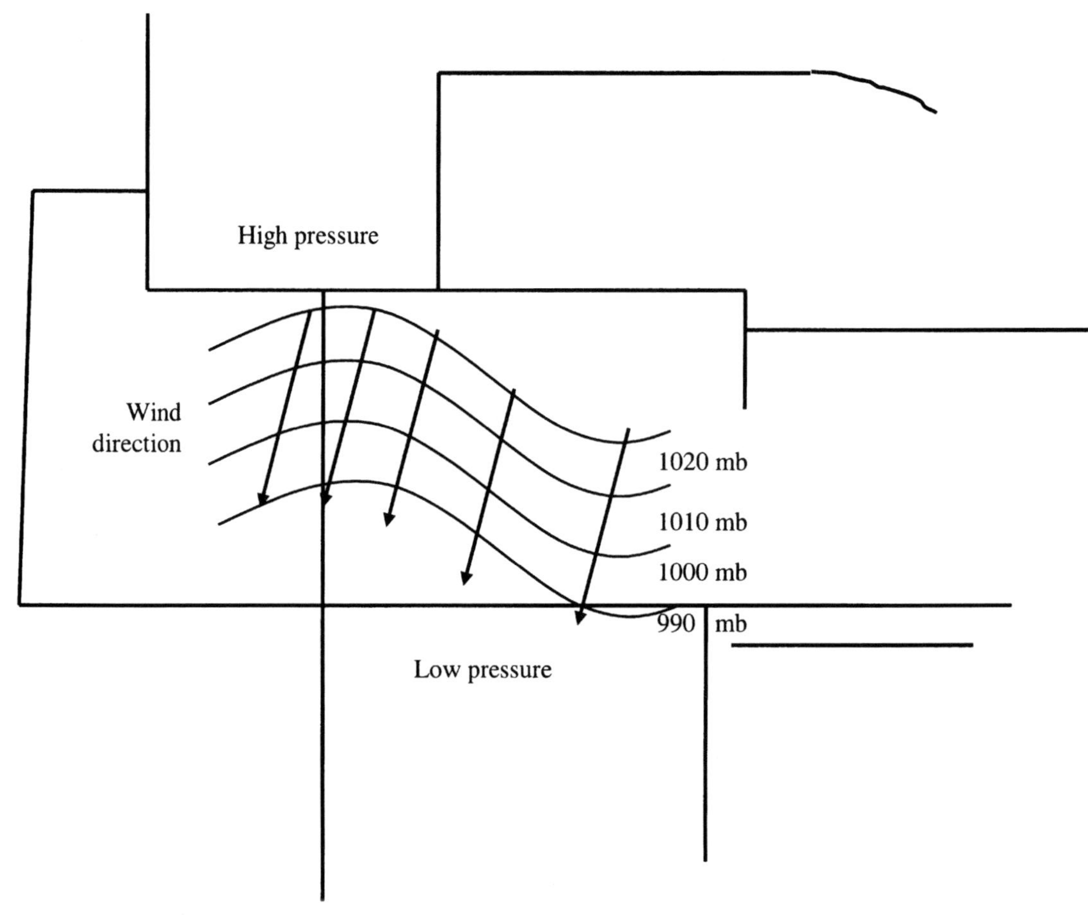

Figure 9.1. Isobars drawn through locations having equal atmospheric pressures. The air motion, or wind direction, is at right angles to the isobars and moves from a region of high pressure to a region of low pressure. Source: Spellman and Whiting (2006).

Localized air circulation gives rise to thermal circulation (a result of the relationship based on a law of physics whereby the pressure and volume of a gas is directly related to its temperature). A change in temperature causes a change in the pressure and/or volume of a gas. With a change in volume comes a change in density, since P = m/V, so regions of the atmosphere with different temperatures may have different air pressures and densities. As a result, localized heating sets up air motion and gives rise to thermal circulation. To gain an understanding of this phenomenon, consider figure 9.2.

Once the air has been set into motion, secondary forces (velocity-dependent forces) act. These secondary forces are (1) Earth's rotation (Coriolis force) and (2) contact with the rotating Earth (friction). The Coriolis force, named after its discoverer, French mathematician Gaspard Coriolis (1772–1843), is the effect of rotation on the atmosphere and on all objects on the Earth's surface. In the Northern Hemi-

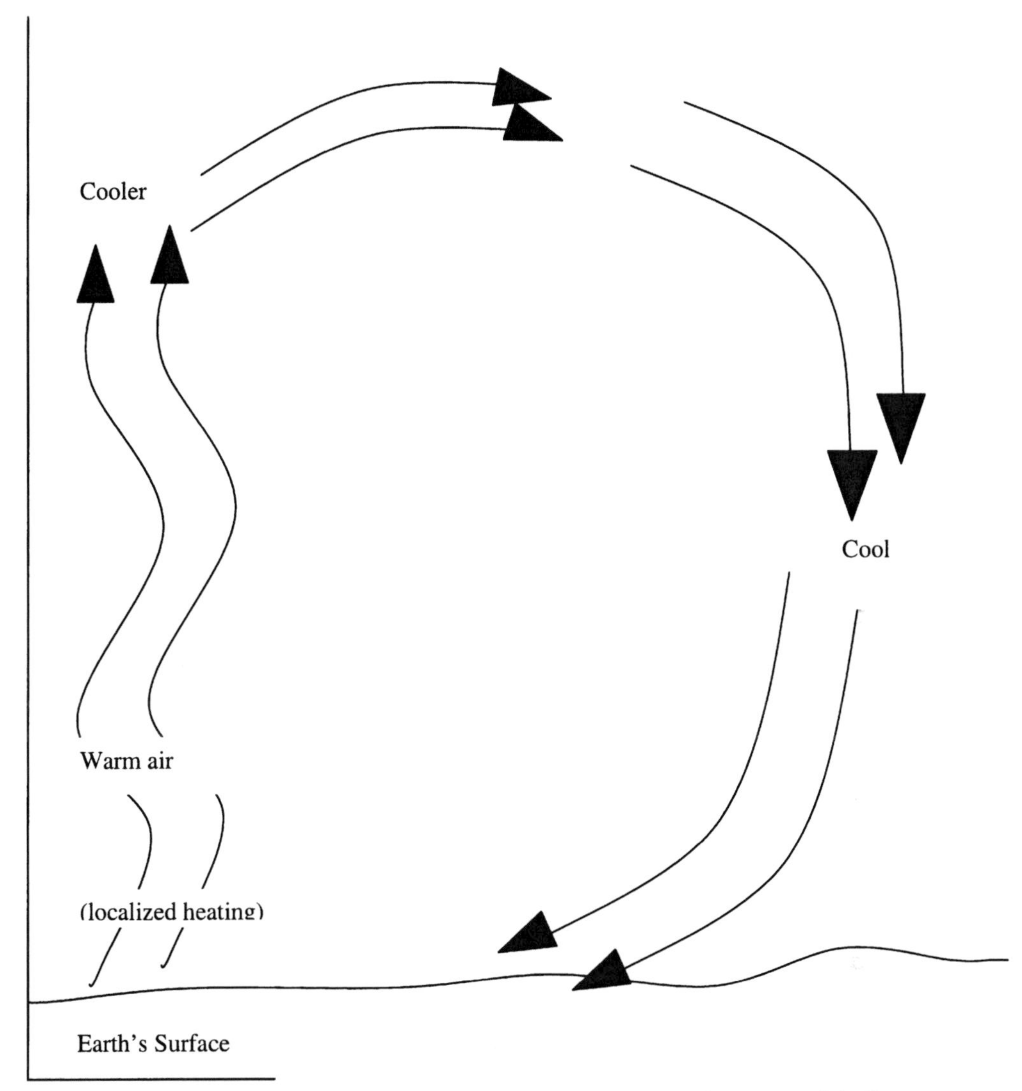

Figure 9.2. Thermal circulation of air. Localized heating, which causes air in the region to rise, initiates the circulation. As the warm air rises and cools, cool air near the surface moves horizontally into the region vacated by the rising air. The upper, still cooler, air then descends to occupy the region vacated by the cool air. Source: Spellman and Whiting (2006).

sphere, it causes moving objects and currents to be deflected to the right; in the Southern Hemisphere, it causes deflection to the left, because of the Earth's rotation. Air, in large-scale north or south movements, appears to be deflected from its expected path—that is, air moving poleward in the Northern Hemisphere appears to be deflected toward the east; air moving southward appears to be deflected toward the west.

Figure 9.3 illustrates the Coriolis effect on a propelled particle (analogous to the

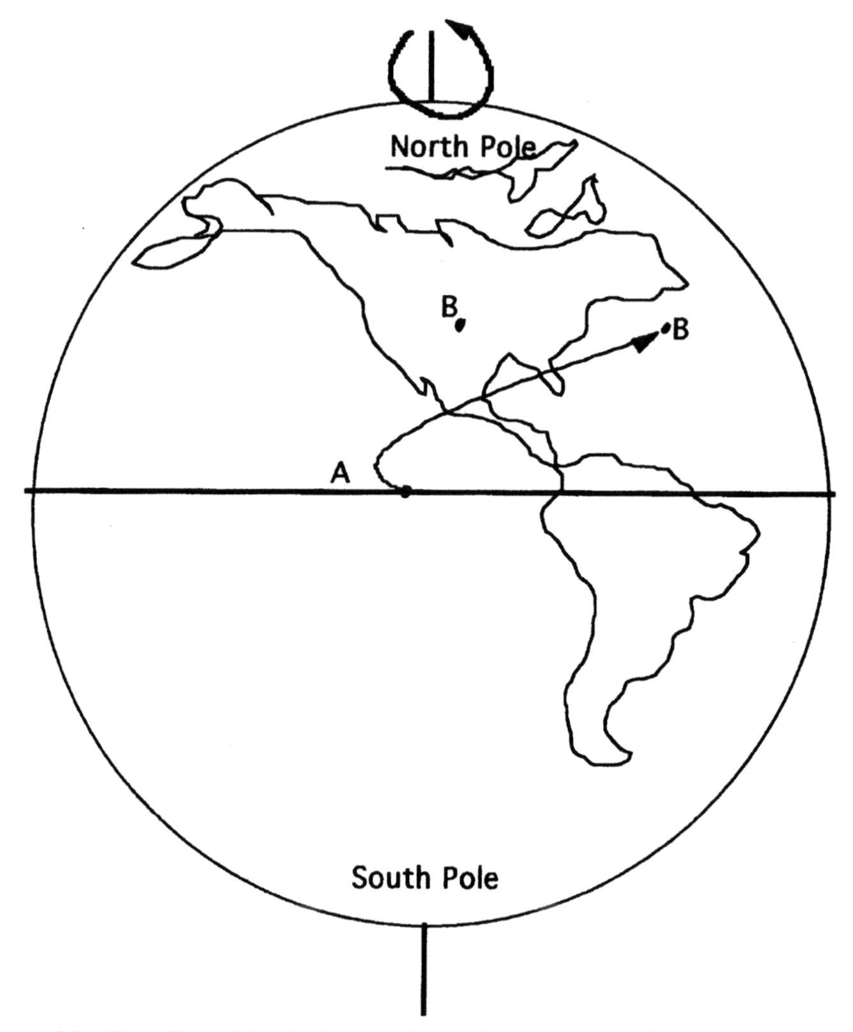

Figure 9.3. The effect of the Earth's rotation on the trajectory of a propelled particle. Source: Spellman and Whiting (2006).

apparent effect of an air mass flowing from point A to point B). From figure 9.3, the action of the Earth's rotation on the air particle as it travels north over the Earth's surface as the Earth rotates beneath it from east to west can be seen. Projected from point A to point B, the particle will actually reach point B because, as it is moving in a straight line (deflected), the Earth rotates east to west beneath it.

Friction (drag) can also cause the deflection of air movements. This friction (resistance) is both internal and external. The friction of the air molecules generates internal friction. Friction is also generated when air molecules run into each other. External friction is caused by contact with terrestrial surfaces. The magnitude of the frictional force along a surface is dependent on the air's magnitude and speed, and the opposing frictional force is in the opposite direction of the air motion.

LOCAL AND WORLD AIR CIRCULATION

Air moves in all directions, and these movements are essential for those of us on Earth: Vertical air motion is essential in cloud formation and precipitation. Horizontal air movement near the Earth's surface produces winds.

Wind is an important factor in human comfort, especially affecting how cold we feel. A brisk wind at moderately low temperatures can quickly make us uncomfortably cold. Wind promotes the loss of body heat, which aggravates the chilling effect, expressed through windchill factors in the winter (see table 9.5) and the heat index in the summer. These two scales describe the cooling effects of wind on exposed flesh at various temperatures.

Local winds are the result of atmospheric pressure differences involved with thermal circulations due to geographic features. Land areas heat up more quickly than do water areas, giving rise to a convection cycle. As a result, during the day, when land is warmer than the water, we experience a lake or sea breeze.

At night, the cycle reverses. Land loses its heat more quickly than water, so the air over the water is warmer. The convection cycle sets to work in the opposite direction, and a land breeze blows.

In the upper troposphere (above eleven to fourteen kilometers, west to east flows) are very narrow, fast-moving bands of air called jet streams. Jet streams have significant effects on surface airflows. When jet streams accelerate, divergence of air occurs at that altitude. This promotes convergence near the surface and the formation of cyclonic motion. Deceleration causes convergency aloft and subsidence near the surface, causing an intensification of high-pressure systems.

Jet streams are thought to result from the general circulation structure in the regions where great high- and low-pressure areas meet.

Table 9.5. Windchill Chart

Wind	*Temperature (degrees Fahrenheit)*											
MPH	*30*	*25*	*20*	*15*	*10*	*5*	*0*	*−5*	*−10*	*−15*	*−20*	*−25*
5	25	19	13	7	1	−5	−11	−16	−22	−28	−34	−40
10	21	15	9	3	−4	−10	−16	−22	−28	−35	−41	−47
15	19	13	3	0	−7	−13	−19	−26	−32	−39	−45	−51
20	17	11	4	−2	−9	−15	−22	−29	−35	−42	−48	−55
25	16	9	3	−4	−11	−17	−24	−31	−37	−44	−51	−58
30	15	8	1	−5	−12	−19	−26	−33	−39	−46	−53	−60
35	14	7	0	−7	−14	−21	−27	−34	−41	−48	−55	−62
40	13	6	−1	−8	−15	−22	−29	−36	−43	−50	−57	−64
45	12	5	−2	−9	−16	−23	−30	−37	−44	−51	−58	−65
50	12	4	−3	−19	−17	−24	−31	−38	−45	−52	−60	−67
55	11	4	−3	−11	−18	−25	−32	−39	−46	−54	−61	−68
60	10	3	−4	−11	−19	−26	−33	−40	−48	−55	−62	−69

Note: Grey cells indicate that frostbite occurs in fifteen minutes or less.

Source: *USA Today*, "Weather: Wind Chill Is a Guide to Winter Danger," August 13, 2001, www.usatoday-.com/weather/resources/basics/windchill/wind-chill-chart.htm (accessed November 15, 2009).

Weather and Climate

An eminent meteorologist once said, "A butterfly flapping its wings in Brazil can cause a tornado in Texas." What the meteorologist was implying is true to a point (and in line with what some critics might say): because of tiny nuances in Earth's weather patterns, making accurate, long-range weather predictions is extremely difficult.

What is the difference between weather and climate? Some people get these two confused, believing they mean the same thing, but they do not. In this section you will gain a clear understanding of the meaning of and difference between the two and also gain basic understanding of the role weather plays in air pollution.

Meteorology: The Science of Weather

Meteorology is the science concerned with the atmosphere and its phenomena. The atmosphere is the medium into which all air pollution is emitted. The meteorologist observes atmospheric processes such as temperature, density, (air) winds, clouds, precipitation, and other characteristics and endeavors to account for its observed structure and evaluation (weather, in part) in terms of external influence and the basic laws of physics. *Air pollution meteorology* is the study of how these atmospheric processes affect the fate of air pollutants.

Since the atmosphere serves as the medium into which air pollutants are released, the transport and dispersion of these releases are influenced significantly by meteorological parameters. Understanding air pollution meteorology and its influence in pollutant dispersion is essential in air quality planning activities. Planners use this knowledge to help locate air pollution monitoring stations and to develop implementation plans to bring ambient air quality into compliance with standards. Meteorology is used in predicting the ambient impact of a new source of air pollution and to determine the effect on air quality from modifications to existing sources (EPA 2005).

Weather is the state of the atmosphere, mainly with respect to its effect upon life and human activities; as distinguished from *climate* (the long-term manifestations of weather), weather consists of the short-term (minutes or months) variations of the atmosphere. Weather is defined primarily in terms of heat, pressure, wind, and moisture.

At high levels above the Earth, where the atmosphere thins to near vacuum, there is no weather; instead, weather is a near-surface phenomenon. This is evidenced clearly on a day-by-day basis where you see the ever-changing, sometimes dramatic, and often violent weather display.

In the study of air science and, in particular, air quality, the following determining factors are directly related to the dynamics of the atmosphere, resulting in local weather. These factors include strength of winds, the direction they are blowing, temperature, available sunlight (needed to trigger photochemical reactions, which produce smog), and the length of time since the last weather event (strong winds and heavy precipitation) cleared the air.

Weather events (such as strong winds and heavy precipitation) that work to clean the air we breathe are beneficial, obviously. However, few people would categorize the weather events such as tornadoes, hurricanes, and typhoons as beneficial. Other weather events have both a positive and negative effect. One such event is El Niño Southern Oscillation (ENSO), discussed below.

CASE STUDY: EL NIÑO SOUTHERN OSCILLATION

ENSO is a natural phenomenon that occurs every two to nine years on an irregular and unpredictable basis. El Niño is a warming of the surface waters in the tropical eastern Pacific, which causes fish to disperse to cooler waters and, in turn, causes the adult birds to fly off in search of new food sources elsewhere.

Through a complex web of events, El Niño (which means "the child" in Spanish because it usually occurs during the Christmas season off the coasts of Peru and Ecuador) can have a devastating impact on all forms of marine life.

During a normal year, equatorial trade winds pile up warm surface waters in the western Pacific. Thunderheads unleash heat and torrents of rain. This heightens the east-west temperature difference, sustaining the cycle. The jet stream blows from north Asia to California. During an ENSO year, trade winds weaken, allowing warm waters to move east. This decreases the east-west temperature difference. The jet stream is pulled farther south than normal, picks up storms it would usually miss, and carries them to Canada or California. Warm waters eventually reach South America.

One of the first signs of its appearance is a shifting of winds along the equator in the Pacific Ocean. The normal easterly winds reverse direction and drag a large mass of warm water eastward toward the South American coastline. The large mass of warm water basically forms a barrier that prevents the upwelling of nutrient-rich cold water from the ocean bottom to the surface. As a result, the growth of microscopic algae that normally flourish in the nutrient-rich upwelling areas diminishes sharply, and that decrease has further repercussions. For example, ENSO has been linked to patterns of subsequent droughts, floods, typhoons, and other costly weather extremes around the globe. Take a look at ENSO's effect on the west coast of the United States where it has been blamed for west coast hurricanes, floods, and early snowstorms. On the positive side, ENSO typically brings good news to those who live on the east coast of the United States: a reduction in the number and severity of hurricanes.

Note that, in addition to reducing the number and severity of hurricanes, in October 1997 the Associated Press reported that a new study has shown that ENSO also deserves credit for invigorating plants and helping to control the pollutant linked to global warming. Researchers have found that El Niño causes a burst of plant growth throughout the world, and this removes carbon dioxide from the atmosphere.

Atmospheric carbon dioxide (CO_2) has been increasing steadily for decades. The culprits are increased use of fossil fuels and the clearing of tropical rain forests. However, during an ENSO phenomenon, global weather is warmer, there is an increase in new plant growth, and CO_2 levels decrease.

Not only does ENSO have a major regional impact in the Pacific, but also its influence extends to other parts of the world through the interaction of pressure, air flow, and temperature effects.

ENSO is a phenomenon that, although not quite yet completely understood by scientists, causes both positive and negative results, depending upon where you live.

THE SUN: THE WEATHER GENERATOR

The Sun is the driving force behind weather. Without the distribution and reradiation to space of solar energy, we would experience no weather (as we know it) on Earth. The Sun is the source of most of the Earth's heat. Of the gigantic amount of solar energy generated by the Sun, only a small portion bombards Earth. Most of the Sun's solar energy is lost in space. A little over 40 percent of the Sun's radiation reaching Earth hits the surface and is changed to heat. The rest stays in the atmosphere or is reflected back into space.

Like a greenhouse, the Earth's atmosphere admits most of the solar radiation. When solar radiation is absorbed by the Earth's surface, it is reradiated as heat waves, most of which is trapped by carbon dioxide and water vapor in the atmosphere, which work to keep the Earth warm in the same way a greenhouse traps heat.

By now you are aware of the many functions performed by the Earth's atmosphere. You should also know that the atmosphere plays an important role in regulating the Earth's heating supply. The atmosphere protects the Earth from too much solar radiation during the day and prevents most of the heat from escaping at night. Without the filtering and insulating properties of the atmosphere, the Earth would experience severe temperatures similar to other planets.

On bright, clear nights, the Earth cools more rapidly than on cloudy nights because cloud cover reflects a large amount of heat back to Earth, where it is reabsorbed.

The Earth's air is heated primarily by contact with the warm Earth. When air is warmed, it expands and becomes lighter. Air warmed by contact with Earth rises and is replaced by cold air, which flows in and under it. When this cold air is warmed, it too rises and is replaced by cold air. This cycle continues and generates a circulation of warm and cold air, which is called *convection.*

At the Earth's equator, the air receives much more heat than the air at the poles. This warm air at the equator is replaced by colder air flowing in from north and south. The warm, light air rises and moves poleward high above the Earth. As it cools, it sinks, replacing the cool surface air that has moved toward the equator.

The circulating movement of warm and cold air (convection) and the differences in heating cause local winds and breezes. Different amounts of heat are absorbed by different land and water surfaces. Soil that is dark and freshly plowed absorbs much more than grassy fields. Land warms faster than water during the day and cools faster at night. Consequently, the air above such surfaces is warmed and cooled, resulting in production of local winds.

Winds should not be confused with air currents. Wind is primarily oriented toward horizontal flow. Air currents, on the other hand, are created by air moving upward and downward. Wind and air currents have direct impacts on air pollution. Air pollutants are carried and dispersed by wind. An important factor in determining the areas most affected by an air pollution source is wind direction. Since air pollution is a global problem, wind direction on a global scale is important.

Along with wind, another constituent associated with Earth's atmosphere is water. Water is always present in the air. It evaporates from the Earth, two-thirds of which is covered by water. In the air, water exists in three states: solid, liquid, and invisible vapor.

As mentioned, the amount of water in the air is called humidity. Again, the *relative humidity* is the ratio of the actual amount of moisture in the air to the amount needed for saturation at the same temperature. Warm air can hold more water than cold. When air with a given amount of water vapor cools, its relative humidity increases; when the air is warmed, its relative humidity decreases.

AIR MASSES

An air mass is a vast body of air (a macroscale phenomenon that can have global implications) in which the condition of temperature and moisture are much the same at all points in a horizontal direction. An air mass takes on the temperature and moisture characteristics of the surface over which it forms and travels, though its original characteristics tend to persist. The processes of radiation, convection, condensation, and evaporation condition the air in an air mass as it travels. Also, pollutants released into an air mass travel and disperse within the air mass. Air masses develop more commonly in some regions than in others. Table 9.6 summarizes air masses and their properties.

When two different air masses collide, a *front* is formed. A front is not a sharp wall but rather a zone of transition that is often several miles wide. Four frontal patterns—warm, cold, occluded, and stationary—can be formed by air of different temperatures. A *cold front* marks the line of advance of a cold air mass from below, as it displaces a warm air mass. A *warm front* marks the advance of a warm air mass as it rises up over a cold one.

When cold and warm fronts merge (the cold front overtaking the warm front) *occluded fronts* form. Occluded fronts can be called cold front or warm front occlusions. But, in either case, a colder air mass takes over an air mass that is not as cold.

The last type of front is the stationary front. As the name implies, the air masses around this front are not in motion. A stationary front can cause bad weather conditions that persist for several days.

Table 9.6. Classification of Air Masses

Name	*Origin*	*Properties*	*Symbol*
Arctic	Polar regions	Low temperatures, low specific but high summer relative humidity, the coldest of the winter air masses	A
Polar continental*	Subpolar continental areas	Low temperatures (increasingc with southward movement), low humidity, remaining constant	cP
Polar maritime	Subpolar area and arctic region	Low temperatures increasing with movement, higher humidity	mP
Tropical continental	Subtropical high-pressure land areas	High temperatures, low moisture content	cT
Tropical maritime	Southern borders of oceanic, subtropical, high-pressure areas	Moderate high temperatures, high relative and specific humidity	mT

* The name of an air mass, such as Polar continental, can be reversed to continental Polar, but the symbol, cP, is the same for either name.

Source: Environmental Protection Agency, "Basic Air Pollution Meteorology," APTI Course SI:409, 2005, www.epa.gov/apti (accessed January 9, 2008).

Thermal Inversions and Air Pollution

Earlier, it was pointed out that during the day the Sun warms the air near the Earth's surface. Normally, this heated air expands and rises during the day, diluting low-lying pollutants and carrying them higher into the atmosphere. Air from surrounding high-pressure areas then moves down into the low-pressure area created when the hot air rises (see figure 9.4a, b). This continual mixing of the air helps keep pollutants from reaching dangerous levels in the air near the ground.

Sometimes, however, a layer of dense, cool air is trapped beneath a layer of less dense, warm air in a valley or urban basin. This is called a *thermal inversion*. In effect, a warm-air lid covers the region and prevents pollutants from escaping in upward-flowing air currents. Usually, these inversions trap air pollutants (i.e., plume dispersion is inhibited) at ground level for a short period of time. However, sometimes they last for several days when a high-pressure air mass stalls over an area, trapping air pollutants at ground level where they accumulate to dangerous levels.

The best-known location in the United States where thermal inversions occur almost on a daily basis is in the Los Angeles basin. The Los Angeles basin is a valley with a warm climate and light winds, surrounded by mountains located near the

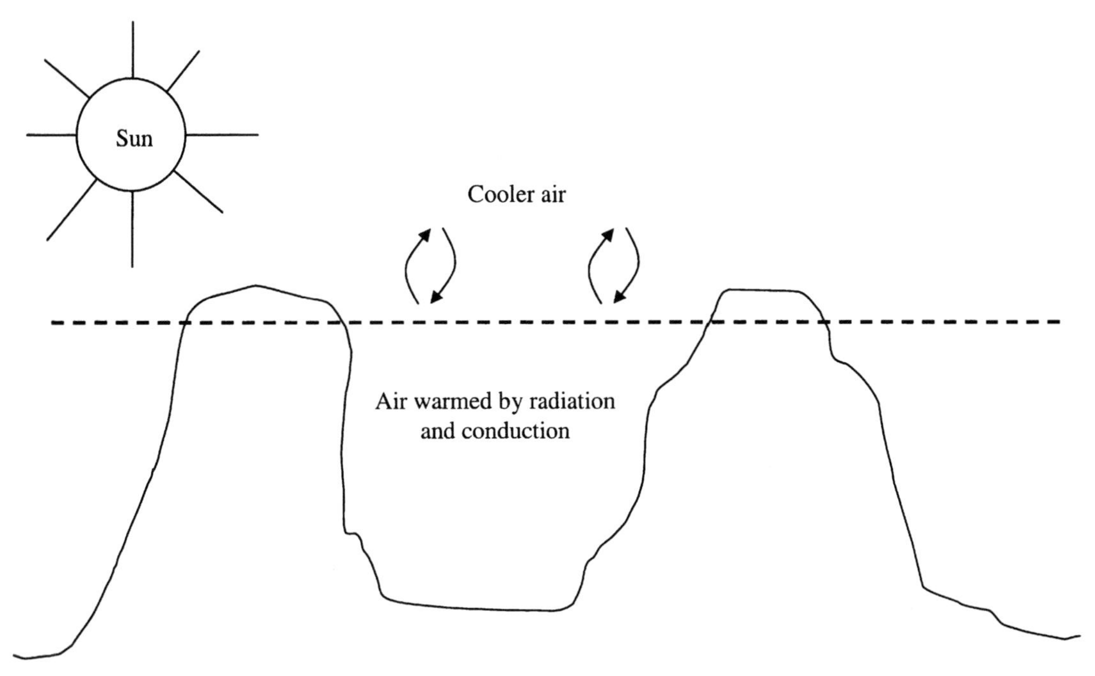

Figure 9.4 (a) Normal conditions. Air at Earth's surface is heated by the sun and rises to mix with the cooler air above it.

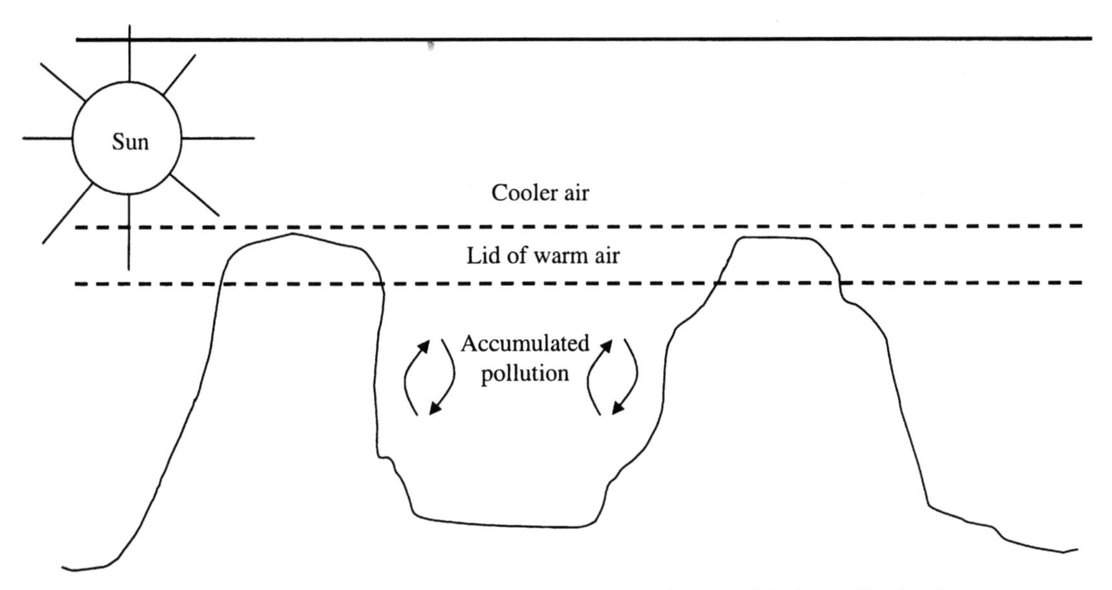

Figure 9.4 (b) Thermal inversion. A layer of warm air forms a lid above the Earth, and the cooler air at the surface is unable to mix with the warm air above. Pollutants are trapped.

Pacific Coast. Los Angeles is a large city with a large population of people and automobiles and possesses the ideal conditions for smog, which is worsened by frequent thermal inversions.

Microclimates

When we think about climate, we are generally referring to overall or generalized weather conditions at a particular place or region over a period of time. In addition to precipitation and temperature, climates have been classified into zones by vegetation, moisture index, and even measures of human discomfort. Using the general climate zone names allows us to differentiate between a particular climate (with its specific climatic conditions) in respect to another climate with differing conditions. When geographical patterns in the weather occur again and again over a long period, they can all be used to define the climate of a region. Some climate zones are known as the hot climates, which include desert, tropical continental, tropical monsoon, tropical marine, or equatorial types. Warm climates include west coast (Mediterranean) and warm east coast. Another category includes the cool climates such as cold desert, west coast (cool), cool temperature interior, and cool, temperate east coast types. Finally, there are the mountain and the cold climate categories of cold continental and polar or tundra. Each of these different climate types is differentiated from the others. However, they all have one major feature in common: they are large-scale regional climates (with variations), occurring at various places throughout the world. They only consider the broad similarities between a particular climate at various locations worldwide; local differences are ignored and boundaries are approximate.

What factors determine the variations of climate over the surface of the Earth? The primary factors are as follows: (1) the effect of latitude and the tilt of the Earth's axis to the plane of the orbit about the Sun; (2) the large-scale movements of different wind belts over the Earth's surface; (3) the temperature difference between land and sea; (4) the contours of Earth's surface; and (5) the location of the area in relation to ocean currents.

What factors determine how climates are distributed? When considering climate distribution, remember that the world does not fall into compartments. The globe is a mosaic of many different types of climate, just as it is a mosaic of numerous types of ecosystems. The complexity of the distribution of land and sea and the consequent complexity of the general circulation of the atmosphere have a direct effect on the distribution of the climate.

WHAT IS A MICROCLIMATE?

In answering this question, scale must first be talked about. For example, let's take a look at the flow of air within a very small environment: the emission of smoke from a chimney. This flow represents one of the smallest spatial subdivisions of atmospheric motion, or microscale weather. On a more realistic but still relatively small scale, we must consider the geographical, biological, and man-made features that make local climate different from the general climate. This local climatic pattern is called a *microclimate.*

What are the elements or conditions that cause local or microclimates? Location, location, location—and local conditions—are the main ingredients making up a microclimate. Let's look at one example.

Large inland lakes moderate temperature extremes and climatic differences between the windward and lee sides. For example, Seattle, on the windward side of Lake Washington, and Bellevue, on the lee side only about nine miles east, have microclimatic differences (although modest) between the two cities. These microclimatic differences exist as temperature fluctuations, precipitation levels, wind speed, and relative humidity.

Even more dramatic differences can be seen in such parameters when a comparison is made between a city such as Milwaukee, on the windward side of Lake Michigan, and Grand Haven, on the lee side, only eighty-five miles east.

These and other examples of microclimates are listed below:

- near the ground
- over open land areas,
- in woodlands or forested areas,
- in valley regions,
- in hillside regions,
- in urban areas, and
- in seaside locations.

In the following sections we take a closer look at these microclimates: at their nature, causative factors, and geographical/topographical locations.

MICROCLIMATES NEAR THE GROUND

Nowhere in the atmosphere are climatic differences as distinct as they are near the ground. For instance, when you go to the beach on a warm summer day, you no doubt have noticed that the grass and water are much cooler to your feet than the sand. So, you may ask, what is it about this area near the ground that produces a microclimate with such major differences?

It's the interface (or activity zone) between the atmosphere and the ground surface (sandy shore) that causes the stark difference in temperature variability. Energy is reaching the sandy beach from the Sun and from the atmosphere (though to a much lesser extent). The energy is either reflected and then returned to the atmosphere in a different form or is absorbed and stored in the sandy surface as heat.

Ground-level energy absorption is very sensitive to the nature of the ground surface. Ground-surface color, wetness, cover (vegetation), and topography are conditions that all affect the interaction between the ground and the atmosphere. Consider a snow-covered ground, for example. Clean snow reflects solar radiation, so the surface remains cool and the snow fails to melt. However, dirty snow absorbs more radiation, heats up, and is likely to melt. If the snowy area is shielded by vegetation, the vegetation, too, may protect the snow from the heat of the Sun.

A surface cover such as clean snow has the ability to reflect solar radiation because of its high albedo. *Albedo* (the ratio between the light reflected from a surface and the total light falling on it) always has a value less than or equal to one. An object with a high albedo, near one, is very bright, while a body with a low albedo, near zero, is dark. For example, freshly fallen snow typically has an albedo that is between 75 percent and 95 percent; that is, 75 percent to 95 percent of the solar radiation that is incident on snow is reflected. At the other extreme, the albedo of a rough, dark surface, such as a green forest, may be as low as 5 percent. The albedos of some common surfaces are listed in table 9.7. The portion of insolation not reflected is absorbed by the Earth's surface, warming it. This means Earth's albedo plays an important part in the Earth's radiation balance and influences the mean annual temperature and the climate on both local and global scales.

MICROCLIMATES OVER OPEN LAND AREAS

Many different properties of ground layer or soil type influence conditions in the thin layer of atmosphere just above it. Light-colored soils do not absorb energy as efficiently as do organically rich, darker soils. Another important factor is soil moisture. Wet soils are normally dark, but moist soil (because water has a large heat capacity) requires a great deal of energy to raise its temperature. A moist soil warms up more slowly than a dry one.

Soil is a heterogeneous mixture of various particles. In between the soil particles is a large amount of air—air that is a poor conductor of heat. The larger the amount of air between the soil particles, the slower the heat transfers through the soil. As demonstrated in our example of the sandy beach, on a hot, sunny day the heat is trapped in the upper layers, so the surface layers warm up more rapidly and become

Table 9.7. The albedo of some surface types

Surface	*Albedo (in % reflected)*
Water (low Sun)	10–100
Water (high Sun)	3–10
Grass	16–26
Glacier ice	20–40
Deciduous forest	15–20
Coniferous forest	5–15
Old snow	40–70
Fresh snow	75–95
Sea ice	30–40
Blacktopped tarmac	5–10
Desert	25–30
Crops	5–25

extremely hot. Water conducts heat more readily than air, so soils that contain some moisture are able to transmit warmth away from the surface more easily than dry soils. This is not always the case, however. If the soil contains too much water, the large heat capacity of the water will prevent the soil warming despite heat being conducted from the surface.

MICROCLIMATES IN WOODLANDS OR FORESTED AREAS

When making microclimate comparisons between open land areas and forested areas (commonly referred to as a forest climate), the differences are quite apparent. Forested areas, for example, are generally warmer in winter than the open areas, while open land is warmer in summer than forested areas. The forest climate has reduced wind speeds, while the open land area has higher wind speeds. The forest climate has higher relative humidity, while the open area has lower relative humidity. In the forest climate, water storage capacity is higher and evaporation rates are lower, while in the open land area water storage capacity is lower and evaporation is higher.

MICROCLIMATES IN VALLEY AND HILLSIDE REGIONS

Heavy, cold air flows downhill, forming cold pockets in valleys. Frost is much more common there, so orchards of apples and oranges and vines of grapes are planted on hillsides to ensure frost drainage when cold spells come.

Probably the best way in which to describe the microclimate in a typical valley region is to compare and contrast it with a hillside environment.

In a typical valley region, the daily minimum temperature is much lower than that in a hillside area. The daily and annual temperature range for a valley is much larger than that of a hillside area. In a valley region, more frost occurs than in a hillside region. Windspeed at night is lower in a valley than on a hillside, and morning fog is more prevalent and lasts longer in a valley region.

MICROCLIMATES IN URBAN AREAS

The microclimate in an urban area as compared to that of the countryside is usually quite obvious. A city, for example, is usually characterized by having haze and smog, higher temperatures, lower wind speed, and reduced radiation. The countryside, on the other hand, is characterized by clean, clear air, lower temperatures, and high wind speeds and radiation.

These different microclimatic conditions should come as no surprise to anyone, especially when you consider what happens when a city is built. Instead of a mixture of soil or vegetation, the surface layer is covered with concrete, brick, glass, and stone surfaces ranging to heights of several hundred feet. These materials have vastly different physical properties from soil and trees. They shed and carry away water, absorb heat, block and channel the passage of winds, and present albedo levels significantly different from those of the natural world. All of these factors (and more) work to alter the climate conditions in the area.

Did You Know?

Urban areas have added roughness features and different thermal characteristics due to the presence of man-made elements. The thermal influence dominates the influence of the frictional components. Building materials such as brick and concrete absorb and hold heat more efficiently than soil and vegetation found in rural areas. After the sun sets, the urban area continues to radiate heat from buildings, paved surfaces, and so on. Air warmed by the urban complex rises to create a dome over the city. It is called the *heat island effect.* The city emits heat all night. Just when the urban area begins to cool, the sun rises and begins to heat the urban complex again. Generally, city areas never revert to stable conditions because of the continual heating that occurs. (EPA 2005)

MICROCLIMATES IN SEASIDE LOCATIONS

The major climatic feature associated with seaside locations is the sea breeze. Sea breezes are formed by the different responses to heating of water and land. For example, if we have a bright, sunny morning with little wind, the ground surface warms rapidly as it absorbs shortwave radiation. Most of this heat is retained at the surface, although some will be transferred through the soil. As a result, the temperature of the ground surface increases and some of the heat warms the air above. When the Sun sets, the surface starts to cool rapidly, because there is little store of heat in the soil. Thus, we find that land surfaces are characterized by high day (and summer) temperatures and low night (and winter) temperatures.

Now let's take a look at the response of the sea, which is very different. Solar energy (sunshine) is able to penetrate through the water to a certain level. Much solar energy has to be absorbed to raise its temperature. Through wave action and convection, the warm surface water is mixed with cooler, deeper water. With enough solar energy and time, the top several feet of water forms an active layer where temperature change is slow. Slight warming occurs during the day and slight cooling at night. This means that the sea is normally cooler than the land by day and warmer by night.

The higher temperature over the land by day generates a weak low-pressure area. As this intensifies during daytime heating, a flow of cool, more humid air spreads inland from the sea, gradually changing in strength and direction during the day. At night the reverse occurs, with circulation of air from the cooler land to the warmer sea, though as the temperature difference is usually less, the land breeze is weak. Even large lakes can show a breeze system of this nature.

Chapter Review Questions

9.1. Defined as the average weather conditions of a place, usually measured annually: ________.
9.2. The part of the atmosphere where all living things exist: ________.
9.3. The ________ contains life-protecting ozone.
9.4. Contains as little as 0.1 percent water vapor.
9.5. Cloud type meaning tufts of hair: ________
9.6. Refers to the mass of the water vapor within a given mass of dry air: ________.
9.7. ________ indicates how much moisture the air can hold above its stated level.
9.8. Particles on which droplets form are called ________.
9.9. ________ is the spontaneous rising of moist air due to instability.
9.10. The "sea of darkness": ________.
9.11. The science of weather: ________.
9.12. The weather generator: ________.
9.13. A (an) ________ is a layer of dense, cool air trapped beneath a layer of less dense, warm air in a valley.
9.14. A parameter used to measure degree of light reflectivity on the surface of Earth: ________.

References and Recommended Reading

Ahrens, D. 1994. *Meteorology today: An introduction to weather, climate and the environment*, 5th ed. Minneapolis: West Publishing.
Anthes, R. A. 1996. *Meteorology*, 7th ed. Upper Saddle River, N.J.: Prentice Hall.
Anthes, R. A., J. J. Cahir, A. B. Fraizer, and H. A. Panofsky. 1984. *The atmosphere*, 3rd ed. Columbus, Ohio: Charles E. Merrill Publishing.
Environmental Protection Agency (EPA). 2005. Basic air pollution meteorology. APTI Course SI:409. www.epa.gov/apti (accessed January 9, 2008).
———. 2007. Air pollution control: Atmosphere. www.epa.gov/apti/course422/ap1.html (accessed December 28, 2007).
Ingersoll, A. P. 1983. The atmosphere. *Scientific American* 249 (33): 162–74.
Lutgens, F. K., and E. J. Tarbuck. 1982. *The atmosphere: An introduction to meteorology*. Englewood Cliffs, N.J.: Prentice Hall.
Miller, G. R., Jr. 2004. *Environmental science*, 10th ed. Sydney Australia: Thomson-Brooks/Cole.
Moran, J. M., M. D. Morgan, and J. H. Wiersma. 1986. *Introduction to environmental science*, 2nd ed. New York: W. H. Freeman.
National Academy of Sciences. 1962. *Water balance in the U.S.* Washington, D.C.: National Research Council Publication 100-B.

National Aeronautics and Space Administration (NASA). 2008a. *Observing cloud type*. Washington, D.C.: Author.

———. 2008b. Precipitation, storms, and other weather phenomena. http://rst.gsfc.nasa.gov/Sect14/Sect14_1d.html (accessed November 15, 2009).

National Oceanic and Atmospheric Administration (NOAA). 2007. Cloud types. www.gfdl.-NOAA.gov/~01/weather/clouds.html (accessed December 29, 2007).

Shipman, J. T., J. L. Adams, and J. D. Wilson. 1987. *An introduction to physical science*, 5th ed. Lexington, Mass.: D. C. Heath.

Spellman, F. R. 2007. *The science of water*, 2nd ed. Boca Raton, Fla.: CRC Press.

Spellman, F. R., and N. E. Whiting. 2006. *Environmental science and technology: Concepts and applications*, 2nd ed. Lanham, Md.: Government Institutes.

U.S. Geological Survey (USGS). 2008. The water cycle: Evapotranspiration. http://ga.water.usgs.gov/edu/watercycleevapotranspiration.html (accessed January 7, 2008).

IV

SOIL ASPECTS

Soil-forming fungi. Appalachian Trail, Virginia. Photo by Frank R. Spellman.

Soil-forming fungi. Appalachian Trail, Virginia. Photo by Frank R. Spellman.

Life sprouting from magma flow. Craters of the Moon, Idaho. Photo by Frank R. Spellman.

Plant life making inroads into soil of Badlands, South Dakota. Photo by Frank R. Spellman.

CHAPTER 10

Soil

> We stand on soil, not on earth.
>
> —Illich et al. (1991)

> Because soil is a dynamic living medium that is a critical element of our environment, we can't study even at the nearest margin of geographical science without studying soil. Soils link the natural environment with climate and vegetation. Simply, soils have a profound effect on humankind's activities through their relative fertility.
>
> —F. R. Spellman (2009)

Soil is part of a complete ecological unit; it is a critical natural system.[1] Yet, we take soil for granted. It's always been there, with the implied corollary that it will always be there. But where does soil come from?

Of course, soil was formed, and in a never-ending process, it is still being formed. However, soil formation is a slow process—one at work over the course of millennia as mountains are worn away to dust through erosion, weathering, and bare rock succession.

Any activity, human or natural, that exposes rock to air begins the process. As a glacier recedes up the valley, land that had been buried for thousands of years is suddenly uncovered. All the soil and plants have long ago been scraped away, but soon the bare rocks will support a rich growth of plants and animals. Through the agents of physical and chemical weathering, through extremes of heat and cold, and through storms and earthquake and entropy, bare rock is gradually broken, reduced (see figures 10.1 and 10.2), and worn away. As its exterior structures are exposed and weakened, plant life appears to speed the process along.

About one year after bare rocks have been exposed, lichens cover the bare rock first, growing on the rock's surface, etching it with mild acids and collecting a thin film of soil that is trapped against the rock and clings. This changes the conditions of growth so much that the lichens can no longer survive and are replaced by mosses.

The mosses establish themselves in the soil, trapped and enriched by the lichens, and collect even more soil (see figure 10.3). They hold moisture to the surface of the rock, setting up another change in environmental conditions.

Well-established mosses hold enough soil to allow herbaceous plant seeds to invade the rock. Grasses and small shrubs and flowering plants move in, sending out fine root systems that hold more soil and moisture and work their way into minute

Figure 10.1. Devils Tower, Wyoming. Photo by Frank R. Spellman.

fissures in the rock's surface. Some of these plants are fireweed, blueberry bushes, and lupine. More and more organisms join the increasingly complex community, including bees and spiders. These small creatures bring seeds, caught on their bodies, to the area so that new plants begin to take root.

Weedy shrubs are the next invaders, with heavier root systems that find their way into every crevice. Each stage of succession affects the decay of the rock's surface and adds its own organic material to the mix. Over the course of time, mountains are worn away, eaten away to soil, as time, plants, weather, and extremes of weather work on them.

The parent material, the rock, becomes smaller and weaker as the years, decades, centuries, and millennia go by, creating the rich, varied, and valuable mineral resource we call soil.

All about Soil

Before we begin a journey that takes us through the territory that is soil and examine soil from micro- to macrolevels, we need to stop for a moment and discuss why, beyond the obvious reason, soil is so important to us, to our environment, to our very survival.

Figure 10.2. Fallen column from Devils Tower. The column and tower fragments shown here are in early stages of bare-rock-succession; lichens and mosses are beginning to grow and to convert the rock column to soil. Photo by Frank R. Spellman.

FUNCTIONS OF SOIL

We normally relate soil to our backyards, to farms, to forests, or to a regional watershed. We think of soil as the substance upon which plants grow. Soils play other roles, though. They have six main functions important to us: (1) soil is a medium for plant growth; (2) soils regulate our water supplies; (3) soils are recyclers of raw materials; (4) soils provide a habitat for organisms; (5) soils are used as an engineering medium; and (6) soils provide materials. Let's take a closer look at each of the functions of soils.

Soil: A Plant Growth Medium

We are all aware of the primary function of soil: soil serves as a plant growth medium, a function that becomes more important with each passing day as Earth's population continues to grow. However, while it is true that soil is a medium for plant growth, soil is actually alive as well. Soil exists in paradox: we depend on soil for life, and at the same time, soil depends on life. Its very origin, its maintenance, and its true nature are intimately tied to living plants and animals. What does this mean? Let's take a look at how one renowned environmental writer, whose elegant prose brought this point to the forefront, explained this paradox. "The soil community . . . consists of a

Figure 10.3. Rock face with lichen and then moss growth. Falling Springs Trail, Virginia. Photo by Frank R. Spellman.

web of interwoven lives, each in some way related to the others—the living creatures depending on the soil, but the soil in turn is a vital element of the earth only so long as this community within it flourishes" (Carson 1962, 56).

The soil might say to us if it could, "Don't kill off the life within me, and I will do the best I can to provide life that will help to sustain your life." What we have here is a tradeoff—one vitally important both to soil and to ourselves. Remember that most of Earth's people are tillers of the soil—and the soil is their source of livelihood—and those soil tillers provide food for us all.

As a plant growth medium, soil provides vital resources and performs important functions for the plant. To grow in soil, plants must have water and nutrients—soil provides these. To grow and to sustain growth, a plant must have a root system—soil provides pore spaces for roots. To grow and maintain growth, a plant's roots must have oxygen for respiration and carbon dioxide exchange and ultimate diffusion out of the soil—soil provides the air and pore spaces (the soil's ventilation system) for this. To continue to grow, a plant must have support—soil provides this support.

If a plant seed is planted in a soil and is exposed to the proper amount of sunlight, for growth to occur, the soil must provide nutrients through a root system that has space to grow, a continuous stream of water (it requires about five hundred grams of water to produce one gram of dry plant material) for root nutrient transport and plant cooling, and a pathway for both oxygen and carbon dioxide transfer. Just as important, soil water provides the plant with its normal fullness or tension (turgor) it needs to stand—the structural support it needs to face the Sun for photosynthesis to occur.

As well as the functions stated above, soil is also an important moderator of temperature fluctuations. If you have ever dug in a garden on a hot, summer day, you probably noticed that the soil was warmer (even hot) on the surface but much cooler just a few inches below the surface.

Soil: Regulator of Water Supplies

When we walk on land, few of us probably realize that we are actually walking across a bridge. This bridge (in many areas) transports us across a veritable ocean of water below us, deep—or not so deep—under the surface of the Earth.

Consider what happens to rain. Where does the rainwater go? Some, falling directly over water bodies, become part of the water body again, but an enormous amount falls on land. Some of the water, obviously, runs off—always following the path of least resistance. In modern communities, stormwater runoff is a hot topic. Cities have taken giant steps to try to control runoff—to send it where it can be properly handled, to prevent flooding.

Let's take a closer look at precipitation and the "sinks" it "pours" into, then relate this usually natural operation to soil water. We begin with surface water, then move on to that ocean of water below the soil's surface: groundwater.

Surface water (water on the Earth's surface as opposed to subsurface water—groundwater) is mostly a product of precipitation: rain, snow, sleet, or hail. Surface water is exposed or open to the atmosphere and results from the movement of water on and just under the Earth's surface (overland flow). This overland flow is the same thing as surface runoff, which is the amount of rainfall that passes over the Earth's surface. Specific sources of surface water include rivers, streams, lakes, impoundments, shallow wells, rain catchments, and tundra ponds or meskegs (peat bogs).

Most surface water is the result of surface runoff. The amount and flow rate of surface runoff is highly variable. This variability stems from two main factors: (1) human interference (influences) and (2) natural conditions. In some cases, surface water runs quickly off land. Generally, this is undesirable (from a water resources standpoint) because it does not provide enough time for water to infiltrate into the ground and recharge groundwater aquifers. Other problems associated with quick surface water runoff are erosion and flooding. Probably the only good thing that can be said about surface water that quickly runs off land is that it does not have enough time (normally) to become contaminated with high mineral content. Surface water running slowly off land may be expected to have all the opposite effects.

Surface water travels over the land to what amounts to a predetermined destination. What factors influence how surface water moves? Surface water's journey over the face of the Earth typically begins at its drainage basin, sometimes referred to as its drainage area, catchment, or watershed. For a groundwater source, this is known as the recharge area—the area from which precipitation flows into an underground water source.

A surface water drainage basin is usually an area measured in square miles, acres, or sections, and if a city takes water from a surface water source, how large (and what lies within) the drainage basin is essential information for the assessment of water quality.

We all know that water doesn't run uphill. Instead, surface water runoff (like the flow of electricity) follows along the path of least resistance. Generally speaking, water within a drainage basin will naturally (by the geological formation of the area) be shunted toward one primary watercourse (a river, stream, creek, or brook) unless some man-made distribution system diverts the flow.

Various factors directly influence the surface water's flow over land. The principal factors are as follows:

1. Rainfall duration: Length of the rainstorm affects the amount of runoff. Even a light, gentle rain will eventually saturate the soil if it lasts long enough. Once the saturated soil can absorb no more water, rainfall builds up on the surface and begins to flow as runoff.
2. Rainfall intensity: The harder and faster it rains, the more quickly soil becomes saturated. With hard rains, the surface inches of soil quickly become inundated, and with short, hard storms, most of the rainfall may end up as surface runoff because the moisture is carried away before significant amounts of water are absorbed into the Earth.
3. Soil moisture: Obviously, if the soil is already laden with water from previous rains, the saturation point will be reached sooner than if the soil were dry. Frozen soil also inhibits water absorption: up to 100 percent of snowmelt or rainfall on frozen soil will end up as runoff because frozen ground is impervious.
4. Soil composition: Runoff amount is directly affected by soil composition. Hard rock surfaces will shed all rainfall, obviously, but so will soils with heavy clay composition. Clay soils possess small void spaces that swell when wet. When the void spaces close, they form a barrier that does not allow additional absorption or infiltration. On the opposite end of the spectrum, coarse sand allows easy water flow-through, even in a torrential downpour.
5. Vegetation cover: Runoff is limited by ground cover. Roots of vegetation and pine needles, pinecones, leaves, and branches create a porous layer (sheet of decaying, natural, organic substances) above the soil. This porous "organic" sheet (ground cover) readily allows water into the soil. Vegetation and organic waste also act as a cover to protect the soil from hard, driving rains. Hard rains can compact bare soils, close off void spaces, and increase runoff. Vegetation and ground cover work to maintain the soil's infiltration and water-holding capacity. Note that vegetation and ground cover reduce evaporation of soil moisture as well.
6. Ground Slope: Flatland water flow is usually so slow that large amounts of rainfall can infiltrate the ground. Gravity works against infiltration on steeply sloping ground where up to 80 percent of rainfall may become surface runoff.
7. Human influences: Various human activities have a definite impact on surface water runoff. Most human activities tend to increase the rate of water flow. For

example, canals and ditches are usually constructed to provide steady flow, and agricultural activities generally remove ground cover that would work to retard the runoff rate. On the opposite extreme, man-made dams are generally built to retard the flow of runoff.

Soil: Recycler of Raw Materials

Can you imagine what it would be like to step out into the open air and be hit by a stench that would not only offend your olfactory sense but could almost reach out and grab you? You look out upon the cluttered fields in front of your domicile and see nothing but stack upon stack of the sources of horrible, putrefied, foul, decaying, gagging, choking, retching stench. We are talking about plant and animal remains and waste (mountains of it), reaching toward the sky—surrounded by colonies of flies of all varieties. "Impossible," you say. Well, thankfully (in most cases) you are right. However, if it were not for the power of the soil to recycle waste products, then this scene or something like it is imaginable and even possible, building a mountain toward the Moon—but of course it would be impossible because under these described conditions there would be no life to die and to stack up anywhere, for that matter.

Soil is a recycler—probably the premier recycler on Earth. The simple fact is that if it were not for soil's incredible recycling ability, plants and animals would have run out of nourishment long ago. Soil recycles in other ways. For example, consider the geochemical cycles (i.e., the chemical interactions between soil, water, air, and life on Earth) in which soil plays a major role.

Soil possesses the incomparable ability and capacity to assimilate great quantities of organic wastes and turn them into beneficial organic matter (humus), then to convert the nutrients in the wastes to forms that can be utilized by plants and animals. In turn, the soil returns carbon to the atmosphere as carbon dioxide, where it again will eventually become part of living organisms through photosynthesis. Soil performs several different recycling functions—most of them good, some of them not so good.

Consider one recycling function of soil that may not be so good. Soils have the capacity to accumulate large amounts of carbon as organic soil matter, which can have a major impact on global change such as greenhouse effect. Moreover, it is important that wastes be applied in appropriate amounts and not contain toxic and environmentally harmful elements or compounds that could poison soils, wastes, and plants.

Soil: Habitat for Soil Organisms

"Life not only formed the soil, but other living things of incredible abundance and diversity now exist within it; if this were not so the soil would be a dead and sterile thing" (Carson 1962, 53). One thing is certain, most soils are not dead and sterile things. The fact is, a handful of soil is an ecosystem. It may contain up to billions of organisms, belonging to thousands of species. Table 10.1 lists a few (very few) of these organisms. Obviously, communities of living organisms inhabit the soil. What is not

Table 10.1. Soil Organisms (a representative sample)

Microorganisms (protists)
Bacteria
Fungi
Actinomycetes
Algae
Protozoa
Nonarthropod animals
Nematodes
Earthworms and potworms
Arthropod animals
Springtails
Mites
Millipedes and centipedes
Harvestman
Ants
Diplopoda
Diptera
Crustacea
Vertebrates
Mice, moles, and voles
Rabbits, gophers, and squirrels

so obvious is that they are as complex and intrinsically valuable as are those organisms that roam the land surface and waters of Earth.

Soil: An Engineering Medium

We usually think of soil as being firm and solid (solid ground, terra firma, etc.). As solid ground, soil is usually a good substrate upon which to build highways and structures. However, not all soils are firm and solid—some are not as stable as others. While construction of buildings and highways may be suitable in one location on one type of soil, it may be unsuitable in another location with different soil. To construct structurally sound, stable highways and buildings, construction on soils and with soil materials requires knowledge of the diversity of soil properties.

Note that working with manufactured building materials that have been "engineered" to withstand certain stresses and forces is much different than working with natural soil materials, even though engineers have the same concerns about soils as they do with man-made building materials (concrete and steel). It is much more difficult to make these predictions or determinations for soil's ability to resist compression, its ability to remain in place, its bearing strength, its shear strength, and its stability than it is to make the same determinations for manufactured building materials.

Soil: Source of Materials

In addition to providing valuable minerals for various purposes, soil is commonly used to provide road-building and dam-construction materials.

CONCURRENT SOIL FUNCTIONS

According to the U.S. Department of Agriculture (USDA 2009), soils perform specific critical functions no matter where they are located, and they perform more that one function at the same time, as described below.

- Soils act like *sponges*, soaking up rainwater and limiting runoff. Soils also impact groundwater recharge and flood-control potentials in urban areas.
- Soils act like *faucets*, storing and releasing water and air for plants and animals to use.
- Soils act like *supermarkets*, providing valuable nutrients and air and water to plants and animals. Soils also store carbon and prevent its loss into the atmosphere.
- Soils act like *strainers* or *filters*, filtering and purifying water and air that flow through them.
- Soils buffer, degrade, immobilize, detoxify, and trap pollutants, such as oil, pesticides, herbicides, and heavy metals, and keep them from entering groundwater supplies. Soils also store nutrients for future use by plants and animals above ground and by microbes within soils.

Soil Basics

Any fundamental discussion about soil should begin with a definition of soil. The word *soil* is derived through Old French from the Latin *solum*, which means floor or ground. John Steinbeck referenced soil by its scars, crusts, and crusting. Charles F. Ramuz (Swiss writer) referred to soil as that soft stuff under the feet. A student of Hippocrates talks about soil as an immense quantity of forces. A more current and concise definition is made difficult by the great diversity of soils throughout the globe. However, here is a generalized definition from the Soil Science Society of America (2008, 1):

> Soil is unconsolidated mineral matter on the surface of the earth that has been subjected to and influenced by genetic and environmental factors of: parent material, climate, macro- and microorganisms, and topography, all acting over a period of time and producing a product—soil—that differs from the material from which it is derived in many physical, chemical, and biological properties, and characteristics.

Engineers might define soil by saying that soil occupies the unconsolidated mantle of weathered rock making up the loose materials on the Earth's surface, commonly known as the regolith. Soil can be described as a three-phase system, composed of a solid, liquid, and gaseous phase. (Note that this phase relationship is important in dealing with soil pollution because each of the three phases of soil are in equilibrium with the atmosphere, and with rivers, lakes, and the oceans. Thus, the fate and transport of pollutants are influenced by each of these components.)

Soil is also commonly described as a mixture of air, water, mineral matter, and organic matter; the relative proportions of these four components greatly influence the productivity of soils. The interface (where the regolith meets the atmosphere) of these materials that make up soil is what concerns us here.

Keep in mind that the four major ingredients that make up soil are not mixed or blended like cake batter. Instead, pore spaces (vital to air and water circulation, providing space for roots to grow and microscopic organisms to live) are a major and critically important constituent of soil. Without sufficient pore space, soil would be too compacted to be productive. Ideally, the pore space will be divided roughly equally between water and air, with about one-quarter of the soil volume consisting of air and one-quarter consisting of water. The relative proportions of air and water in a soil typically fluctuate significantly as water is added and lost. Compared to surface soils, subsoils tend to contain less total pore space, less organic matter, and a larger proportion of micropores, which tend to be filled with water.

Let's take a closer look at the four major components that make up soil.

Soil air circulates through soil pores in the same way air circulates through a ventilation system. Only when the pores (the ventilation ducts) become blocked by water or other substances does the air fail to circulate. Though soil pores normally connect to interface with the atmosphere, soil air is not the same as atmospheric air. It differs in composition from place to place. Soil air also normally has a higher moisture content than the atmosphere. The content of carbon dioxide (CO_2) is usually higher as well, and that of oxygen (O_2) lower than accumulations of these gases found in the atmosphere.

Earlier, we stated that only when soil pores are occupied by water or other substances does air fail to circulate in the soil. For proper plant growth, this is of particular importance because in soil pore spaces that are water dominated, air oxygen content is low and carbon dioxide levels are high, which restricts plant growth.

The presence of water in soil (often reflective of climatic factors) is essential for the survival and growth of plant and other soil organisms. Soil moisture is a major determinant of the productivity of terrestrial ecosystems and agricultural systems. Water moving through soil materials is a major force behind soil formation. Along with air, water, and dissolved nutrients, soil moisture is critical to the quality and quantity of local and regional water resources.

Mineral matter varies in size and is a major constituent of nonorganic soils. Mineral matter consists of large particles (rock fragments) including stones, gravel, and coarse sand. Many of the smaller mineral matter components are made of a single mineral. Minerals in the soil (for plant life) are the primary source of most of the chemical elements essential for plant growth.

Soil organic matter consists primarily of living organisms and the remains of plants, animals, and microorganisms that are continuously broken down (biodegraded) in the soil into new substances that are synthesized by other microorganisms. These other microorganisms continually use this organic matter and reduce it to carbon dioxide (via respiration) until it is depleted, making repeated additions of new plant and animal residues necessary to maintain organic soil matter (Brady and Weil 2007).

Now that we have defined soil, let's take a closer look at a few of the basics pertaining to soil and some of the common terms used in any discussion related to soil basics.

Soil is the layer of bonded particles of sand, silt, and clay that covers the land surface of the Earth. Most soils develop in multiple layers. The topmost layer (topsoil) is the layer of soil moved in cultivation and in which plants grow. This topmost layer is actually an ecosystem composed of both biotic and abiotic components—inorganic chemicals, air, water, and decaying organic material that provides vital nutrients for plant photosynthesis and living organisms. Below the topmost layer is the subsoil, the part of the soil below the plow level, usually no more than a meter in thickness. Subsoil is much less productive, partly because it contains much less organic matter. Below that is the parent material, the unconsolidated (and more or less chemically weathered) bedrock or other geologic material from which the soil is ultimately formed. The general rule of thumb is that it takes about thirty years to form one inch of topsoil from subsoil; it takes much longer than that for subsoil to be formed from parent material—the length of time depending on the nature of the underlying matter (Franck and Brownstone 1992).

SOIL: PHYSICAL PROPERTIES

From the soil pollution technologist's point of view (regarding land conservation and methodologies for contaminated soil remediation through reuse and recycling), five major physical properties of soil are of interest. They are soil texture, slope, structure, organic matter, and color. Soil texture is determined by the size of the rock particles (sand, silt, and clay particles) or the soil separates within the soil. The largest soil particles are gravel, which consist of fragments larger than two millimeters in diameter.

Particles between 0.05 and 2 millimeters are classified as sand. Silt particles range from 0.002 to 0.05 millimeters in diameter, and the smallest particles (clay particles) are less than 0.002 millimeters in diameter. Though clays are composed of the smallest particles, those particles have stronger bonds than silt or sand, though once broken apart, they erode more readily. Particle size has a direct impact on erodibility. Rarely does a soil consist of only one single size of particle—most are a mixture of various sizes.

The slope (or steepness of the soil layer) is another given, important because the erosive power of runoff increases with the steepness of the slope. Slope also allows runoff to exert increased force on soil particles, which breaks them apart more readily and carries them farther away.

Soil structure (tilth) should not be confused with soil texture—they are different. In fact, in the field, the properties determined by soil texture may be considerably modified by soil structure. Soil structure refers to the combination or arrangement of primary soil particles into secondary particles (units or peds). Simply stated, soil structure refers to the way various soil particles clump together. The size, shape, and arrangement of clusters of soil particles called aggregates form naturally larger clumps called peds. Sand particles do not clump because sandy soils lack structure. Clay soils tend to stick together in large clumps. Good soil develops small, friable (easily crumbled) clumps. Soil develops a unique, fairly stable structure in undisturbed landscapes, but agricultural practices break down the aggregates and peds, lessening erosion resistance.

The presence of decomposed or decomposing remains of plants and animals (organic matter) in soil helps not only fertility but also soil structure—especially the soil's ability to store water. Live organisms such as protozoa, nematodes, earthworms, insects, fungi, and bacteria are typical inhabitants of soil. These organisms work to either control the population of organisms in the soil or aid in the recycling of dead organic matter. All soil organisms, in one way or another, work to release nutrients from the organic matter, changing complex organic materials into products that can be used by plants.

Just about anyone who has looked at soil has probably noticed that soil color is often different from one location to another. Soil colors range from very bright to dull grays, to a wide range of reds, browns, blacks, whites, yellows, and even greens. Soil color is dependent primarily on the quantity of humus and the chemical form of iron oxides present.

Soil scientists use a set of standardized color charts (the *Munsell Color Book*) to describe soil colors. They consider three properties of color—hue, value, and chroma—in combination to come up with a large number of color chips to which soil scientists can compare the color of the soil being investigated.

SOIL SEPARATES

As pointed out in the previous section, soil particles have been divided into groups based on their size, termed *soil separates*—sand, silt, and clay—by the International Soil Science Society System, the U.S. Public Roads Administration, and the USDA. In this text, we use the classification established by the USDA. The size ranges in these separates reflect major changes in how the particles behave and in the physical properties they impart to soils.

In table 10.2, the names of the separates are given, together with their diameters and the number of particles in one gram of soil (according to the USDA).

Sand ranges in diameter from 2 to 0.05 millimeters and is divided into five classes (see table 10.2). Sand grains are more or less spherical (rounded) in shape, with variable angularity, depending on the extent to which they have been worn down by abrasive processes such as rolling around by flowing water during soil formation.

Sand forms the framework of soil and gives it stability when in a mixture of finer

Table 10.2. Characteristics of Soil Separates (USDA)

Separate	*Diameter (mm)*	*Number of Particles/Gram*
Very coarse sand	2.00–1.00	90
Coarse sand	1.00–0.50	720
Medium sand	0.50–0.25	5,700
Fine sand	0.25–0.10	46,000
Very fine sand	0.10–0.05	722,000
Silt	0.05–0.002	5,776,000
Clay	Below 0.002	90,260,853,000

particles. Sand particles are relatively large, which allows voids that form between each grain to also be relatively large. This promotes free drainage of water and the entry of air into the soil. Sand is usually composed of a high percentage of quartz because it is most resistant to weathering and its breakdown is extremely slow. Many other minerals are found in sand, depending upon the rocks from which the sand was derived. In the short term (on an annual basis), sand contributes little to plant nutrition in the soil. However, in the long term (thousands of years of soil formation), soils with a lot of weatherable minerals in their sand fraction develop a higher state of fertility.

Silt (essentially microsand), though spherically and mineralogically similar to sand, is smaller—too small to be seen with the naked eye. It weathers faster and releases soluble nutrients for plant growth faster than sand. Too fine to be gritty, silt imparts a smooth feel (like flour) without stickiness. The pores between silt particles are much smaller than those in sand (sand and silt are just progressively finer pieces of the original crystals in the parent rocks). In flowing water, silt is suspended until it drops out when flow is reduced. On the land surface, silt, if disturbed by strong winds, can be carried great distances and is deposited as loess.

The clay soil separate is (for the most part) much different from sand and silt. Clay is composed of secondary minerals that were formed by the drastic alteration of the original forms or by the recrystallization of the products of their weathering. Because clay crystals are platelike (sheeted) in shape, they have a tremendous surface area–to-volume ratio, giving clay a tremendous capacity to adsorb water and other substances on its surfaces. Clay actually acts as a storage reservoir for both water and nutrients. There are many kinds of clay, each with different internal arrangements of chemical elements, which give them individual characteristics.

Soil Formation

Everywhere on Earth's land surface is either rock formation or exposed soil. When rocks formed deep in the Earth are thrust upward and exposed to the Earth's atmosphere, they adjust to the new environment and soil formation begins. Soil is formed as a result of physical, chemical, and biological interactions in specific locations. Just as vegetation varies among biomes, so do the soil types that support that vegetation.

The vegetation of the tundra and rain forest differ vastly from each other and from vegetation of the prairie and coniferous forest; soils differ in similar ways.

In the soil-forming process, two related—but fundamentally different—processes are occurring simultaneously. The first is the formation of parent soil materials by weathering of rocks, rock fragments, and sediments. This set of processes is carried out in the zone of weathering. The end point is to produce parent material in which the soil can develop and is referred to as C horizon material (see figure 10.4). It applies in the same way for glacial deposits as for rocks. The second set of processes is the formation of the soil profile by soil-forming processes, which gradually changes the C horizon material into A, E, and B horizons. Figure 10.4 illustrates two soil profiles, one on hard granite and one on a glacial deposit.

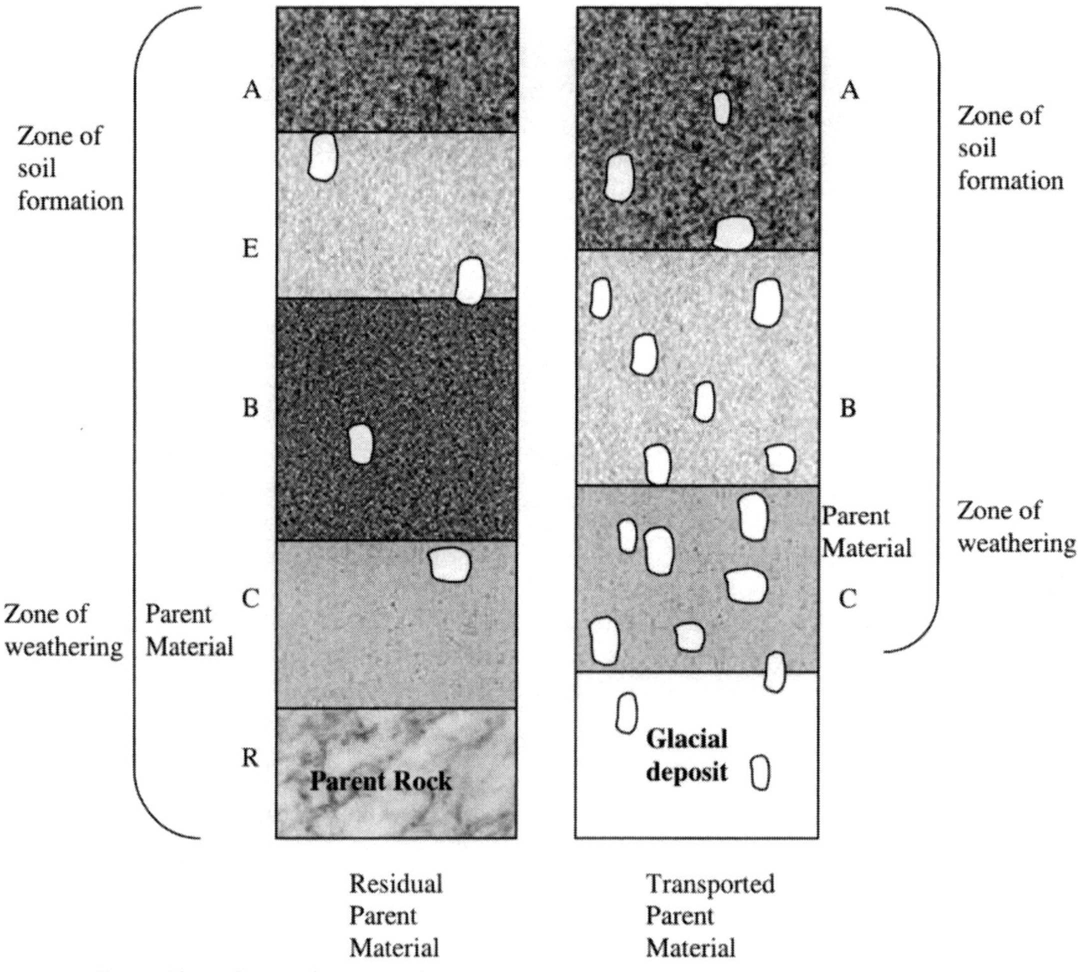

Figure 10.4. Soil profiles on residual and transported parent materials.

WEATHERING

Soil development takes time and is the result of two major processes: weathering and morphogenesis (morphogenesis was described earlier as bare rock succession). *Weathering* (the breaking down of bedrock and other sediments that have been deposited on the bedrock by wind, water, volcanic eruptions, or melting glaciers) happens physically, chemically, or by a combination of both. Weathering is the first step in the erosion process; again, it causes the breakdown of rocks, to either form new minerals that are stable on the surface of Earth or break the rocks down to smaller particles. Simply, weathering (which projects itself on all surface material above the water table) is the general term used for all the ways in which a rock may be broken down.

Soil Characterization

When people become ill, they may go to a doctor to seek a diagnosis of what is causing the illness and, hopefully, prognosis of how long before they feel well again.

What do diagnosis and prognosis have to do with soil? Actually, quite a lot. The diagnostic techniques used by a physician to identify the causative factors leading to a particular illness are analogous to the soil practitioner using diagnostic techniques to identify a particular soil. Sound far-fetched? It shouldn't because it isn't. Soil scientists must be able to determine the type of soil they study or work with.

Determining the type of soil makes sense, but what does prognosis have to do with all this? Soil practitioners need to be able to not only identify or classify a soil type but also correctly predict how a particular pollutant will react or respond when spilled in that type of soil. The fate of the pollutant is important in determining the possible damage incurred to the environment—soil, groundwater, and air—because, ultimately, a spill could easily affect all three. Thus, the soil practitioner must not only use diagnostic tools in determining soil type but also must be familiar with the soil type, to judge how a particular pollutant or contaminant will respond when spilled in the soil type.

Let's take a closer look at the genesis of soil classification. From the time humans first advanced from hunter-gatherer status to cultivators of crops they noticed differences in productive soils and unproductive soils. The ancient Chinese, Egyptians, Romans, and Greeks all recognized and acknowledged the differences in soils as media for plant growth. These early soil classification practices were based primarily upon texture, color, and wetness.

Soil classification as a scientific practice did not gain a foothold until the later eighteenth and early nineteenth centuries when the science of geology was born. Such terms (with an obvious geological connotation) as limestone soils and lake-laid soils as well as clayey and sandy soils came into being. The Russian scientist Vasily Dokuchaev was the first to suggest a generic classification of soils—that soils were natural bodies. Dokuchaev's classification work was then further developed by Europeans and Americans. The system is based on the theory that each soil has a definite form and structure (morphology), related to a particular combination of soil-forming factors. This system was used until 1960, when the USDA published *Soil Classification: A Comprehensive System* (USDA 1960). This classification system places major emphasis on soil morphology and gives less emphasis to genesis or the soil-forming factors as compared to previous systems. In 1975, *Soil Classification: A Comprehensive System* was replaced by *Soil Taxonomy* (USDA 1999), which classifies objects according to their natural relationships. Soils are classified based on measurable properties of soil profiles.

Note that no clear delineation or line of demarcation can be drawn between the properties of one soil and those of another. Instead, a gradation (sometimes quite subtle—like from one shade of white to another) occurs in soil properties as one moves from one soil to another. Brady and Weil (2007) point out that "the gradation in soil properties can be compared to the gradation in the wavelengths of light as you move from one color to another. The changing is gradual, and yet we identify a boundary that differentiates what we call 'green' from what we call 'blue'" (58).

To properly characterize the primary characteristics of a soil, a soil must be identified down to the smallest three-dimensional characteristic sample possible. However, to accurately perform a particular soil sample characterization, a sampling unit must be large enough so that the nature of its horizons can be studied and the range of its properties identified. The pedon (rhymes with head-on) is this unit. The pedon is roughly polygonal in shape and designates the smallest characteristic unit that can still be called a soil.

Because pedons occupy a very small space (from approximately one to ten square meters), they cannot, obviously, be used as the basic unit for a workable field soil classification system. To solve this problem, a group of pedons, termed a polypedon, is of sufficient size to serve as a basic classification unit (or as commonly called a soil individual). In the United States, these groupings have been called a soil series.

There is a difference between *a soil* and *the soil.* This difference is important in the soil classification scheme. A soil is characterized by a sampling unit (pedon), which as a group (polypedons) form a soil individual. The soil, on the other hand, is a collection of all these natural ingredients and is distinguishable from other bodies such as water, air, solid rock, and other parts of the Earth's crust. By incorporating the difference between a soil and the soil, a classification system has been developed that is effective and widely used.

DIAGNOSTIC HORIZONS, TEMPERATURE, AND MOISTURE REGIMES

Soil taxonomy uses a strict definition of soil horizons called diagnostic horizons, which are used to define most of the orders. Two kinds of diagnostic horizons are recognized: surface and subsurface. The surface diagnostic horizons are called epipedons (Greek *epi,* over; *pedon,* soil). The epipedon includes the dark (organic rich), upper part of the soil, or the upper eluvial horizons, and sometimes both. Those soils beneath the epipedons are called subsurface diagnostic horizons. Each of these layers is used to characterize different soils in soil taxonomy.

In addition to using diagnostic horizons to strictly define soil horizons, soil moisture regime classes can also be used. A soil moisture regime refers to the presence of plant-available water or groundwater at a sufficiently high level. The control section of the soil (ranging from ten to thirty centimeters for clay and from thirty to ninety centimeters for sandy soils) designates that section of the soil where water is present or absent during given periods in a year. The control section is divided into sections: upper and lower portions. The upper portion is defined as the depth to which 2.5 centimeters of water will penetrate within twenty-four hours. The lower portion is the depth that 7.5 centimeters of water will penetrate.

Six soil moisture regimes are identified:

- Aridic—characteristic of soils in arid regions;
- Xeric—characteristic of having long periods of drought in the summer;
- Ustic—in which soil moisture is generally high enough to meet plant needs during the growing season;
- Udic—common soil in humid climatic regions;
- Perudic—an extremely wet moisture regime annually; and
- Aquic—soil saturated with water and free of gaseous oxygen.

Table 10.3 lists the moisture regime classes and the percentage distribution of areas with different soil moisture regimes.

In soil taxonomy, several soil temperature regimes are also used to define classes of soils. Based on mean annual soil temperature, mean summer temperature, and the difference between mean summer and winter temperatures, soil temperature regimes are shown in table 10.4.

The diagnostic horizons and moisture/temperature regimes just discussed are the main criteria used to define the various categories in soil taxonomy.

SOIL TAXONOMY

The U.S. Soil Conservation Service's soil classification system, soil taxonomy (which is based on measurable properties of soil profiles), places soils in categories (see table 10.5). Let's take a closer look at each one of these categories.

Table 10.3. Soil Moisture Regimes (percent of global area occupied by each)

Moisture Regime	*Percent of Soils*
Aridic	35.9
Xeric	3.5
Ustic	18.0
Udic	33.1
Perudic	1.0
Aquic	8.3

Source: Adapted from H. Eswaran, "Assessment of Global Resources: Current Status and Future Needs," *Pedologie* 43 (1993): 19–39.

Table 10.4. Soil Temperature Regimes (percent of global areas occupied by each)

Soil Temperature Regimes/ Mean Annual Temperature (° C)	*Percent*
Pergelic (0° C)	10.9
Cryic (0–8° C)	3.5
Frigid (0–8° C)	1.2
Mesic (8–15° C)	12.5
Thermic (15–22° C)	11.4
Hyperthermic (>22° C)	18.5
Isofrigid (0–8° C)	0.1
Isomesic (8–15° C)	0.3
Isothermic (15–22° C)	2.4
Isohyperthermic (>22° C)	26.0
Water (NA)	1.2
Ice (NA)	1.4

Source: Adapted from H. Eswaran, "Assessment of Global Resources: Current Status and Future Needs," *Pedologie* 43 (1993): 19–39.

- *Order*: soils not too dissimilar in their genesis. There are twelve soil orders in soil taxonomy. The names and major characteristics of each soil order are shown in table 10.6.
- *Suborder*: sixty-four subdivisions of order that emphasize properties that suggest some common features of soil genesis.
- *Great group*: diagnostic horizons are the major bases for differentiating approximately three hundred great groups.
- *Subgroup*: approximately 1,200 subdivisions of the great groups.
- *Family*: approximately 7,500 soils with subgroups having similar physical and chemical properties.
- *Series*: a subdivision of the family and the most specific unit of the classification system. More than eighteen thousand soil series are recognized in the United States.

SOIL ORDERS

As stated earlier, twelve soil orders are recognized; they constitute the first category of the classification. The classification of the orders is illustrated in table 10.6.

Table 10.5. Subdivision of Soil Taxonomy Classification System (in hierarchical order)

Category	*Number of Taxa*
Order	12
Suborder	55
Great group	Approximately 230
Subgroup	Approximately 1,200
Family	Approximately 7,500
Series	Approximately 18,500 in United States

Table 10.6. Soil Orders (with simplified definitions)

Alfisol	Mild forest soil with gray to brown surface horizon, medium to high base supply (refers to amount of interchangeable cations that remain in soil), and a subsurface horizon of clay accumulation.
Andisol	Formed on volcanic ash and cinders and lightly weathered.
Aridsol	Dry soil with pedogenic (soil-forming) horizon; low in organic matter.
Entisol	Recent soil without pedogenic horizons.
Gelisols	Soils of very cold climates; defined as containing permafrost.
Histosol	Organic (peat or bog) soil.
Inceptisol	Soil at the beginning of the weathering process with weakly differentiated horizons.
Mollisol	Soft soil with a nearly black, organic-rich surface horizon and high base supply.
Oxisol	Oxide-rich soil; principally a mixture of kaolin, hydrated oxides, and quartz.
Spodosol	Soil that has an accumulation of amorphous materials in the subsurface horizons.
Ultisol	Soil with a horizon of silicate clay accumulation and low base supply.
Vertisol	Soil with high-activity clays (cracking clay soil).

Source: U.S. Department of Agriculture, Soil Survey Staff, *Keys to Soil Taxonomy* (Washington, D.C.: Natural Resources Conservation Service, 1994).

SOIL SUBORDERS

Soil orders are further divided into sixty-four suborders, based primarily on the chemical and physical properties that reflect either the presence or absence of water logging or genetic differences caused by climate and vegetation—to give the class the greatest genetic homogeneity. Thus, the aqualfs (formed under wet conditions) are "wet" (aqu for aqua); alfisols become saturated with water sometime during the year. The suborder names all have two syllables, with the first syllable indicating the order, such as *alf* in alfisol and *oll* for Mollisol.

SOIL GREAT GROUPS AND SUBGROUPS

Suborders are divided into great groups. They are defined largely by the presence or absence of diagnostic horizons and the arrangements of those horizons. Great group names are coined by prefixing one or more additional formative elements to the appropriate suborder name. More than three hundred great groups are identified.

Subgroups are subdivisions of great groups. Subgroup names indicate to what extent the central concept of the great group is expressed. A Typic fragiaqualf is a soil that is typical for the fragiaqualf great group.

SOIL FAMILIES AND SERIES

The family category of classification is based on features that are important to plant growth such as texture, particle size, mineralogical class, and depth. Terms such as clayey, sandy, loamy, and others are used to identify textural classes. Terms used to describe mineralogical classes include *mixed*, *oxidic*, *carbonatic*, and others. For temperature classes, terms such as *hypothermic*, *frigid*, *cryic*, and others are used.

The soil series (subdivided from soil family) gets down to the individual soil, and the name is that of a natural feature or place near where the soil was first recognized. Familiar series names include Amarillo (Texas), Carlsbad (New Mexico), and Fresno (California). In the United States, there are more than eighteen thousand soil series.

Chapter Review Questions

10.1. Discuss and describe the stages of bare rock succession.
10.2. Describe and discuss soil's six chief functions.
10.3. Runoff is limited by ________.
10.4. Soil is a mixture of ________.
10.5. Tilth relates to the ________ of soil.

Note

1. Much of the material in this chapter is adapted from F. R. Spellman, *The Science of Environmental Pollution*, 2nd ed. (Boca Raton, Fla.: CRC Press, 2009).

References and Recommended Reading

American Society for Testing Materials (ASTM). 1969. *Manual on water*. Philadelphia: Author.

Blumberg, L., and R. Gottlieg. 1989. *War on waste: Can America win its battle with garbage?* Washington, D.C.: Island Press.

Brady, N. C., and R. R. Weil. 2007. *The nature and properties of soils*, 11th ed. Upper Saddle River, N.J.: Prentice Hall.

Carson, R. 1962. *Silent spring*. Boston: Houghton Mifflin.

Christian Science Monitor. 1994. Environmental justice. March 15.

Ciardi, J. 1997. From Stoneworks. In *The Collected Poems of John Ciardi*, ed. E. M. Cifelli. Fayettville: University of Arkansas Press.

Davis, G. H., and G. L. Pollock. 2003. Geology of Bryce Canyon National Park, Utah. In *Geology of Utah's Parks and Monuments*, ed. D. A. Sprinkel et al., 2nd ed. Salt Lake City: Utah Geological Association.

Environmental Protection Agency (EPA). 1988. *The solid waste dilemma: An agenda for action—background document*. Washington, D.C.: EPA/530-SW-88–054A.

———. 1989. *Decision-makers guide to solid waste management*. Washington, D.C.: EPA/530-SW89–072.

———. 1993. *Characterization of municipal solid waste in U.S.: 1992 update.* Washington, D.C.: EPA/530-5-92-019.

———. 2007. Municipal solid waste generation, recycling, and disposal in the United States: Facts and figures for 2007. www.epa.gov/osw (accessed March 20, 2009).

Eswaran, H. 1993. Assessment of global resources: Current status and future needs. *Pedologie* 43:19–39.

Foth, H. D. 1978. *Fundamentals of soil science*, 6th ed. New York: Wiley.

Franck, I., and D. Brownstone. 1992. *The green encyclopedia.* New York: Prentice Hall.

GLOBE Program. 2003. Teacher's guide: Soil investigation chapter. http://archive.globe.gov/tctg/globetg.jsp (accessed March 20, 2009).

Illich, I., S. Groeneveld, and L. Hoinacki. 1991. *Declaration of Soil.* Kassel, Germany: University of Kassel.

Kemmer, F. N. 1979. *Water: The universal solvent.* Oak Ridge, Ill.: NALCO Chemical Company.

Konigsburg, E. M. 1996. *The view from Saturday.* New York: Scholastic.

MacKay, D., W. Y. Shiu, and K-C. Ma. 1997. *Illustrated handbook of physical-chemical properties and environmental fate for organic chemicals.* Boca Raton, Fla.: CRC Press/Lewis Publishers.

Morris, D. 1991. As if materials mattered. *Amicus Journal* 13 (4): 17–21.

Mowet, F. 1957. *The dog who wouldn't be.* New York: Willow Books.

National Park Service (NPS). 2008. *The hoodoo* (park newsletter). Bryce, Utah: Bryce Canyon National Park.

O'Reilly, J. T. 1994. *State and local government solid waste management.* Deerfield, Ill.: Clark, Boardman, Callahan.

Peterson, C. 1987. Mounting garbage problem. *Washington Post*, April 5.

Richards, B. 1988. Burning issue. *Wall Street Journal*, June 16.

Robotham, M., and J. Hart. 2003. Soil sampling for home gardens and small acreages. Extension service bulletin, Oregon State University. http://extension.oregonstate.edu/catalog/html/ec/ec628/ (accessed March 20, 2009).

Soil Science of America. 2008. *Soil Science Glossary.* Madison, WI: Soil Science of America.

Spellman, F. R. 2008. *The science of water.* Boca Raton, Fla.: CRC Press.

———. 2009. *The science of environmental pollution*, 2nd ed. Boca Raton, Fla.: CRC Press.

Spellman, F. R., and N. E. Whiting. 2006. *Environmental science and technology*, 2nd ed. Lanham, Md.: Government Institutes.

Tchobanoglous, G., H. Theisen, and S. Vigil. 1993. *Integrated solid waste management: Engineering principles and management issues.* New York: McGraw-Hill.

Tomera, A. N. 1989. *Understanding basic ecological concepts.* Portland, Maine: J. Weston Walch, Publisher.

Tonge, P. 1987. All that trash. *Christian Science Monitor*, July 6.

U.S. Department of Agriculture (USDA). 1960. *Soil classification: A comprehensive system*, soil survey staff. Washington, D.C.: Natural Resources Conservation Service.

———. 1994. *Keys to soil taxonomy*, soil survey staff. Washington, D.C.: Natural Resources Conservation Service.

———. 1999. *Soil taxonomy: A basic system of soil classification for making and interpreting soil surveys*, 2nd ed. Washington, D.C.: Natural Resources Conservation Service.

———. 2009. *Urban soil primer.* Washington, D.C.: Author.

The way nature works. 1992. New York: Macmillan.

Wolf, N., and E. Feldman. 1990. *Plastics: America's packaging dilemma.* Washington, D.C.: Island Press.

V

ECOLOGICAL ASPECTS

Ecology is about interrelationships between organisms.

CHAPTER 11

Ecology

> The "control of nature" is a phrase conceived in arrogance, born of the Neanderthal age of biology and the convenience of man.
>
> —Rachel Carson (1962, 297)

Setting the Stage

Geography and ecology? Absolutely. Although ecology is a science all by itself, it is also a branch of the larger sciences of biology and geography. As a science all by itself, ecology can be defined in various and numerous ways. For example, ecology, or ecological science, is commonly defined in the literature as the scientific study of the distribution and abundance of living organisms and how the distribution and abundance are affected by interactions between the organisms and their environment. The term ecology was coined in 1866 by the German biologist Ernst Haeckel, and it loosely means "the study of the household [of nature]." Eugene P. Odum (1983) explains that the word "ecology" is derived from the Greek *oikos*, meaning home. Ecology, then, means the study of organisms at home. In a broader sense, ecology is the study of the relation of organisms or groups to their environment.

Did You Know?

No ecosystem can be studied in isolation. If we were to describe ourselves, our histories, and what made us the way we are, we could not leave the world around us out of our description! So it is with streams: they are directly tied in with the world around them. They take their chemistry from the rocks and dirt beneath them as well as for a great distance around them. (Spellman 1996)

Charles Darwin explained ecology in a famous passage in the *Origin*, a passage that helped establish the science of ecology. A "web of complex relations" binds all living things in any region, Darwin writes. Adding or subtracting even a single species causes waves of change that race through the web, "onwards in ever-increasing circles of complexity." The simple act of adding cats to an English village would reduce the number of field mice. Killing mice would benefit the bumblebees, whose nest and honeycombs the mice often devour. Increasing the number of bumblebees would benefit the heartsease and red clover, which are fertilized almost exclusively by bumblebees. So adding cats to the village could end by adding flowers. For Darwin the whole of the Galapagos archipelago argues this fundamental lesson. The volcanoes are much more diverse in their ecology than their biology. The contrast suggests that, in the struggle for existence, species are shaped at least as much by the local flora and fauna as by the local soil and climate. "Why else would the plants and animals differ radically among islands that have the same geological nature, the same height, and climate" (Darwin 1998, 189).

Probably the best way to understand ecology—to get a really good "feel" for it—or to get to the heart of what ecology is all about is to read the following by Carson (1962):

> We poison the caddis flies in a stream and the salmon runs dwindle and die. We poison the gnats in a lake and the poison travels from link to link of the food chain and soon the birds of the lake margins become victims. We spray our elms and the following springs are silent of robin song, not because we sprayed the robins directly but because the poison traveled, step by step, through the now familiar elm leaf-earthworm-robin cycle. These are matters of record, observable, part of the visible world around us. They reflect the web of life—or death—that scientists know as ecology.

As Carson points out, what we do to any part of our environment has an impact upon other parts. In other words, there is an interrelationship between the parts that make up our environment. Probably the best way to state this interrelationship is to define ecology—that is, to define it as it is used in this text: "Ecology is the science that deals with the specific interactions that exist between organisms and their living and nonliving environment" (Tomera 1989).

When environment was mentioned in the preceding and as it is discussed throughout this text, it (the environment) includes everything important to the organism in its surroundings. The organism's environment can be divided into four parts:

1. Habitat and distribution—its place to live
2. Other organisms—whether friendly or hostile
3. Food
4. Weather—light, moisture, temperature, soil, and so on

There are four major subdivisions of ecology:

- Behavioral ecology
- Population ecology (autecology)
- Community ecology (synecology)
- Ecosystem ecology

Behavioral ecology is the study of the ecological and evolutionary basis for animal behavior. *Population ecology* (or autecology) is the study of the individual organism or a species. It emphasizes life history, adaptations, and behavior. It is the study of communities, ecosystems, and biosphere. An example of autecology would be when biologists spend their entire lifetime studying the ecology of the salmon. *Community ecology* (or synecology), on the other hand, is the study of groups of organisms associated together as a unit and deals with the environmental problems caused by mankind. For example, the effect of discharging phosphorous-laden effluent into a stream involves several organisms. The activities of human beings have become a major component of many natural areas. As a result, it is important to realize that the study of ecology must involve people. *Ecosystem ecology* is the study of how energy flow and matter interact with biotic elements of ecosystems (Odum 1971).

Did You Know?

Ecology is generally categorized according to complexity; the primary kinds of organism under study (plant, animal, and insect ecology); the biomes principally studied (forest, desert, benthic, grassland, etc.); the climatic or geographic area (e.g., arctic or tropics); and the spatial scale (macro or micro) under consideration.

Key Ecological Terms

- *Abiotic factor*—the nonliving part of the environment composed of sunlight, soil, mineral elements, moisture, temperature, topography, minerals, humidity, tide, wave action, wind, and elevation.
- *Atmosphere*—the gaseous mantle enveloping the hydrosphere and lithosphere, of which 78 percent is nitrogen by volume.
- *Autotroph*—(green plants) fixes energy of the Sun and manufactures food from simple, inorganic substances.
- *Biogeochemical cycles*—cyclic mechanisms in all ecosystems by which biotic and abiotic materials are constantly exchanged.
- *Biotic factor (community)*—the living part of the environment composed of organisms that share the same area, are mutually sustaining, interdependent, and constantly fixing, utilizing and dissipating energy.
- *Carbon*—an essential part of all organic compounds; photosynthesis is a source of carbon. Photosynthesis is the chemical process by which solar energy is stored as chemical energy.
- *Community*—in an ecological sense, includes all the populations occupying a given area.
- *Consumers and decomposers*—dissipate energy fixed by the producers through food

chains or webs. The available energy decreases by 80–90 percent during transfer from one trophic level to another.

- *Ecological pyramids*—three types of ecological pyramids: pyramids of numbers, productivity, and energy. All of these pyramids are based on the fact that, due to energy loss, fewer animals can be supported at each additional trophic level, which is the number of energy transfers an organism is from the rest of the pyramid.
- *Ecology*—the study of the interrelationship of an organism or a group of organisms and their environment.
- *Ecosystem*—the community and the nonliving environment functioning together as an ecological system.
- *Energy*—the ability or capacity to do work. Energy is degraded from a higher to a lower state.
- *Environment*—everything that is important to an organism in its surroundings.
- *First Law of Thermodynamics*—states energy is transformed from one form to another, but is neither created nor destroyed. Given this principle, we should be able to account for all the energy in a system in an energy budget, a diagrammatic representation of the energy flows through an ecosystem.
- *Heterotrophs*—(animals) use food stored by the autotroph, rearrange it, and finally decompose complex materials into simple, inorganic compounds. Heterotrophs may be carnivorous (meat eaters), herbivorous (plant eaters), or omnivorous (plant and meat eaters).
- *Homeostasis*—a natural occurrence during which an individual population or an entire ecosystem regulates itself against negative factors and maintains an overall stable condition.
- *Hydrosphere*—the water covering the Earth's surface, of which 80 percent is salt water and 19 percent is groundwater.
- *Lithosphere*—the solid components of the Earth's surface such as rocks and weathered soil.
- *Niche*—the role that an organism plays in its natural ecosystem, including its activities, resource use, and interaction with other organisms.
- *Nitrogen*—required for construction of proteins and nucleic acids; the major source is the atmosphere.
- *Organisms*—require twenty to forty elements for survival.
- *Phosphorus cycle*—very inefficient; the greatest source is the lithosphere. Humans have greatly speeded up this cycle through mining.
- *Pollution*—an adverse alteration to the environment by a pollutant.
- *Second Law of Thermodynamics*—asserts that energy is only available due to degradation of energy from a concentrated to a dispersed form. This indicates that energy becomes more and more dissipated (randomly arranged) as it is transformed from one form to another or moved from one place to another. It also suggests that any transformation of energy will be less than 100 percent efficient (i.e., the transfers of energy from one trophic level to another are not perfect); some energy is dissipated during each transfer.

- *Stability*—ability of a living system to withstand or recover from externally imposed changes or stresses.

Why Is Ecology Important?

Ecology, in its true sense, is a holistic discipline that does not dictate what is right or wrong. Instead, ecology is important to life on Earth simply because it makes us aware, to a certain degree, of what life on Earth is all about. Ecology shows us that each living organism has an ongoing and continual relationship with every other element that makes up our environment. Simply, ecology is all about interrelationships, intraspecific and interspecific, and how important it is to maintain these relationships—to ensure our very survival.

History of Ecology

The chronological development of most sciences is clear and direct. Listing the progressive stages in the development of biology, math, chemistry, and physics is a relatively easy, straightforward process. The science of ecology is different. Having only gained prominence in the later part of the twentieth century, ecology is generally spoken of as a new science. However, ecological thinking at some level has been around for a long time, and the principles of ecology have developed gradually and more like a multistemmed bush than a tree with a single trunk (Smith 1996).

Thomas Smith and Robert Smith (2006) point out that one can argue that ecology goes back to Aristotle or perhaps his friend and associate, Theophrastus, both of whom had interest in the relations between organisms and the environment and in many species of animals. Theophrastus described interrelationships between animals and between animals and their environment as early as the fourth century BC (Ramalay 1940).

Modern ecology has its early roots in plant geography (i.e., plant ecology, which developed earlier than animal ecology) and natural history. The early plant geographers (ecologists) included Carl Ludwig Willdenow (1765–1812) and Friedrich Alexander von Humboldt (1769–1859). Willdenow was one of the first phytogeographers; he was also a mentor to von Humboldt. Willdenow, for whom the perennial vine Willdenow's spikemoss (*Selaginella willdenowil*) is named, developed the notion, among many others, that plant distribution patterns changed over time. Von Humboldt, considered by many to be the father of ecology, further developed many of Willdenow's notions, including the notion that barriers to plant dispersion were not absolute (Smith 2007).

Another scientist who is considered a founder of plant ecology was Johannes E. B. Warming (1841–1924). Warming studied the tropical vegetation of Brazil. He is best known for working on the relations between living plants and their surroundings. He is also recognized for his flagship text on plant ecology, *Plantesamfund* (1895). He also wrote *A Handbook of Systematic Botany* (1878).

Meanwhile, other naturalists were assuming important roles in the development of ecology. First and foremost among the naturalists was Darwin. While working on his *Origin of Species*, Darwin (1998) came across the writings of Thomas Malthus (1766–1834). Malthus advanced the principle that populations grow in a geometric fashion, doubling at regular intervals until they outstrip the food supply—ultimately resulting in death and thus restraining population growth (Smith and Smith 2006). Darwin, in his autobiography (1876), stated,

> In October 1838, that is, fifteen months after I had begun my systematic inquiry, I happened to read for amusement Malthus on *Population* and being well prepared to appreciate the struggle for existence which everywhere goes on from long-continued observation of the habits of animals and plants it at once struck me that under these circumstances favorable variations would tend to be preserved, and unfavorable ones to be destroyed. The results of this would be the formation of a new species. Here, then I had at last got a theory by which to work.

During the period Darwin was formulating his *Origin of Species*, Gregor Mendel (1822–1884) was studying the transmission of characteristics from one generation of pea plants to another. Mendel's plus Darwin's work provided the foundation for population genetics, the study of evolution and adaptation.

Time marched on and the preceding work of chemists Antoine-Laurent Lavoisier (he lost his head during the French Revolution) and Horace B. de Saussere and the Austrian geologist Eduard Suess, who proposed the term *biosphere* in 1875, all set the foundations of the advanced work that followed.

Did You Know?

The Russian geologist, Vladimir Vernadsky, detailed the idea of the biosphere in 1926.

Several forward strides in animal ecology, independent of plant ecology, were made during the nineteenth century that enabled the twentieth-century scientists Richard Hesse, Charles Elton, Charles Adams, and Victor Shelford to refine the discipline.

Smith and Smith (2006, 39) point out that many early plant ecologists were "concerned with observing the patterns of organisms in nature, attempting to understand how patterns were formed and maintained by interactions with the physical environment." Instead of looking for patterns, Frederic E. Clements (1874–1945) sought a system of organizing nature. Conducting his studies on vegetation in Nebraska, he postulated that the plant community behaves as a complex organism that grows and develops through stages, resembling the development of an individual organism, to a mature (climax) stage. Clements's theory of vegetation was criticized significantly by Arthur Tansley, a British ecologist, and others.

New World Encyclopedia (2007) points out that, in 1935, Tansley coined the term *ecosystem*—the interactive system established between the group of living creatures (biocoenosis) and the environment in which they live (biotype). Tansley's ecosystem concept was adopted by the well-known and influential biology educator Eugene Odum. Along with his brother, Howard Odum, Eugene Odum wrote a textbook that (starting in 1953) educated multiple generations of biologists and ecologists in North America (including the author of this text). Eugene Odum is often called the "father of modern ecosystem ecology."

A new direction in ecology was given a boost in 1913 when Victor Shelford stressed the interrelationship of plants and animals. He conducted early studies on succession in the Indiana dunes and on experimental *physiological ecology*. Because of his work, ecology became a science of communities. His *Animal Communities in Temperate America* (1913) was one of the first books to treat ecology as a separate science. Eugene Odum was one of Shelford's students.

Human ecology began in the 1920s. About the same time, the study of populations split into two fields, *population ecology* and *evolutionary ecology*. Closely associated with population ecology and evolutionary ecology is *community ecology*. At the same time, physiological ecology arose. Later, natural history observations spawned behavioral ecology (Smith and Smith 2006).

The history of ecology has been tied to advances in biology, physics, and chemistry that have spawned new areas of study in ecology, such as landscape, conservation, restoration, and global ecology. At the same time, ecology was rife with conflicts and opposing camps. Smith (1996) notes that the first major split in ecology was between plant ecology and animal ecology, which even led to a controversy over the term ecology, with botanists dropping the initial "o" from oecology, the spelling in use at the time, and zoologists refusing to use the term ecology at all because of its perceived affiliation with botany. Other historical schisms were between organismal and individualist ecology, holism versus reductionism, and theoretical versus applied ecology (New World Encyclopedia 2007).

Holism versus Reductionism

In ecology, holism is the most important and leading approach because it tries to include biological, chemical, physical, and economic views in a given area. However, at the area level the complexity grows. Thus, it is necessary to reduce the characteristic of the view in other ways, for instance, to a specific time of duration. The early conservationist John Muir stated that "when we try to pick out anything by itself we find it hitched to everything else in the universe."

(Gifford 2006, 104)

Levels of Organization

Eugene Odum explains that "the best way to delimit modern ecology is to consider the concept of levels of organization" (Odum 1983, 3). Levels of organization can be simplified as follows:

Organs → Organism → Population → Communities → Ecosystem → Biosphere

In this relationship, organs form an organism; organisms of a particular species form a population; populations occupying a particular area form a community; communities, interacting with nonliving or abiotic factors, separate in a natural unit to create a stable system known as the ecosystem (the major ecological unit); and the part of Earth in which an ecosystem operates is known as the biosphere. Audrey Tomera (1989, 41) points out that "every community is influenced by a particular set of abiotic factors." The abiotic part of the ecosystem is represented by inorganic substances such as oxygen, carbon dioxide, several other inorganic substances, and some organic substances.

The physical and biological environment in which an organism lives is referred to as its habitat. For example, the habitat of two common aquatic insects, the "backswimmer" (Notonecta) and the "water boatman" (Corixa), is the littoral zone of ponds and lakes (shallow, vegetation-choked areas; see figure 11.1; Odum 1983).

Within each level of organization of a particular habitat, each organism has a special role. The role the organism plays in the environment is referred to as its niche. A niche might be that the organism is food for some other organism or is a predator of other organisms. Odum refers to an organism's niche as its "profession" (Odum 1975). In other words, each organism has a job or role to fulfill in its environment. Although two different species might occupy the same habitat, "niche separation based on food habits" differentiates between two species. Such niche separation can be seen by comparing the niches of the water backswimmer and the water boatman. The backswimmer is an active predator, while the water boatman feeds largely on decaying vegetation (Odum 1983).

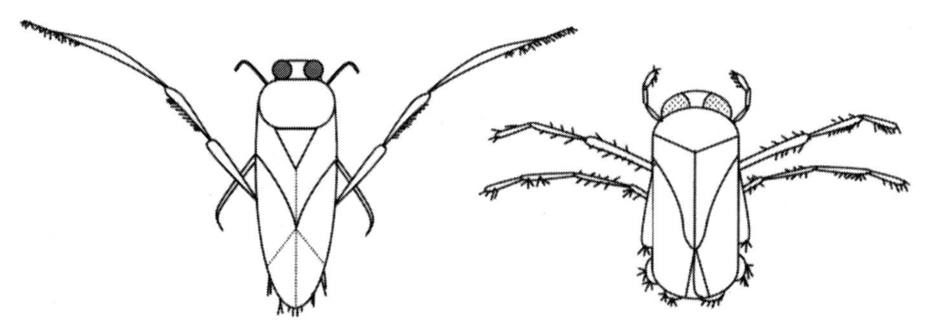

Figure 11.1. Notonecta (left) and Corixa (right). Adapted from Odum (1983, 402).

Did You Know?

In order for an ecosystem to exist, a dynamic balance must be maintained among all biotic and abiotic factors—a concept known as homeostasis.

Ecosystem

As mentioned, *ecosystem*, a contraction of "ecological" and "system," is a term introduced by Tansley to denote an area that includes all organisms therein and their physical environment. Specifically, an ecosystem is defined as the geographic area including all the living organisms, their physical surroundings, and the natural cycles that sustain them. All of these elements are interconnected (USFWS 2007). Simply, the ecosystem is the major ecological unit in nature. Elements of an ecosystem may include flora, fauna, lower life-forms, water, and soil (Wikipedia 2007). "There is a constant interchange of the most various kinds within each system, not only between the organisms but between the organic and the inorganic" (Tansley 1935, 299). Living organisms and their nonliving environment are inseparably interrelated and interact upon each other to create a self-regulating and self-maintaining system. To create a self-regulating and self-maintaining system, ecosystems are homeostatic, that is, they resist any change through natural controls. These natural controls are important in ecology. This is especially the case since it is people through their complex activities who tend to disrupt natural controls.

Tansley regarded the ecosystem as not only the organism complex but also the whole complex of physical factors forming what we call the environment. It was first applied to levels of biological organization represented by units such as community and the biome. Odum (1952) and Evans (1956) expanded the extent of the concept to include other levels of organization (USDA 1982).

As stated earlier, an ecosystem encompasses both the living and nonliving factors in a particular environment. The living or biotic part of the ecosystem is formed by two components: autotrophic and heterotrophic. The *autotrophic* (self-nourishing) component does not require food from its environment but can manufacture food from inorganic substances. For example, some autotrophic components (plants) manufacture needed energy through photosynthesis. *Heterotrophic* components, on the other hand, depend upon autotrophic components for food (Porteous 1992).

The nonliving or abiotic part of the ecosystem is formed by three components: inorganic substances, organic compounds (link biotic and abiotic parts), and climate regime. Figure 11.2 is a simplified diagram showing a few of the living and nonliving components of an ecosystem found in a freshwater pond.

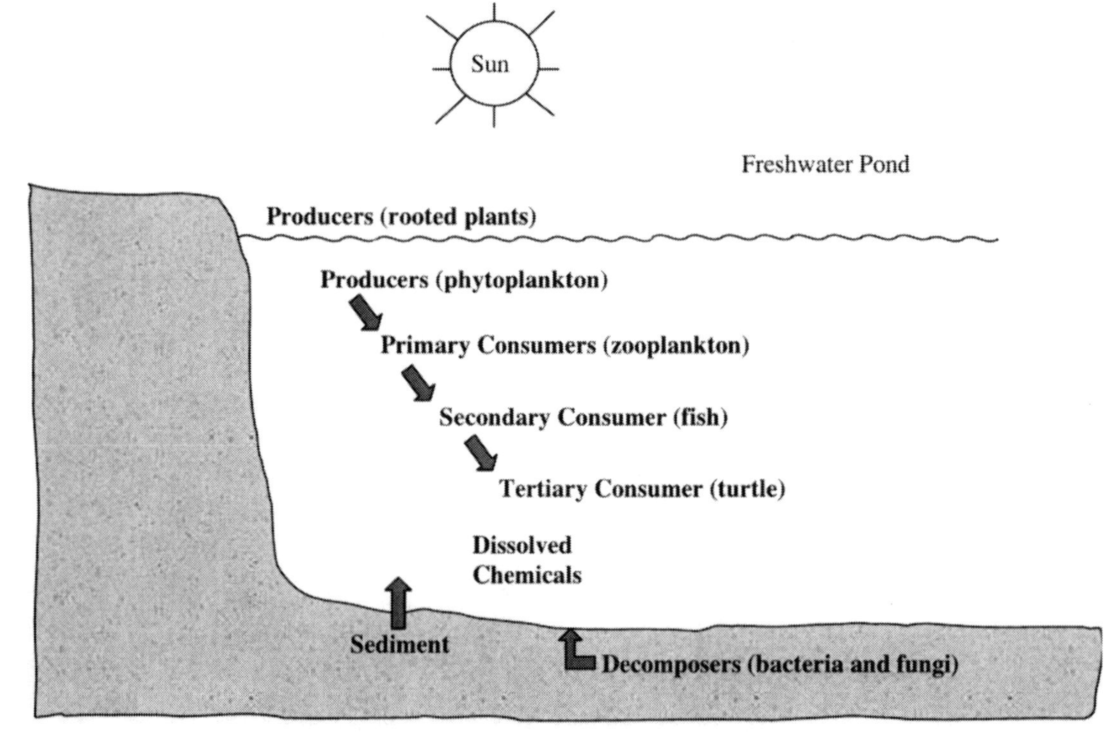

Figure 11.2. Major components of a freshwater pond ecosystem.

Did You Know?

According to the *Concise Oxford Dictionary of Ecology* (1994), "Natural" is commonly defined as being "present in or produced by nature . . . with relatively little modification by humans."

Modern usage of the term ecosystem derives from the work done by Raymond Lindeman. Lindeman's central concepts were that of functional organization and ecological energy efficiency ratios. This approach is connected to ecological energetics and might also be thought of as environmental rationalism. It was subsequently applied by H. T. Odum in founding the transdiscipline known as *systems ecology* (Lindeman 1942).

An ecosystem is a cyclic mechanism in which biotic and abiotic materials are constantly exchanged through biogeochemical cycles. Biogeochemical cycles are defined as follows: *bio* refers to living organisms and *geo* to water, air, rocks, or solids. *Chemical* is concerned with the chemical composition of the Earth. Biogeochemical cycles are driven by energy, directly or indirectly from the Sun. They will be discussed later.

The simplified freshwater pond shown in figure 11.2 depicts an ecosystem where biotic and abiotic materials are constantly exchanged. Producers construct organic substances through photosynthesis and chemosynthesis. Consumers and decomposers use organic matter as their food and convert it into abiotic components, that is, they dissipate energy fixed by producers through food chains. The abiotic part of the pond in figure 11.2 is formed of inorganic and organic compounds dissolved and in sediments such as carbon, oxygen, nitrogen, sulfur, calcium, hydrogen, and humic acids. The biotic part is represented by producers such as rooted plants and phytoplanktons. Fish, crustaceans, and insect larvae make up the consumers. Detrivores, which feed on organic detritus, are represented by mayfly nymphs. Decomposers make up the final biotic part. They include aquatic bacteria and fungi, which are distributed throughout the pond.

As stated earlier, an ecosystem is a cyclic mechanism. From a functional viewpoint, an ecosystem can be analyzed in terms of several factors. The factors important in this study include biogeochemical cycles, energy, and food chains; these factors are discussed in detail later.

TYPES OF ECOSYSTEMS

Individual ecosystems consist of physical, chemical, and biological components. As mentioned, the physical and chemical components are known as abiotic (not living environment) factors that influence living organisms in both terrestrial and aquatic ecosystems. The abiotic factors are as follows:

For terrestrial ecosystems

- Sunlight
- Temperature
- Precipitation
- Wind
- Latitude
- Altitude
- Fire frequency
- Soil

For aquatic ecosystems

- Light penetration
- Water currents
- Dissolved nutrient concentrations
- Suspended solids
- Salinity

The biotic (living environment) factors making up an ecosystem include the producers, consumers, and decomposers.

Abiotic and biotic factors combine to make up the following types of terrestrial and aquatic ecosystems:

- Estuaries
- Swamps and marshes
- Tropical rain forest
- Temperate forest
- Northern coniferous forest (taiga)

- Savanna
- Agricultural land
- Woodland and shrubland
- Temperate grassland
- Lake and streams
- Continental shelf
- Open ocean
- Tundra (arctic and alpine)
- Desert scrub
- Extreme desert

Biogeochemical/Nutrient Cycles

All matter (i.e., carbon, nitrogen, oxygen, or molecules [water]) cycles; it is neither created nor destroyed. Because the Earth is essentially a closed system with respect to matter, it can be said that all matter on Earth cycles. Ecologists study the flow of nutrients in ecosystems.

Ecosystem elements, like streams—and their larger cousins, rivers—are complex ecosystems that take part in the physical and chemical cycles (biogeochemical cycles) that shape our planet and allow life to exist. A *biogeochemical cycle* is composed of bioelements (chemical elements that cycle through living organisms) and occurs when there is an interaction between the biological and physical exchanges of bioelements. In a biogeochemical cycle, nutrient cycling and recycling through ecosystems results from the actions of geology, meteorology, and living things. Various nutrient biogeochemical cycles include the following (Spellman 1996):

- water cycle (discussed earlier)
- carbon cycle
- oxygen cycle
- nitrogen cycle
- phosphorus cycle
- sulfur cycle
- *Important note*: Contrary to an incorrect assumption, energy does not cycle through an ecosystem; chemicals do. The inorganic nutrients cycle through more than the organisms, however; they also enter into the oceans, atmosphere, and even rocks. Since these chemicals cycle through both the biological and the geological world, we call the overall cycles biogeochemical cycles.

Each chemical has its own unique cycle, but all of the cycles do have some things in common. Reservoirs are those parts of the cycle where the chemical is held in large quantities for long periods of times (e.g., the oceans for water and rocks for phosphorous). In exchange pools, on the other hand, the chemical is held for only a short time (e.g., the atmosphere, a cloud). The length of time a chemical is held in an exchange pool or a reservoir is termed its residence time. The biotic community includes all living organisms. This community may serve as an exchange pool (although for some chemicals like carbon, bound in certain tree species for a thousand years, it may seem more like a reservoir) and also serve to move chemicals (bioelements) from one stage of the cycle to another. For instance, the trees of the tropical rain forest bring water up from the forest floor to be transpired into the atmosphere. Likewise, coral organisms take carbon from the water and turn it into limestone rock. The energy for most of the transportation of chemicals from one place to another is provided either by the Sun or by the heat released from the mantle and core of the Earth (Spellman 1996).

- *Important point*: Water is exchanged between the hydrosphere, lithosphere, atmosphere, and biosphere. The oceans are large reservoirs that store water; they ensure thermal and climatic stability.

In addition to these exchanges of nutrients (losses and gains), Stephen Abedon (1997) points out examples of other losses and gains:

- Minerals can be lost from ecosystems by the action of rain.
- Nutrients can also be carried into ecosystems by the action of wind or migrating animals.
- Movement of salmon up rivers is an example of how nutrients might be delivered into an upstream ecosystem (e.g., from the oceans back to terrestrial forests).
- A consequence of ecosystem disruption is an impaired ability to recycle nutrients that leads to nutrient loss and long-term ecosystem impoverishment.
- In general, a disturbed habitat probably loses (rather than recycles) nutrients to a much greater degree than an undisturbed habitat where the action of human activities are not necessarily rapidly or readily reversible. A common consequence of human disturbance of ecosystems and the associated irreversible loss of nutrients is desertification.

In the case of chemical elements that cycle through living things, that is, bioelements, the following occurs (ISWS 2005):

- All bioelements reside in compartments or defined spaces in nature.
- A compartment contains a certain quantity, or pool, of bioelements.
- Compartments exchange bioelements. The rate of movement of bioelements between two compartments is called the flux rate.
- The average length of time a bioelement remains in a compartment is called the mean residence time (MRT).
- The flux rate and pools of bioelements together define the nutrient cycle in an ecosystem.
- Ecosystems are not isolated from one another, and bioelements come into an ecosystem through meteorological, geological, or biological transport mechanisms—
  - meteorological (e.g., deposition in rain and snow, and atmospheric gases);
  - geological (e.g., surface and subsurface drainage); and
  - biological (e.g., movement of organisms between ecosystems).

As a result, biogeochemical cycles can be

- local or
- global.

Smith (1974) categorizes biogeochemical cycles into two types, the gaseous and the sedimentary. Gaseous cycles include the carbon and nitrogen cycles. The main pool (or sink) of nutrients in the gaseous cycle is the atmosphere and the ocean. The sedimentary cycles include the sulfur and phosphorous cycles. The main sink for sedimentary cycles is soil and rocks of the Earth's crust.

Between twenty to forty elements of the Earth's ninety-two naturally occurring elements are ingredients that make up living organisms. The chemical elements carbon, hydrogen, oxygen, nitrogen, and phosphorus are critical in maintaining life as we know it on Earth. Odum (1971) points out that, of the elements needed by living organisms to survive, oxygen, hydrogen, carbon, and nitrogen are needed in larger quantities than are some of the other elements. The point is—no matter what particular elements are needed to sustain life, these elements exhibit definite biogeochemical cycles. These biogeochemical cycles will be discussed in detail later. For now, it is important to cover the life-sustaining elements in greater detail.

The elements needed to sustain life are products of the global environment. The global environment consists of three main subdivisions (see figure 11.3):

1. Hydrosphere—includes all the components formed of water bodies on the Earth's surface.

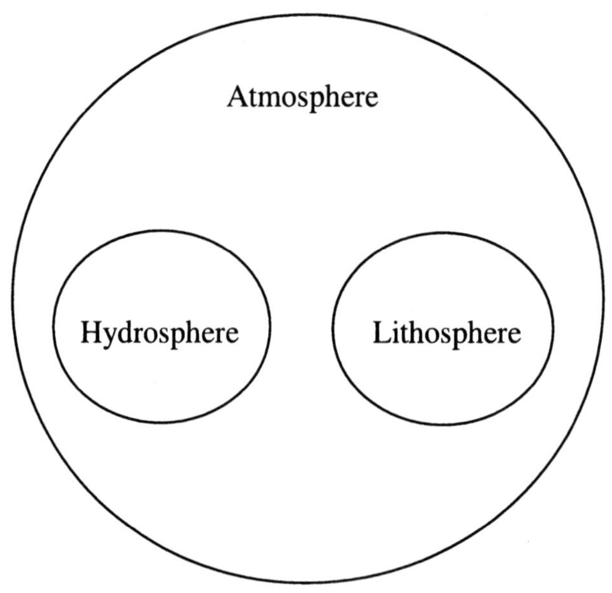

Figure 11.3. The global environment.

2. Lithosphere—comprises the solid components, such as rocks.
3. Atmosphere—is the gaseous mantle that envelopes the hydrosphere and lithosphere.

To survive, organisms require inorganic metabolites from all three parts of the biosphere. For example, the hydrosphere supplies water as the exclusive source of needed hydrogen. Essential elements such as calcium, sulfur, and phosphorus are provided by the lithosphere. Finally, oxygen, nitrogen, and carbon dioxide are provided by the atmosphere.

Within the biogeochemical cycles, all the essential elements circulate from the environment to organisms and back to the environment. Because of the critical importance of elements in sustaining life, it may be easily understood why the biogeochemical cycles are readily and realistically labeled nutrient cycles.

Through these biogeochemical or nutrient cycles, nature processes and reprocesses the critical life-sustaining elements in definite inorganic-organic cycles. Some cycles, such as carbon, are more perfect than others, that is, there is no loss of material for long periods of time. One major point to keep in mind is that energy (to be explained later) flows "through" an ecosystem, but nutrients are cycled and recycled.

Humans need most of these recycled elements to survive. Because we need almost all the elements in our complex culture, we have speeded up the movement of many materials so that the cycles tend to become imperfect or what Odum calls acyclic. Odum goes on to explain that our environmental impact on phosphorus demonstrates one example of a somewhat imperfect cycle.

> We mine and process phosphate rock with such careless abandon that severe local pollution results near mines and phosphate mills. Then, with equally acute myopia we increase the input of phosphate fertilizers in agricultural systems without controlling in any way the inevitable increase in run-off output that severely stresses our waterways and reduces water quality through eutrophication. (Odum 1971, 87)

As related above, in agricultural ecosystems, we often supply necessary nutrients in the form of fertilizer to increase plant growth and yield. In natural ecosystems, however, these nutrients are recycled naturally through each trophic level. For an example, the elemental forms are taken up by plants. The consumers ingest these elements in the form of organic plant material. Eventually, the nutrients are degraded to the inorganic form again. The following pages present and discuss the nutrient cycles for carbon, nitrogen, phosphorus, and sulfur.

CARBON CYCLE

Carbon, which is an essential ingredient of all living things, is one of the most abundant elements in the solar system; it is the basic building block of the large, organic molecules necessary for life. Inorganic forms of carbon (carbon dioxide, bicarbonate, and carbonate) strongly affect the acidity of soils and natural waters, the heat insulating capability of the atmosphere, and the rates of such key natural processes as photosynthesis, weathering, and biomineralization. In reduced form, carbon provides the elemental backbone for myriad organic molecules that comprise living organisms, soil humus, and fossil fuels (NSF 2000). Carbon is cycled into food chains from the atmosphere, as shown in figure 11.4 and the carbon cycle box model (figure 11.5) below.

CARBON CYCLE BOX MODEL

The carbon cycle (see figure 11.4) is based on carbon dioxide (CO_2), which makes up only a small percentage of the atmosphere. From figure 11.4 it can be seen that green plants obtain carbon dioxide (CO_2) from the air and, through photosynthesis, described by Isaac Asimov as the "most important chemical process on Earth," it produces the food and oxygen on which all organisms live (Asimov 1989, 20). Part of the carbon produced remains in living matter, and the other part is released as CO_2 in cellular respiration. G. Tyler Miller points out that the carbon dioxide released by cellular respiration in all living organisms is returned to the atmosphere (Miller 1988).

Did You Know?

About a tenth of the estimated seven hundred billion tons of carbon dioxide in the atmosphere is fixed annually by photosynthetic plants. A further trillion tons are dissolved in the ocean, more than half in the photosynthetic layer.

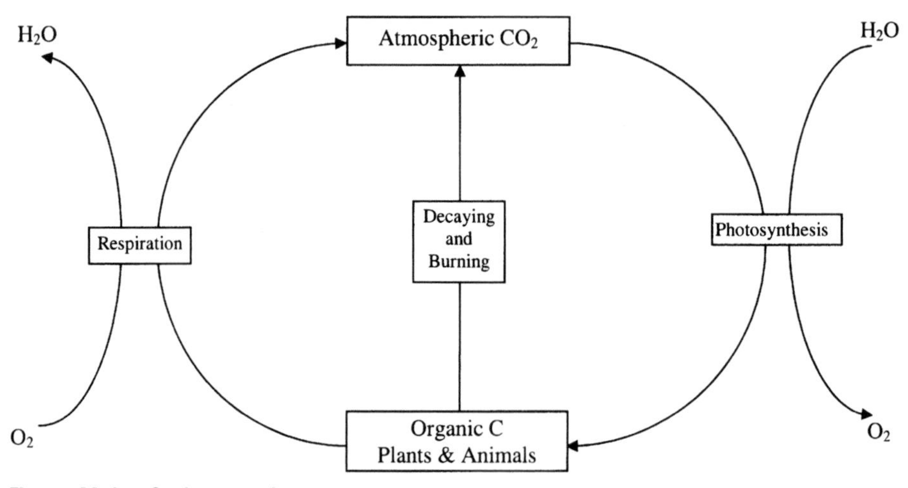

Figure 11.4. Carbon cycle.

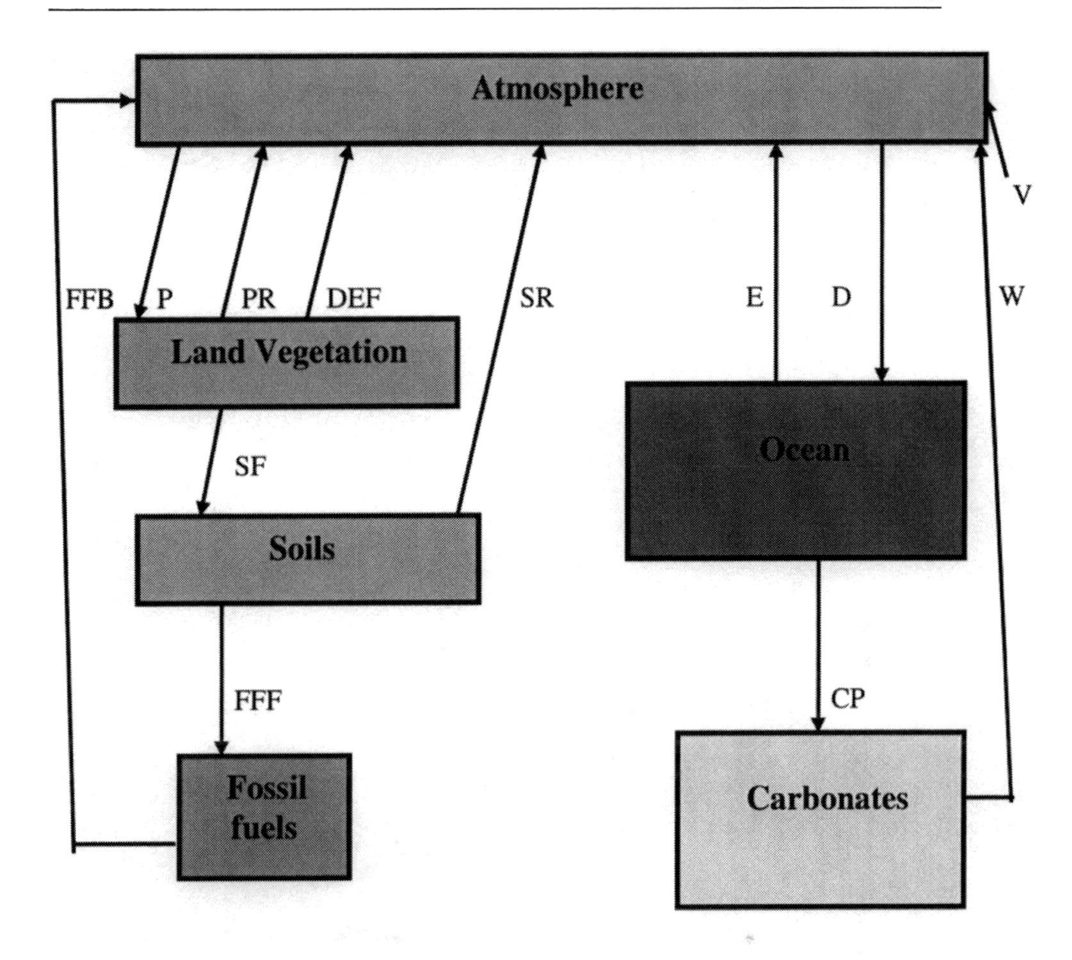

P	Photosynthesis
PR	Plant respiration
SR	Soil respiration
SF	Plants to soils
FFF	Fossil fuel formation
FFB	Fossil fuel burning
DEF	Deforestation
D	Dissolving
E	Exsolving
CP	Carbonate formation
W	Weathering
V	Volcanoes

Figure 11.5. Carbon cycle box model.

Some carbon is contained in buried dead animal and plant materials. Much of these buried plant and animal materials were transformed into fossil fuels. Fossil fuels, coal, oil, and natural gas contain large amounts of carbon. When fossil fuels are burned, stored carbon combines with oxygen in the air to form carbon dioxide, which enters the atmosphere (Moran, Morgan, and Wiersma 1986).

In the atmosphere, carbon dioxide acts as a beneficial heat screen as it does not allow the radiation of Earth's heat into space. This balance is important. The problem is that, as more carbon dioxide from burning is released into the atmosphere, the balance can and is being altered. Odum (1983, 4) warns that the recent increases in consumption of fossil fuels "coupled with the decrease in the 'removal capacity' of the green belt is beginning to exceed the delicate balance." Massive increases of carbon dioxide into the atmosphere tend to increase the possibility of global warming. The consequences of global warming "would be catastrophic . . . and the resulting climatic change would be irreversible" (Abrahamson 1988, 202).

NITROGEN CYCLE

Nitrogen is important to all life because it is a necessary nutrient. Nitrogen in the atmosphere or in the soil can go through many complex chemical and biological changes, be combined into living and nonliving material, and return back to the soil or air in a continuing cycle. This is called the nitrogen cycle (Killpack and Buchholz 1993). The nitrogen cycle consists of the following processes and various states:

Nitrogen Cycle

Processes	*States*
Fertilizer	N_2—elemental nitrogen is a gaseous form of nitrogen
Volatilization	NH_3—ammonia is a gaseous form of nitrogen
Animal wastes	NO—nitric oxide is a gaseous form of nitrogen
Organic matter	NH_4+—ammonium is attracted to soil particles
Immobilization	N_2O—nitrous oxide is a gaseous form of nitrogen
Nitrification	NO_3—nitrate is not attracted to soil particles
Biological fixation	
Mineralization	
Denitrification	
Crop uptake	

Important Nitrogen Cycle Terminology

- *Limiting nutrient*—amount of an element necessary for plant life is in short supply.
- *Nitrogen fixation*—chemical conversion from N_2 to NH_3 (ammonia) or NO_3 (nitrate).
- *Denitrification*—chemical conversion from nitrate (NO_3) back to N_2.

Today, we readily apply nitrogen in the form of fertilizer to plants to ensure proper growth. Before fertilizer containing nitrogen was commercially available, how did agriculture survive? Early farmers had to rely on natural regeneration of fixed nitrogen:

- annual floods—bring fresh sediments (e.g., Nile Valley).
- slash/burn agriculture—once the soil nutrients are depleted, move on to a new place.
- crop rotation—certain crops (e.g., soybeans) are good at fixing nitrogen, while others (e.g., corn) use it up; plant in alternate years.

The atmosphere contains 78 percent by volume of nitrogen. Moreover, as stated above, nitrogen is an essential element for all living matter and constitutes 1–3 percent dry weight of cells, yet nitrogen is not a common element on Earth. Although it is an essential ingredient for plant growth, it is chemically very inactive, and before it can be incorporated by the vast majority of the biomass, it must be "fixed" (Porteous 1992).

Peter Price (1984) describes the nitrogen cycle as an example "of a largely complete chemical cycle in ecosystems with little leaching out of the system." From the water/wastewater specialist's point of view, nitrogen and phosphorous are both commonly considered as limiting factors for productivity. Of the two, nitrogen is harder to control but is found in smaller quantities in wastewater.

As stated earlier, nitrogen gas makes up about 78 percent of the volume of the Earth's atmosphere. As such, it is useless to most plants and animals. Fortunately, nitrogen gas is converted into compounds containing nitrate ions, which are taken up by plant roots as part of the nitrogen cycle, shown in simplified form in figure 11.6 and in box model form below.

Aerial nitrogen is converted into nitrates mainly by microorganisms, bacteria, and blue-green algae. Lightning also converts some aerial nitrogen gas into forms that return to the Earth as nitrate ions in rainfall and other types of precipitation. From figure 11.6 it can be seen that ammonia plays a major role in the nitrogen cycle. Excretion by animals and anaerobic decomposition of dead organic matter by bacteria produce ammonia. Ammonia, in turn, is converted by nitrification bacteria into nitrites and then into nitrates. This process is known as nitrification. Nitrification bacteria are aerobic. Bacteria that convert ammonia into nitrites are known as nitrite bacteria (Nitrosococcus and Nitrosomonas); they convert nitrites into nitrates and nitrate bacteria (Nitrobacter). In wastewater treatment, ammonia is produced in the sludge digester, nitrates in the aerobic sewage treatment process.

NITROGEN CYCLE BOX MODEL

Nitrogen reservoirs and quantities in millions of metric tons are as follows:

- Atmosphere: 4,000,000,000
- Land plants: 3,500

- Soils: 9,500
- Oceans: 23,000,000
- Sediments and rocks: 200,000,000,000

Did You Know?

Buried rocks and sediments are the largest pool of nitrogen, but this reservoir is a minor part of the cycle.

Specialized bacteria and lightning are the only natural ways that nitrogen is fixed. Lightning may have been necessary for life to begin:

No life => no bacteria => no bacterial fixation => no usable nitrogen => no life . . .

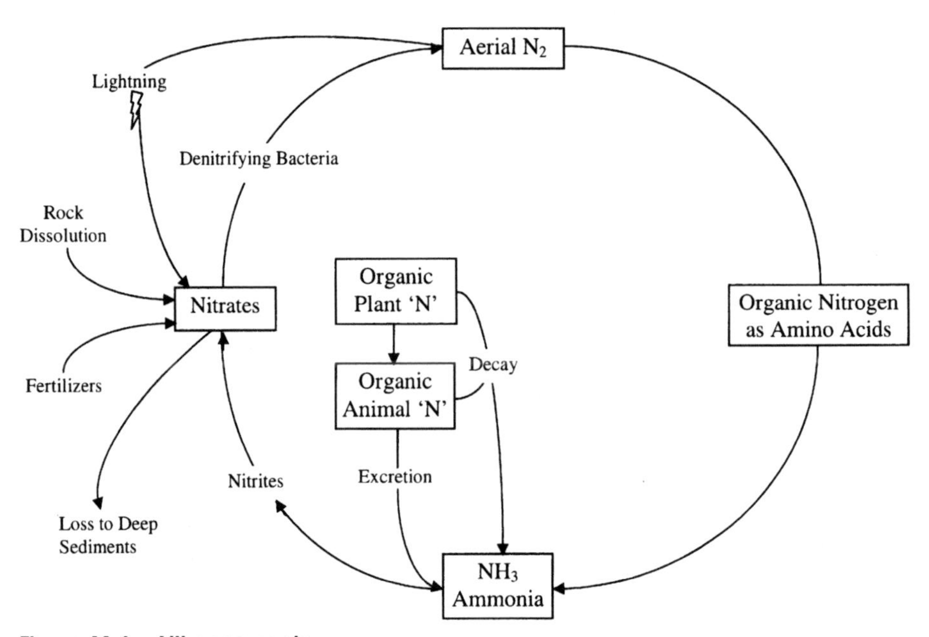

Figure 11.6. Nitrogen cycle.

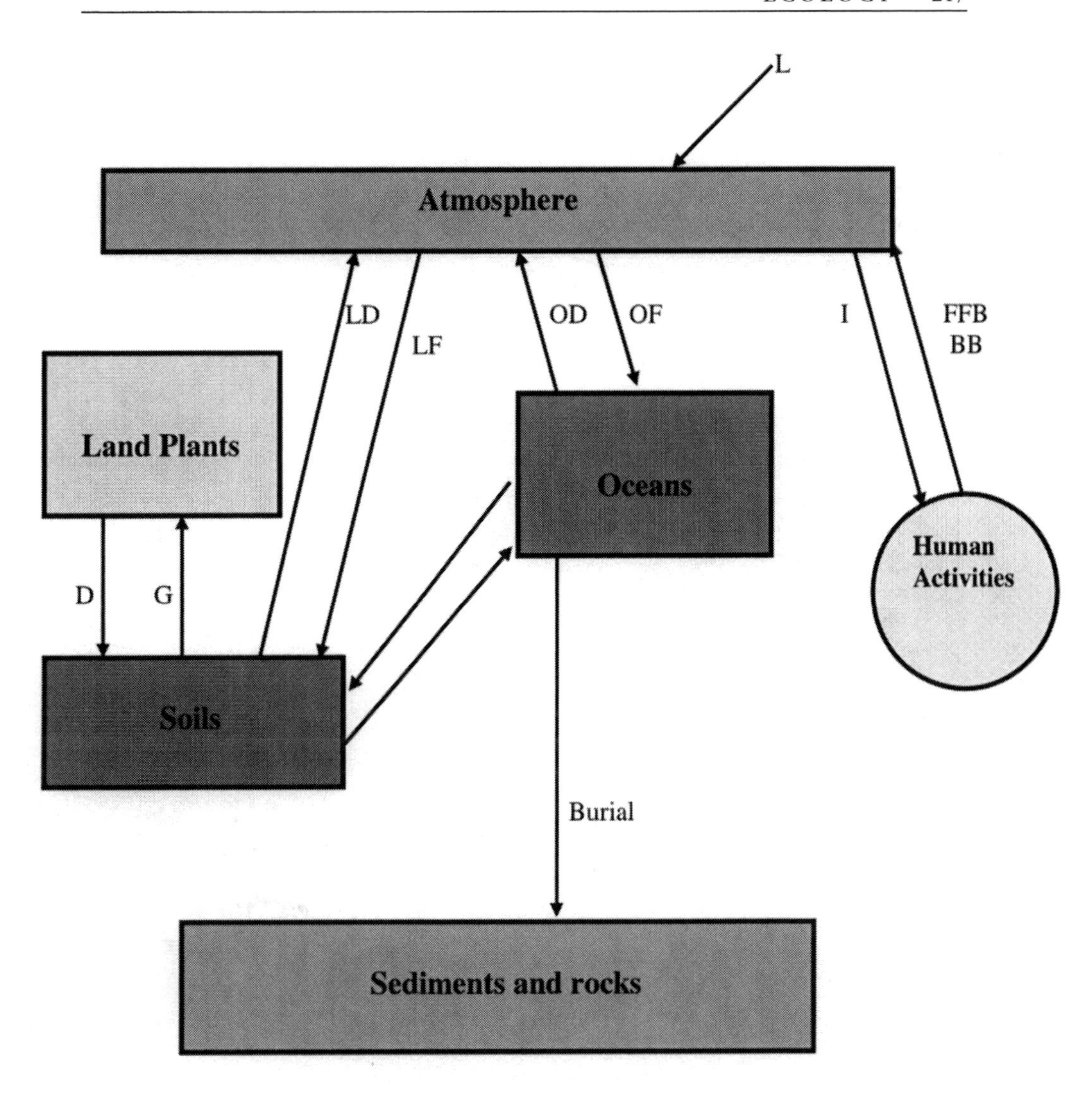

LF	Land Fixation	**O – L**	Ocean-to-Land
LD	Land Denitrification	**L – O**	Land-to-Ocean
OF	Oceanic Fixation		
OD	Oceanic Denitrification		
I	Industrial Fixation		
FFB	Fossil Fuel Burning		
BB	Biomass Burning		
L	Lightning		
D	Decay		
G	Growth		

Figure 11.7. Nitrogen cycle box model.

PHOSPHORUS CYCLE

Phosphorus is another element that is common in the structure of living organisms. However, of all the elements recycled in the biosphere, phosphorus is the scarcest and therefore the one most limiting in any given ecological system. It is indispensable to life, being intimately involved in energy transfer and in the passage of genetic information in the DNA of all cells.

The ultimate source of phosphorus is rock, from which it is released by weathering, leaching, and mining. Phosphorus has no stable gas phase, so addition of phosphorus to land is slow. Phosphorus occurs as phosphate or other minerals formed in past geological ages. These massive deposits are gradually eroding to provide phosphorus to ecosystems. A large amount of eroded phosphorus ends up in deep sediments in the oceans and lesser amounts in shallow sediments. Part of the phosphorus comes to land when marine animals are brought out. Birds also play a role in the recovery of phosphorus. The great guano deposit, bird excreta, of the Peruvian coast is an example. Man has hastened the rate of loss of phosphorus through mining activities and the subsequent production of fertilizers, which is washed away and lost. Even with the increase in human activities, however, there is no immediate cause for concern, since the known reserves of phosphate are quite large.

Did You Know?

Humans have greatly accelerated phosphorus transfer from rocks to plants and soils (about five times faster than weathering).

Phosphorous has become very important in water quality studies, since it is often found to be a limiting factor (i.e., limiting plant nutrient). Control of phosphorus compounds that enter surface waters, and contribute to growth of algal blooms, is of much interest to stream ecologists. Phosphates, upon entering a stream, act as fertilizer, which promotes the growth of undesirable algae populations or algal blooms. As the organic matter decays, dissolved oxygen levels decrease, and fish and other aquatic species die. Figure 11.8 shows the phosphorus cycle. The phosphorus cycle box model is shown below.

PHOSPHORUS CYCLE BOX MODEL

While it is true that phosphorus discharged into streams is a contributing factor to stream pollution (and causes eutrophication), it is also true that phosphorus is not the lone factor (see figure 11.9). Odum (1975) warns against what he calls the one-factor control hypothesis, that is, the one problem/one solution syndrome. He goes on to point out that environmentalists in the past have focused on one or two items, like

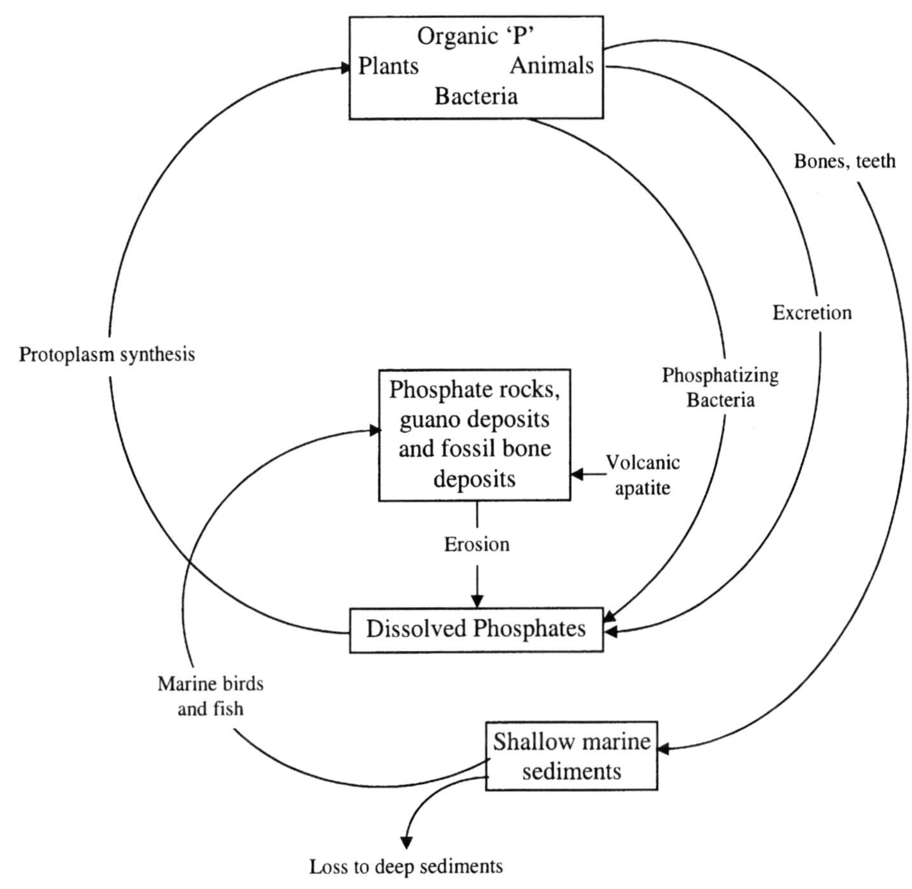

Figure 11.8. Phosphorus cycle.

phosphorous contamination, and "have failed to understand that the strategy for pollution control must involve reducing the input of all enriching and toxic materials" (p. 110)

Did You Know?

Because of its high reactivity, phosphorus exists in combined form with other elements. Microorganisms produce acids that form soluble phosphate from insoluble phosphorus compounds. The phosphates are utilized by algae and terrestrial green plants, which in turn pass into the bodies of animal consumers. Upon death and decay of organism, phosphates are released for recycling (Spellman 1996).

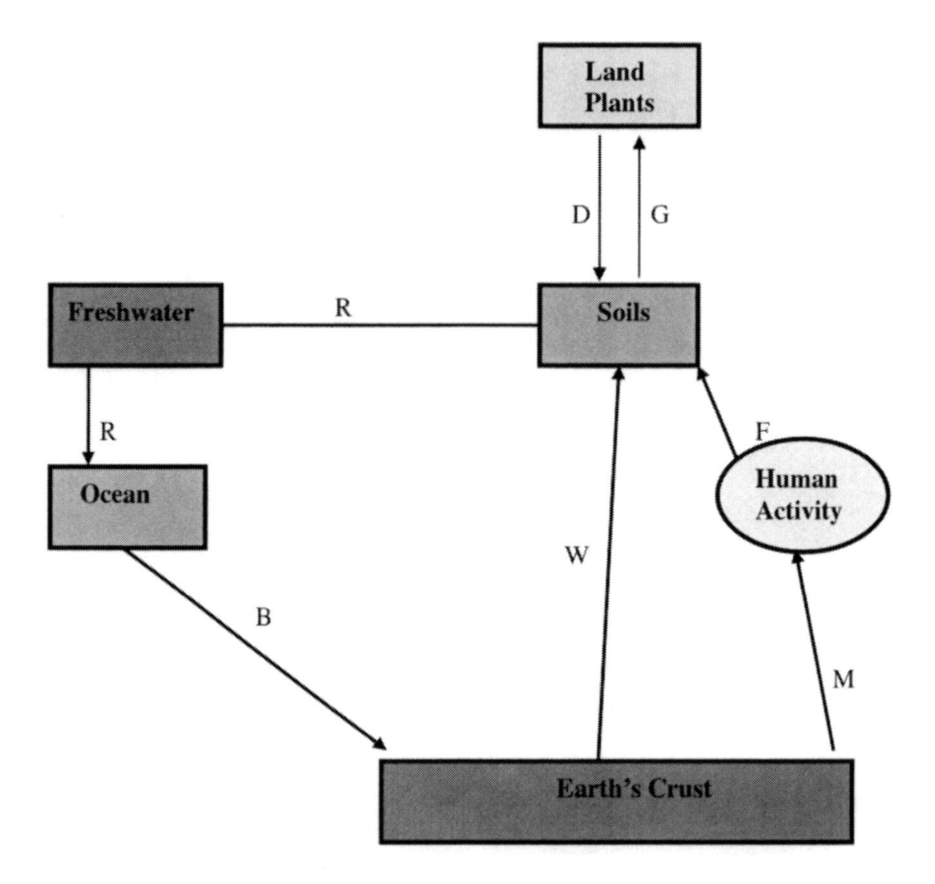

M	Mining
F	Fertilization
W	Weathering
R	Runoff
B	Burial
D	Decay
G	Growth

Figure 11.9. Phosphorus cycle box model.

SULFUR CYCLE

Sulfur, like nitrogen and carbon, is characteristic of organic compounds. However, an important distinction between cycling of sulfur and cycling of nitrogen and carbon is that sulfur is "already fixed," that is, plenty of sulfate anions are available for living organisms to utilize; the largest reservoir is Earth's crust. By contrast, the major biological reservoirs of nitrogen atoms (N_2) and carbon atoms (CO_2) are gases that must be pulled out of the atmosphere. Sulfur is rarely a limiting nutrient for ecosystems or organisms.

Did You Know?

In the sulfur cycle, elementary sulfur of the lithosphere is not available to plants and animals unless converted to sulfates.

The sulfur cycle (see figure 11.10) is both sedimentary and gaseous. George Tchobanoglous and Edward Schroeder (1985, 276) note that "the principal forms of sulfur that are of special significance in water quality management are organic sulfur, hydrogen sulfide, elemental sulfur and sulfate."

Bacteria play a major role in the conversion of sulfur from one form to another. In an anaerobic environment, bacteria break down organic matter producing hydrogen sulfide with its characteristic rotten egg odor. A bacterium called Beggiatoa converts hydrogen sulfide into elemental sulfur. An aerobic sulfur bacterium, Thiobacillus thiooxidans, converts sulfur into sulfates. Other sulfates are contributed by the dissolving of rocks and some sulfur dioxide. Sulfur is incorporated by plants into proteins. Some of these plants are then consumed by organisms. Sulfur from proteins is liberated by many heterotrophic anaerobic bacteria, as hydrogen sulfide.

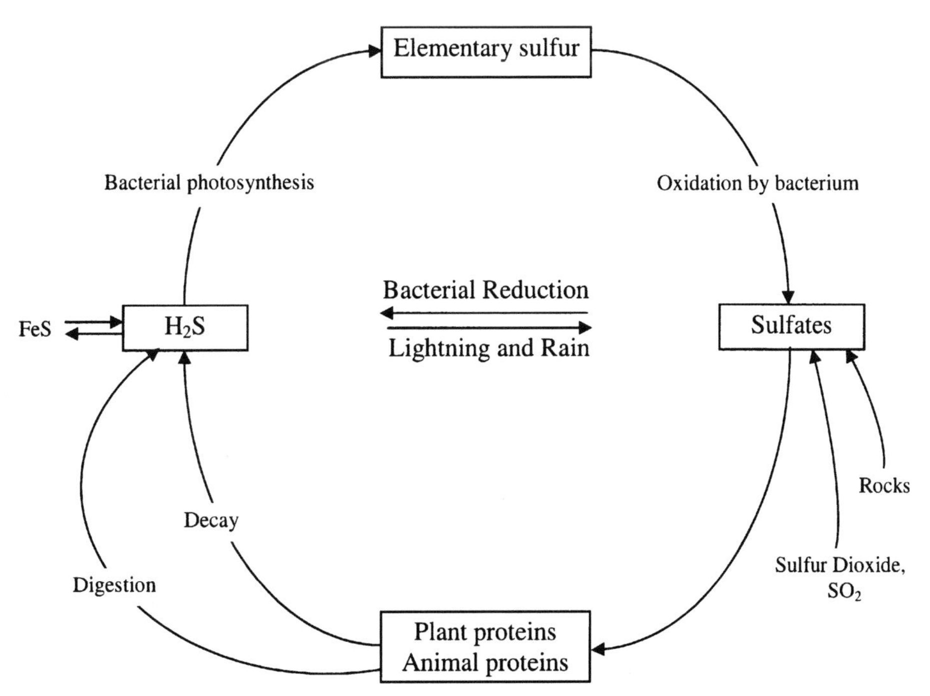

Figure 11.10. Sulfur cycle.

Energy Flow in the Ecosystem

> Three hundred trout are needed to support one man for a year. The trout, in turn, must consume 90,000 frogs, that must consume 27 million grasshoppers that live off of 1,000 tons of grass.
>
> —G. Tyler Miller Jr., American chemist (1988, 62)

The main concept covered in this section is how energy moves through an ecosystem. Again, it is important to point out that energy "moves" through an ecosystem and is not cycled through it. If you can understand this, you are in good shape, because then you have an idea of how ecosystems are balanced, how they may be affected by human activities, and how pollutants will move through an ecosystem.

Simply defined, energy is the ability or capacity to do work. For an ecosystem to exist, it must have energy. All activities of living organisms involve work, which is the expenditure of energy. This means the degradation of a higher state of energy to a lower state. The flow of energy through an ecosystem is governed by two laws: the First and Second Laws of Thermodynamics.

The first law, sometimes called the conservation law, states that energy may not be created or destroyed. The second law states that no energy transformation is 100 percent efficient, that is, in every energy transformation, some energy is dissipated as heat. The term entropy is used as a measure of the nonavailability of energy to a system. Entropy increases with an increase in dissipation. Because of entropy, input of energy in any system is higher than the output or work done; thus, the resultant efficiency is less than 100 percent.

Odum (1975, 61) explains that "the interaction of energy and materials in the ecosystem is of primary concern of ecologists." Earlier, we discussed the biogeochemical nutrient cycles. It is important to remember that it is the flow of energy that drives these cycles. Again, it should be noted that energy does not cycle as do nutrients in biogeochemical cycles. For example, when food passes from one organism to another, energy contained in the food is reduced step by step until all the energy in the system is dissipated as heat. Price (1984, 11) refers to this process as "a unidirectional flow of energy through the system, with no possibility for recycling of energy." When water or nutrients are recycled, energy is required. The energy expended in this recycling is not recyclable. And, as Odum (1975) points out, this is a "fact not understood by those who think that artificial recycling of man's resources is somehow an instant and free solution to shortages."

While there is a slight input of geothermal energy, as pointed out earlier, the principal source of energy for any ecosystem is sunlight. Green plants, through the process of photosynthesis, transform the Sun's light energy into chemical energy: carbohydrates that are consumed by animals. This transfer of energy, as stated previously, is unidirectional—from producers to consumers—it is accomplished by cellular respiration, the process by which organisms (like mammals) break the glucose back down into its constituents, water and carbon dioxide, thus regaining the stored energy the Sun originally gave to the plants. Often this transfer of energy to different organisms is called a *food chain*. It is safe to say that food energy passes through a community in various ways—each separate way is called a food chain. Figure 11.11 shows a simple aquatic food chain.

All organisms, alive or dead, are potential sources of food for other organisms. All organisms that share the same general type of food in a food chain are said to be at the same trophic level (nourishment or feeding level—each level of consumption in a food chain is called a trophic level). Since green plants use sunlight to produce food for animals, they are called the producers, or the first trophic level. The herbivores, which eat plants directly, are called the second trophic level, or the primary consumers. The carnivores are flesh-eating consumers; they include several trophic levels from the third on up. At each transfer, a large amount of energy (about 80 to 90 percent) is lost as heat and wastes. Thus, nature normally limits food chains to four or five links; however, in aquatic ecosystems, "food chains are commonly longer than those on land" (Dasmann 1984, 65). The aquatic food chain is longer because several predatory fish may be feeding on the plant consumers. Even so, the built-in inefficiency of the energy transfer process prevents development of extremely long food chains.

Tomera (1989) describes a simple food chain that can be seen in a prairie dog community.

> The grass in the community manufactures food. The grass is called a food producer. The grass is eaten by a prairie dog. Because the prairie dog lives directly off the grass, it is termed a first-order consumer. A weasel [or other predator] may kill and eat the prairie dog. The weasel is, therefore, a predator and would be termed a second-order consumer. The second-order consumer is twice removed from the green grass. The weasel, in turn, may be eaten by a large hawk or eagle. The bird that kills and eats the weasel would therefore be a third-order consumer, three times removed from the grass. Of course, the hawk would give off waste materials and eventually die itself. Wastes and dead organisms are then acted on by decomposers. (50)

Only a few simple food chains are found in nature. Thus, when attempting to identify the complex food relationships among many animals and plants within a community, it is useful to create a food web. The fact is that most simple food chains are interlocked; this interlocking of food chains forms a food web. A *food web* can be characterized as a map that shows what eats what (Miller 1988). Most ecosystems support a complex food web. A food web involves animals that do not feed on one trophic level. For example, humans feed on both plants and animals. The point is, an

Figure 11.11. Aquatic food chain.

organism in a food web may occupy one or more trophic levels. Trophic level is determined by an organism's role in its particular community, not by its species. Food chains and webs help to explain how energy moves through an ecosystem.

An important trophic level of the food web that has not been discussed thus far is comprised of the decomposers (bacteria, mushrooms, etc.). The decomposers feed on dead plants or animals and play an important role in recycling nutrients in the ecosystem. As Miller (1988, 62) points out, "There is no waste in ecosystems. All organisms, dead or alive, are potential sources of food for other organisms." An example of an aquatic food web is shown in figure 11.12

From the preceding discussion about food chains and food webs, the important point to be gained is that there is a distinct difference between the two. A food chain, for example, is a simple, straight-line process going from producer to first-, second-, and possibly third-order consumers and ending with the decomposers. On the other hand, in a food web, there are a number of second- and third-order consumers.

Did You Know?

The natural world is pretty much a question of competition. Plants compete for water, sunlight, and nutrients. Resources are scarce, organisms compete for them, and survival goes to the strongest. Humans do what we can to help the good species win over the bad.

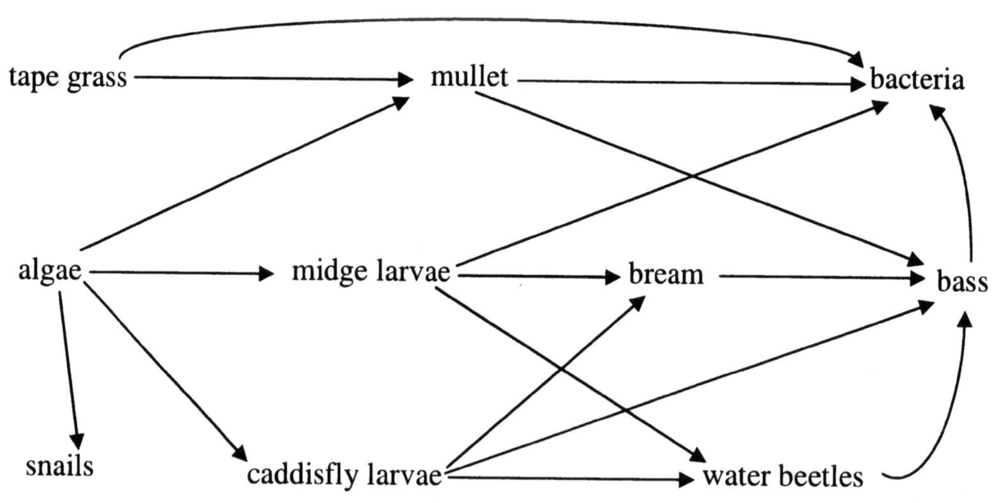

Figure 11.12. Aquatic food web.

Food Chain Efficiency

Earlier, it was pointed out that energy from the Sun is captured (via photosynthesis) by green plants and used to make food. Most of this energy is used to carry on the plant's life activities. The rest of the energy is passed on as food to the next level of the food chain.

Did You Know?

A food chain is the path of food from a given final consumer back to a producer.

It is important to note that nature limits the amount of energy that is accessible to organisms within each food chain. Not all food energy is transferred from one trophic level to the next. For ease of calculation, "ecologists often assume an ecological efficiency of 10 percent (10 percent rule) to estimate the amount of energy transferred through a food chain" (Moran, Morgan, and Wiersma 1986, 37). For example, if we apply the 10 percent rule to the diatoms-copepods-minnows-medium fish–large fish food chain shown in figure 11.13, we can predict that 1,000 grams of diatoms produce 100 grams of copepods, which will produce 10 grams of minnows, which will produce 1 gram of medium fish, which, in turn, will produce 0.1 gram of large fish. Thus, only about 10 percent of the chemical energy available at each trophic level is transferred and stored in usable form at the next level. What happens to the other 90 percent? The other 90 percent is lost to the environment as low-quality heat in accordance with the Second Law of Thermodynamics.

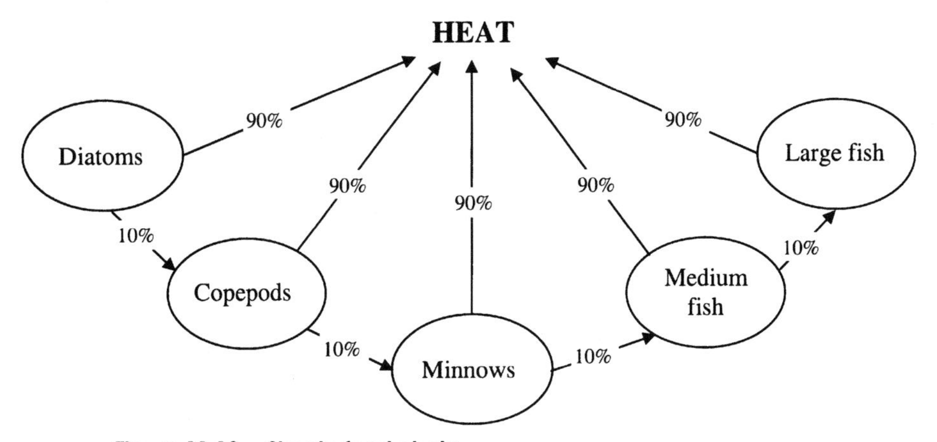

Figure 11.13. Simple food chain.

Did You Know?

The ratio of net production at one level to net production at the next higher level is called the *conversion efficiency.*

Ecological Pyramids

As we proceed in the food chain from the producer to the final consumer, it becomes clear that a particular community in nature often consists of several small organisms associated with a smaller and smaller number of larger organisms. A grassy field, for example, has a larger number of grass and other small plants, a smaller number of herbivores like rabbits, and an even smaller number of carnivores like foxes. The practical significance of this is that we must have several more producers than consumers.

This pound-for-pound relationship, where it takes more producers than consumers, can be demonstrated graphically by building an ecological pyramid. In an ecological pyramid, the number of organisms at various trophic levels in a food chain is represented by separate levels or bars placed one above the other with a base formed by producers and the apex formed by the final consumer. The pyramid shape is formed due to a great amount of energy loss at each trophic level. The same is true if numbers are substituted by the corresponding biomass or energy. Ecologists generally use three types of ecological pyramids: pyramids of number, biomass, and energy. Obviously, there will be differences among them. Some generalizations follow:

1. *Energy pyramids* must always be larger at the base than at the top (because of the Second Law of Thermodynamics, which has to do with dissipation of energy as it moves from one trophic level to another). Simply, energy pyramids depict the decrease in the total available energy at each higher trophic level.
2. Likewise, *biomass pyramids* (in which biomass is used as an indicator of production) are usually pyramid shaped. This is particularly true of terrestrial systems and aquatic ones dominated by large plants (marshes), in which consumption by heterotrophs is low and organic matter accumulates with time. A census of the population, multiplied by the weight of an average individual in it, gives an estimate of the weight of the population. This is called the *biomass* (or standing crop). However, it is important to point out that biomass pyramids can sometimes be inverted. This is especially common in aquatic ecosystems, in which the primary producers are microscopic planktonic organisms that multiply very rapidly, have very short life spans, and undergo heavy grazing by herbivores. At any single point in time, the amount of biomass in primary producers is less than that in larger, long-lived animals that consume primary producers.
3. *Numbers pyramids* can have various shapes (and not be pyramids at all, actually)

depending on the sizes of the organisms that make up the trophic levels. In forests, the primary producers are large trees, and the herbivore level usually consists of insects, so the base of the pyramid is smaller than the herbivore level above it, that is, the pyramid is inverted. In grasslands, the number of primary producers (grasses) is much larger than that of the herbivores above (large grazing animals). (Spellman 1996)

To get a better idea of how an ecological pyramid looks and how it provides its information, we need to look at an example. The example used here is the energy pyramid. According to Odum (1983, 154), the energy pyramid is a fitting example because, among the "three types of ecological pyramids, the energy pyramid gives by far the best overall picture of the functional nature of communities."

In an experiment conducted in Silver Springs, Florida, Odum (1983) measured the energy for each trophic level in terms of kilocalories. A kilocalorie is the amount of energy needed to raise one cubic centimeter of water one degree centigrade. When an energy pyramid is constructed to show Odum's findings, it takes on the typical upright form (as it must because of the Second Law of Thermodynamics) as shown in figure 11.14.

Simply put, as reflected in Figure 11.14 and according to the Second Law of Thermodynamics, no energy transformation process is 100 percent efficient. This fact is demonstrated, for example, when a horse eats hay. The horse cannot obtain, for its own body, 100 percent of the energy available in the hay. For this reason, the energy productivity of the producers must be greater than the energy production of the primary consumers. When human beings are substituted for the horse, it is interesting to note that, according to the Second Law of Thermodynamics, only a small population could be supported. But this is not the case. Humans also feed on plant matter, which allows a larger population. Therefore, if meat supplies become scarce, we must eat more plant matter. This is the situation we see today in countries where meat is scarce. Consider this: if we all ate soybean, there would be at least enough food for ten times as many of us as compared to a world where we all eat beef (or pork, fish, chicken, etc.). Here is another way of looking at this: every time we eat meat, we are taking food out of the mouths of nine other people, who could be fed with the plant material that was fed to the animal we are eating (Spellman 1996). Food-energy relationships are often referred to as eater-eaten relationships. It's not quite that simple, of course, but you probably get the general idea.

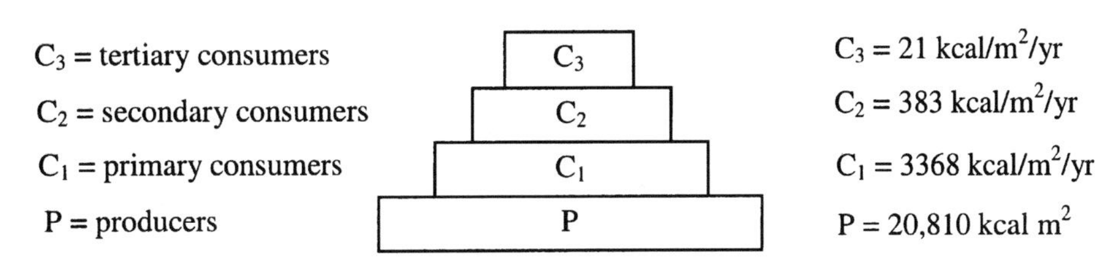

Figure 11.14. Energy flow pyramid. Adapted from Odum (1971).

Important Relationships in Living Communities

In addition to the pyramid-shaped relationships, there are other important relationships in living communities. Some of these involve food energy and some do not. In this section, several of these relationships are described.

Symbiosis is a close (intimate) ecological relationship (organisms live together in close proximity) between the individuals of two (or more) different species. Sometimes a symbiotic relationship benefits both species, sometimes one species benefits at the other's expense, and in other cases neither species benefits. One thing is certain; the relationship is *obligate*, meaning at least one of the species must be involved in the relationship to survive.

Ecologists use a different term for each type of symbiotic relationship:

- Mutualism—both organisms benefit.
- Commensalism—one organism benefits; the other is unaffected.
- Parasitism—one organism benefits; the other is harmed.
- Competition—neither organism benefits.
- Neutralism—both organisms are unaffected.

The effects of these interactions on population growth can be positive, negative, or neutral (see table 11.1).

Productivity

As mentioned previously, the flow of energy through an ecosystem starts with the fixation of sunlight by plants through photosynthesis. In evaluating an ecosystem, the measurement of photosynthesis is important. Ecosystems may be classified into highly productive or less productive. Therefore, the study of ecosystems must involve some measure of the productivity of that ecosystem.

Smith (1974, 38) defines production (or, more specifically, primary production because it is the basic form of energy storage in an ecosystem) as being "the energy accumulated by plants." Stated differently, primary production is the rate at which

Table 11.1. Population Interactions, Two-Species System

	Response	
Type of Interaction	*A*	*B*
Neutral	0	0
Mutualism	+	+
Commensalism	+	0
Parasitism	+	−
Competition	−	−

the ecosystem's primary producers capture and store a given amount of energy, in a specified time interval. In even simpler terms, primary productivity is a measure of the rate at which photosynthesis occurs. Odum (1971) lists four successive steps in the production process as follows:

1. *Gross primary productivity*—the total rate of photosynthesis in an ecosystem during a specified interval at a given trophic level.
2. *Net primary productivity*—the rate of energy storage in plant tissues in excess of the rate of aerobic respiration by primary producers.
3. *Net community productivity*—the rate of storage of organic matter not used.
4. *Secondary productivity*—the rate of energy storage at consumer levels.

When attempting to comprehend the significance of the term productivity as it relates to ecosystems, it is wise to consider an example. Consider the productivity of an agricultural ecosystem such as a wheat field. Often its productivity is expressed as the number of bushels produced per acre. This is an example of the harvest method for measuring productivity. For a natural ecosystem, several one-square meter plots are marked off, and the entire area is harvested and weighed to give an estimate of productivity as grams of biomass per square meter per given time interval. From this method, a net primary production (net yield) can be measured.

Productivity, both in the natural and cultured ecosystem, may vary considerably, not only between type of ecosystems but also within the same ecosystem. Several factors influence year-to-year productivity within an ecosystem. Such factors as temperature, availability of nutrients, fire, animal grazing, and human cultivation activities are directly or indirectly related to the productivity of a particular ecosystem.

Table 11.2 shows representative values for the net productivity of a variety of ecosystems—both natural and managed. Keep in mind that these values are only approximations derived from Odum's (1971; 1983) work and are subject to marked fluctuations because of variations in temperature, fertility, and availability of water.

In the aquatic (and any other) ecosystem, pollution can have a profound impact upon the system's productivity. For example, certain kinds of pollution may increase the turbidity of the water. This increase in turbidity causes a decrease in energy delivered by photosynthesis to the ecosystem. Accordingly, this turbidity and its aggregate effects decrease net community productivity on a large scale (Laws 1993).

The ecological trends paint a clear picture. Wherever we look, ecological productivity is limping behind human consumption. Since 1984, the global fish harvest has been dropping and so has the per capita yield of grain crops (Brown 1994). Moreover, stratospheric ozone is being depleted; the release of greenhouse gases has changed the atmospheric chemistry and might lead to climate change; erosion and desertification are reducing nature's biological productivity; irrigation water tables are falling; contamination of soil and water is jeopardizing the quality of food; other natural resources are being consumed faster than they can regenerate; and biological diversity is being lost—to reiterate only a small part of a long list. These trends indicate a decline in the quantity and productivity of nature's assets (Wachernagel 1997).

Table 11.2. Estimated Net Productivity of Certain Ecosystems

Ecosystem	*Kilocaleries/m²/year*
Temperate deciduous forest	5,000
Tropical rain forest	15,000
Tall-grass prairie	2,000
Desert	500
Coastal marsh	12,000
Ocean close to shore	2,500
Open ocean	800
Clear (oligotrophic) lake	800
Lake in advanced state of eutrophication	2,400
Silver Springs, Florida	8,800
Field of alfalfa (Lucerne)	15,000
Corn (maize) field, United States	4,500
Rice paddies, Japan	5,500
Lawn, Washington, D.C.	6,800
Sugar cane, Hawaii	25,000

Biodiversity

The Earth contains a diverse array of organisms whose species diversity, genetic diversity, and ecosystems are together called biodiversity. The U.S. Agency for International Development (USAID; 2007) defines *biodiversity* as the variety and variability of life on Earth. This includes all of the plants and animals that live and grow on the Earth, all of the habitats that they call home, and all of the natural processes of which they are a part. The Earth supports an incredible array of biodiversity with plants and animals of all shapes and sizes. This fantastic variety of life is found in diverse habitats ranging from the hottest desert to tropical rain forests to the arctic tundra. Biodiversity is essential to every aspect of the way that humans live around the world. Plants and animals provide people with food and medicine, trees play an important role in absorbing greenhouse gases and cleaning the air we breathe, and rivers and watersheds provide the clean water that we drink.

Unfortunately, however we define it, the fact is the Earth's biodiversity is disappearing, with an estimated one thousand species per year becoming extinct. Conserving biodiversity is especially crucial in developing countries where people's livelihoods are directly dependent on natural resources such as forests, fisheries, and wildlife.

In its simplest terms, biodiversity is the variety of life at all levels; it includes the array of plants and animals; the genetic differences among individuals; the communities, ecosystems, and landscapes in which they occur; and the variety of processes on which they depend (LaRoe 1995).

Biodiversity is important for several reasons. Its value is often reported in economic terms: for example, Keystone Center (1991) and Edward Wilson (1992) report that about half of all medicinal drugs come from—or were first found in—natural plants and animals, and therefore these resources are critical for their existing and as yet undiscovered medicinal benefits. Moreover, most foods were domesticated from wild stocks, and interbreeding of different, wild genetic stocks is often used to increase crop yield. LaRoe (1995) reports that today we use but a small fraction of the food crops used by native cultures: many of these underused plants may become critical new food sources for the expanding human population or in times of changing environmental conditions.

It should be noted that it is the great variety of life that makes existence on Earth possible, thus pointing out the greater importance of biodiversity. As a case in point, consider that plants convert carbon dioxide to oxygen during the photosynthetic process; animals breathe this fresh air, releasing energy and providing the second level of the food chain. In turn, animals convert oxygen back to carbon dioxide, providing the building blocks for the formation of sugars during photosynthesis by plants. Decomposers (microbes: fungi, bacteria, and protozoans) break down the carcasses of dead organisms, recycling the minerals to make them available for new life; along with some algae and lichens, they create soils and improve soil fertility (LaRoe 1995).

Additionally, biodiversity provides the reservoir for change in our life-support systems, allowing life to adapt to changing conditions. This diversity is the basis for not only short-term adaptation to changing conditions but also long-term evolution.

Humans cannot survive in the absence of nature. We depend on the diversity of life on Earth for about 25 percent of our fuel (wood and manure in Africa, India, and much of Asia); more than 50 percent of our fiber (for clothes and construction); almost 50 percent of our medicines; and, of course, for all our food (Miller et al. 1985).

Some people believe that because extinction is a natural process we therefore should not worry about endangered species or the loss of biodiversity (LaRoe 1995).

Loss of Biodiversity

According to the U.S. Geological Survey (USGS; 1995), loss of biodiversity is real. Biologists have alerted each other and much of the general public to the contemporary mass extinction of species. Less recognized is loss of biodiversity at the ecosystem level, which occurs when distinct habitats, species assemblages, and natural processes are diminished or degraded in quality. Tropical forests, apparently the most species-rich terrestrial habitats on Earth, are the most widely appreciated, endangered ecosystems; they almost certainly are experiencing the highest rates of species extinction today (Myers 1984, 1988; Wilson 1988). However, biodiversity is being lost more widely than just in the tropics. Peter Moyle and Jack Williams (1990) point out that some temperate habitats, such as freshwaters in California and old-growth forests in the Pacific Northwest (Norse 1990) to name but two, are being destroyed faster than most tropical rain forests and stand to lose as great a proportion of their species. Because so much of the temperate zone has been settled and exploited by humans, losses of biodiversity at the ecosystem level have been greatest there so far.

Ecosystems can be lost or impoverished in basically two ways. USGS (1995) reports that the most obvious kind of loss is quantitative—the conversion of a native prairie to a cornfield or to a parking lot. Quantitative losses, in principle, can be measured easily by a decline to a real extent of a discrete ecosystem type (i.e., one that can be mapped). The second kind of loss is qualitative and involves a change or degradation in the structure, function, or composition of an ecosystem (Franklin et al. 1981; Noss 1990a). At some level of degradation, an ecosystem ceases to be natural, for example, a ponderosa pine forest may be high graded by removing the largest, healthiest, and frequently, the genetically superior trees; a sagebrush steppe may be grazed so heavily that native perennial grasses are replaced by exotic annuals; or a stream may become dominated by trophic generalist and exotic fishes. Qualitative changes may be expressed quantitatively, for instance, by reporting that 99 percent of the sagebrush steppe is affected by livestock grazing, but such estimates are usually less precise than estimates of habitat conversion. In some cases, as in the conversion of an old-growth forest to a tree farm, the qualitative changes in structure and function are sufficiently severe to qualify as outright habitat loss.

Several biologists (Diamond 1984; Ehrlich and Ehrlich 1981; Ehrlich and Wilson 1991; Soule 1991; Wilcox and Murphy 1985; Wilson 1985) agree that the major proximate causes of biotic impoverishment today are habitat loss, degradation, and fragmentation. Hence, modern conservation is strongly oriented toward habitat protection. The stated goal of the Endangered Species Act of 1973 is "to provide a means whereby the ecosystems upon which endangered species and threatened species depend may be conserved" (P.L. 94–325, as amended). The mission of the Nature Conservancy, the largest, private land-protection organization in the United States, is to save "the last of the least and the best of the rest" by protecting natural areas that harbor rare species and communities and high-quality samples of all natural communities (Jenkins 1985, 21).

USGS (1995) reports that, despite the many important accomplishments of natural-area programs in the United States, areas selected under conventional inventories tend to be small. As predicted by island biogeographic theory (MacArthur and Wilson 1967) and, more generally, by species-area relationships, smaller areas tend to have fewer species. All else being equal, smaller areas hold smaller populations, each of which is more vulnerable to extinction than larger populations (Soule 1987). Recognizing that small, natural areas that are embedded in intensely used landscapes seldom maintain their diversities for long, scientists called for habitat protection and management at broad, spatial scales such as landscapes and regions (Harris 1984; Noss 1983, 1987, 1992; Scott, Csuti, and Caicco 1991; Scott et al. 1991). In practice, however, most modern conservation continues to focus on local habitats of individual species and not directly on communities, ecosystems, or landscapes (Noss and Harris 1986).

Ecosystem conservation is a complement to—not a substitute for—species-level conservation. Protecting and restoring ecosystems serve to protect species about which little is known and to provide the opportunity to protect species while they are still common. Yet, ecosystems remain less tangible than species (Noss 1991a). And as USGS (1995) points out that the logic behind habitat protection as a means of conserving biodiversity is difficult to refute, conservationists face a major hurdle: convincing policy makers that significantly more and different kinds of habitat must be designated as reserves or otherwise managed for natural values. Scientists cannot yet say with accuracy how much land or what percentage of an ecosystem type must be kept in a natural condition to maintain viable populations of a given proportion of the native biota or the ecological processes of an ecosystem. However, few biologists doubt that the current level of protection is inadequate. Estimates of the fraction of major terrestrial ecosystem types that are not represented in protected areas in the United States range from 21 to 52 percent (Shen 1987). Probably a smaller percentage is adequately protected. For example, 60 percent of 261 major terrestrial ecosystems in the United States and in Puerto Rico, defined by the Bailey-Kuchler classification, were represented in designated wilderness areas in 1988 (Davis 1988). Only 19 percent of those ecosystem types, however, were represented in units of one hundred thousand hectares or more and only 2 percent in units of one million hectares or more—all of them in Alaska (Noss 1990b). Because the size of an area has a pronounced effect on the viability of species and on ecological processes, representation of ecosystem types in small units, in most cases, cannot be considered adequate protection.

Biodiversity and Stability

Biodiversity promotes stability. Gary Meffe and C. Ronald Carroll (1997) purport a major benefit of biodiversity is that more diverse ecosystems may be more stable or more predictable through time when compared to species-poor ecosystems. Stability can be defined at the community level as fewer invasions and less extinction, meaning that a more stable community will contain a more stable composition of species. Stated differently, the stability of a system is an inherent property of its component populations and communities, and it is a measure of the ability of that system to accommodate environmental change (Jones 1997). Three main components of stability are as follows:

- *persistence* (inertia): the ability of a community or ecosystem to resist disturbance or alteration.
- *constancy*: the ability to maintain a certain size or maintain its number within limits—system remains unchanged.
- *resilience*: the tendency of a system to return to a previous state after a perturbation.

Biodiversity: Estimated Decline

In this section, the estimated decline (USGS 1995) of biodiversity, with emphasis on the United States, is presented. As noted below, estimated decline includes area loss and degradation.

FIFTY UNITED STATES

85 percent of original primary (virgin) forest destroyed by late 1980s (Postel and Ryan 1991)

90 percent loss of ancient (old-growth) forests (World Resources Institute 1992)

30 percent loss of wetlands from 1780s to 1980s (Dahl 1990)
12 percent loss of forested wetlands from 1940 to 1980 (Abernethy and Turner 1987)
81 percent of fish communities adversely affected by anthropogenic limiting factors (Judy et al. 1982).

FORTY-EIGHT CONTERMINOUS STATES

c. 95–98 percent of virgin forests destroyed by 1990 (Postel and Ryan 1991)
99 percent loss of primary (virgin) eastern deciduous forest (Allen and Jackson 1992)
>70 percent loss of riparian forests since presettlement time (Brinson et al. 1981)
23 percent loss of riparian forest since the 1950s (Abernethy and Turner 1987)
53 percent loss of wetlands from 1780s to 1980s (Dahl 1990)
25 percent loss of wetlands between mid-1970s and mid-1980s (Dahl and Johnson 1991)
98 percent of an estimated 5.2 million kilometers of streams degraded enough to be unworthy of federal designation as wild or scenic rivers (Benke 1990).

Chapter Review Questions

11.1. Another word for environmental interrelationship is ________.
11.2. Define autecology.
11.3. Define synecology.
11.4. Define pollution.
11.5. List eight abiotic factors found in a typical ecosystem:
1. ________________
2. ________________
3. ________________
4. ________________
5. ________________
6. ________________
7. ________________
8. ________________
11.6. Define biogeochemical cycle.
11.7. ________________ are those parts of the cycle where the chemical is held in large quantities for long periods of time.
11.8. The length of time a chemical is held in an exchange pool or a reservoir is termed its ________________ time.
11.9. Name the three transport mechanisms.
11.10. The ________________ cycle is both sedimentary and gaseous.

References and Recommended Reading

Abedon, S. T. 1997. Ecosystems. www.mansfield.ohio-state.edu (accessed February 15, 2007).
Abernethy, Y., and R. E. Turner. 1987. U.S. forested wetlands: 1940–1980. *BioScience* 37:721–27.

Abrahamson, D. E., ed. 1988. *The challenge of global warming*. Washington, D.C.: Island Press.
Allen, E. G., and L. L. Jackson. 1992. The arid west. *Restoration Plans and Management Notes* 10 (1): 56–59.
Asimov, I. 1989. *How did we find out about photosynthesis?*. New York: Walker & Company.
Barlocher, R., and L. Kendrick. 1975. Leaf conditioning by microorganisms. *Oecologia* 20:359–62.
Benfield, E. F. 1996. Leaf breakdown in streams ecosystems. In *Methods in stream ecology*, ed. F. R. Hauer and G. A. Lambertic, 579–90. San Diego: Academic Press.
Benfield, E. F., D. R. Jones, and M. F. Patterson. 1977. Leaf pack processing in a pastureland stream. *Oikos* 29:99–103.
Benjamin, C. L., G. R. Garman, and J. H. Funston. 1997. *Human biology*. New York: McGraw-Hill.
Benke, A. C. 1990. A perspective on America's vanishing streams. *Journal of the North American Benthological Society* 91:77–88.
Brinson, M. M., B. L. Swift, R. C. Plantico, and J. S. Barclay. 1981. *Riparian ecosystems: Their ecology and status*. FWS/OBS-83/17. Washington, D.C.: U.S. Fish and Wildlife Service, Biological Services Program.
Brown, L. R. 1994. Facing food insecurity. In *State of the world*, ed. L. R. Brown. New York: W. W. Norton.
Carson, R. 1962. *Silent spring*. Boston: Houghton Mifflin.
Clements, E. S. 1960. *Adventures in ecology*. New York: Pageant Press.
Crossley, D. A., Jr., G. J. House, R. M. Snider, R. J. Snider, and B. R. Stinner. 1984. The positive interactions in agroecosystems. In *Agricultural ecosystems*, ed. R. Lowrance, B. R. Stinner, and G. J. House. New York: Wiley.
Cummins, K. W. 1974. Structure and function of stream ecosystems. *Bioscience* 24:631–41.
Cummins, K. W., and M. J. Klug. 1979. Feeding ecology of stream invertebrates. *Annual Review of Ecology and Systematics* 10:631–41.
Dahl, T. E. 1990. *Wetland losses in the United States 1780's to 1980's*. Washington, D.C.: U.S. Fish and Wildlife Service.
Dahl, T. E., and C. E. Johnson. 1991. *Wetlands: Status and trends in the conterminous United States mid-1970's to mid-1980's*. Washington, D.C.: U.S. Fish and Wildlife Service.
Darwin, C. 1998. *The origin of species*, ed. G. Suriano. New York: Grammercy.
Dasmann, R. F. 1984. *Environmental conservation*. New York: Wiley.
Davis, G. D. 1988. Preservation of natural diversity: The role of ecosystem representation within wilderness. Paper presented at National Wilderness Colloquium, Tampa, Florida.
Diamond, J. M. 1984. Historic extinctions: A Rosetta stone for understanding prehistoric extinctions. In *Quaternary extinctions: A prehistoric revolution*, ed. P. S. Martin and R. G. Klein, 824–62. Tucson: University of Arizona Press.
Dolloff, C. A., and J. R. Webster. 2000. Particulate organic contributions from forests to streams: Debris isn't so bad. In *Riparian management in forests of the continental eastern United States*, ed. E. S. Verry, J. W. Hornbeck, and C. A. Dolloff. Boca Raton, Fla.: Lewis Publishers.
Ehrlich, P. R., and A. H. Ehrlich. 1981. *Extinction: The causes and consequences of the disappearance of species*. New York: Random House.
Ehrlich, P. R., and E. O. Wilson. 1991. Biodiversity studies: Science and policy. *Science* 253:757–62.
Endangered Species Act of 1973. P.L. 94–325. Amended 108th Cong.
Evans, F. C. 1956. Ecosystem as the basic unit in ecology. *Science* 23:1127–28.
Franklin, J. F., et al. 1981. *Ecological characteristics of old-growth Douglas-fir forests*. General Technical Report PNW-118. Portland, Ore.: U.S. Forest Service, Pacific Northwest Forest and Range Experiment Station.

Gifford, T. 2006. *Reconnecting with John Muir: Essays in post-pastoral practice.* Athens: University of Georgia Press.

Hardin, G. 1968. The tragedy of the commons. *Science* 162 (3859): 1243–48.

Harris, L. D. 1984. *Bottomland hardwoods: Valuable, vanishing, vulnerable.* Gainesville: Florida Cooperative Extension Service, University of Florida.

Illinois State Water Survey (ISWS). 2005. Biogeochemical cycles II: The nitrogen cycle. www.sws.uiuc.edu/ (accessed February 15, 2007).

Jenkins, R. E. 1985. Information methods: Why the heritage programs work. *Nature Conservancy News* 35 (6): 21–23.

Jones, A. M. 1997. *Environmental biology.* New York: Routledge.

Judy, R. D., et al. 1982. *National fisheries survey, vol. I: Technical report: Initial findings.* FWS/OBS-84/06. Washington, D.C.: U.S. Environmental Protection Agency and U.S. Fish and Wildlife Service.

Kelly, K., and T. F. Homer-Dixon. 1995. *Environmental scarcity and violent conflict: The case of Gaza, project on environment, population and security.* Washington, D.C.: American Association for the Advancement of Science.

Keystone Center. 1991. *Biological diversity on federal lands, report of a Keystone policy dialogue.* Keystone, Colo.: Author.

Killpack, S. C., and D. Buchholz. 1993. *Nitrogen in the environment: Nitrogen.* Columbia: University of Missouri Press.

Krebs, C. H. 1972. *Ecology: The experimental analysis of distribution and abundance.* New York: Harper and Row.

LaRoe, E. T. 1995. Biodiversity: A new challenge. In *Our Living Resources.* Washington, D.C.: U.S. Geological Survey.

Laws, E. A. 1993. *Environmental science: An introductory text.* New York: Wiley.

Lindeman, R. L. 1942. The trophic-dynamic aspect of ecology. *Ecology* 23:399–418.

Lowdermilk, W. C. 1938. *Conquest of the land through 7,000 years.* Washington, D.C.: U.S. Department of Agriculture.

MacArthur, R. H., and E. O. Wilson. 1967. *The theory of island biogeography.* Princeton, N.J.: Princeton University Press.

Margulis, L., and D. Sagan. 1997. *Microcosmos: Four billion years of evolution from our microbial ancestors.* Berkeley: University of California Press.

Marshall, P. 1950. *Mr. Jones, meet the master.* New York: Revell.

Meffe, G. K., and C. R. Carroll. 1997. *Principles of conservation ecology.* Sunderland, Mass.: Sinauer Associates.

Metcalf and Eddy, Inc. 1991. *Wastewater engineering: Treatment, disposal, reuse,* 3rd ed. New York: McGraw-Hill.

Miller, G.T., Jr. 1988. *Environmental science: An introduction.* Belmont, Calif.: Wadsworth.

Miller, K. R., et al. 1985. Issues on the preservation of biological diversity. In *The global possible: Resources, development, and the new century,* ed. R. Repetto. New Haven, Conn.: Yale University Press.

Moran, J. M., M. D. Morgan, and J. H. Wiersma. 1986. *Introduction to environmental science.* New York: W. H. Freeman.

Moyle, P. B., and J. E. Williams. 1990. Biodiversity loss in the temperate zone: Decline of the native fish fauna of California. *Conservation Biology* 4:475–84.

Myers, N. 1984. *The primary source: Tropical forests and our future.* New York: W. W. Norton.

———. 1988. Tropical forests and their species. Going, going . . . ? In *Biodiversity,* ed. E. O. Wilson. Washington, D.C.: National Academy Press.

National Science Foundation (NSF). 2000. Report of the Workshop on the Terrestrial Carbon Cycle. www.carboncyclescience.gov/ (accessed February 18, 2007).

New World Encyclopedia. 2007. Ecology. www.newworldencyclopedia.org/entry/Ecology (accessed February 10, 2007).
Norse, E. A. 1990. *Ancient forests of the Pacific Northwest*. Washington, D.C.: Wilderness Society and Island Press.
Noss, R. F. 1983. A regional landscape approach to maintain diversity. *BioScience* 33:700–706.
———. 1987. From plant communities to landscapes in conservation inventories: A look at the Nature Conservancy (USA). *Biological Conservation* 41:11–37.
———. 1990a. Indicators for monitoring biodiversity: A hierarchical approach. *Conservation Biology* 4:355–64.
———. 1990b. What can wilderness do for biodiversity? In *Preparing to manage wilderness in the 21st century*, comp. P. Reed, 49–61. Asheville, N.C.: U.S. Forest Service.
———. 1991a. From endangered species to biodiversity. In *Balancing on the brink of extinction: The Endangered Species Act and lessons for the future*, ed. K. A. Kohm, 227–46. Washington, D.C.: Island Press.
———. 1991b. Sustainability and wilderness. *Conservation Biology* 5:120–21.
———. 1992. The wildlands project: Land conservation strategy. *Wild Earth* (Special Issue): 10–25.
Noss, R. F., and L. D. Harris. 1986. Nodes, networks, and MUMs: Preserving diversity at all scales. *Environmental Management* 10:299–309.
Odum, E. P. 1952. *Fundamentals of ecology*, 1st ed. Philadelphia: Saunders.
———. 1971. *Fundamentals of ecology*, 3rd ed. Philadelphia: Saunders.
———. 1975. *Ecology: The link between the natural and the social sciences*. New York: Holt, Rinehart and Winston.
———. 1983. *Basic ecology*. Philadelphia: Saunders.
———. 1984. Properties of agroecosystems. In *Agricultural ecosystems*, ed. R. Lowrance, B. R. Stinner, and G. J. House. New York: Wiley.
Odum, E. P., and G. W. Barrett. 2005. *Fundamentals of ecology*, 5th ed. Belmont, Calif.: Thomson Brooks/Cole.
Paul, R. W., Jr., E. F. Benfield, and J. Cairns Jr. 1978. Effects of thermal discharge on leaf decomposition in a river ecosystem. *Verhandlugen der Internationalen Vereinigung fur Thoeretsche and Angewandte Limnologie* 20:1759–66.
Peterson, R. C., and K. W. Cummins. 1974. Leaf processing in woodland streams. *Freshwater Biology* 4:345–68.
Porteous, A. 1992. *Dictionary of environmental science and technology*. New York: Wiley.
Postel, S., and J. C. Ryan. 1991. Reforming forestry. In *State of the world 1991: A Worldwatch Institute report on progress toward a sustainable society*, ed. L. Starker, 74–92. New York: W. W. Norton.
Price, P. W. 1984. *Insect ecology*. New York: Wiley.
Ramalay, F. 1940. The growth of a science. *University of Colorado Studies* 26:3–14.
Scott, J. M., B. Csuti, and S. Caicco. 1991. Gap analysis: assessing protection needs. In *Landscape linkages and biodiversity*, ed. W. E. Hudson, 15–26. Washington, D.C.: Defenders of Wildlife and Island Press.
Scott, J. M., B. Csuti, J. D. Jacobi, and J. E. Estes. 1987. Species richness: A geographic approach to protecting future biological diversity. *BioScience* 37:782–88.
Scott, J. M., B. Csuti, K. Smith, J. E. Estes, and S. Caicco. 1991. Gap analysis of species richness and vegetation cover: An integrated biodiversity conservation strategy. In *Balancing on the brink of extinction: The Endangered Species Act and lessons for the future*, ed. K. A. Kohm, 282–97. Washington, D.C.: Island Press.
Shen, S. 1987. Biological diversity and public policy. *BioScience* 37:709–12.
Smith, R. L. 1974. *Ecology and field biology*. New York: Harper and Row.
———. 1996. *Ecology and field biology*. New York: HarperCollins College.

———. 2007. Karl Ludwig Willdenow. www.wku.edu/~smithch/chronob/WILL1765.htm (accessed February 9, 2007).

Smith, T. M., and R. L. Smith. 2006. *Elements of ecology*, 6th ed. San Francisco: Pearson, Benjamin Cummings.

Soule, M. E., ed. 1987. *Viable populations for conservation*. Cambridge: Cambridge University Press.

———. 1991. Conservation: Tactics for a constant crisis. *Science* 253:744–50.

Spellman, F. R. 1996. *Stream ecology and self-purification*. Lancaster, PA: Technomic Publishing.

Suberkoop, K., G. L. Godshalk, and M. J. Klug. 1976. Changes in the chemical composition of leaves during processing in a woodland stream. *Ecology* 57:720–27.

Tansley, A. G. 1935. The use and abuse of vegetational concepts and terms. *Ecology* 16:284–307.

Tchobanoglous, G., and E. D. Schroeder. 1985. *Water supply*. Reading, Mass.: Addison-Wesley.

Tomera, A. N. 1989. *Understanding basic ecological concepts*. Portland, Maine: J. Weston Walch, Publisher.

Townsend, C. R., J. L. Harper, and M. Begon. 2000. *Essentials of ecology*. Blackwell Science.

United Nations Environment Programme (UNEP). 1995. *Global biodiversity assessment*, ed. V. H. Heywood. Cambridge: Cambridge University Press.

U.S. Agency for International Development (USAID). 2007. Environment. www.usaid.gov/our_work/environment/biodiversity/index.html (accessed February 4, 2007).

U.S. Department of Agriculture (USDA). 1982. *Agricultural statistics 1982*. Washington, D.C.: U.S. Government Printing Office.

———. 1999. Autumn colors: How leaves change color. www.na.fs.fed.us/spfo/pubs/misc/autumn/autumn_colors.htm (accessed February 8, 2007).

———. 2007. Agricultural ecosystems and agricultural ecology. http://nrcs.usda.gov/technical/ECS/agecol/ecosystem.html (accessed February 11, 2007).

U.S. Fish and Wildlife Service (USFWS). 2007. Ecosystem conservation. www.fws.gov/ecosystems (accessed February 11, 2007).

U.S. Geological Survey (USGS). 1995. *Endangered ecosystem of the United States: A preliminary assessment of loss and degradation*. Washington, D.C.: Author.

———. 2005. The water cycle: Evapotranspiration. http://ga.water.usgs.gov/edu/watercycle evapotranspiration.html (accessed February 16, 2007).

Wachernagel, M. 1997. Framing the sustainability crisis: Getting from concerns to action. www.ires.ubc.ca/ (accessed February 26, 2007).

Wessells, N. K., and J. L. Hopson. 1988. *Biology*. New York: Random House.

Wikipedia. 2007. Ecosystem. http://en.wikipedia.org/wiki/Ecosystem (accessed February 11, 2007).

Wilcox, B. A., and D. D. Murphy. 1985. Conservation strategy: The effects of fragmentation on extinction. *American Naturalist* 125:879–87.

Wilson, E. O. 1985. The biological diversity crisis. *Bio-Science* 35:700–706.

———. 1988. *Biodiversity*. Washington, D.C.: National Academy Press.

———. 1992. *The diversity of life*. Cambridge, Mass.: Belknap Press.

World Resources Institute. 1992. *The 1992 information please environmental almanac*. Washington, D.C.: Author.

VI

HUMAN AND CULTURAL GEOGRAPHY

Food snacks. Bangkok, Thailand.

CHAPTER 12

Human Geography: Thematic Units

> Whether we look or whether we listen,
> We hear life murmur or see it glisten.
>
> James Russell Lowell, 1819–1891

Human geography is the branch of geography that encompasses the human, political, cultural, social, and economic aspects of geographical studies. Moreover, it focuses on the study of patterns and processes that shape human interaction with the environment with particular reference to the causes and consequences of the spatial distribution of human activity on the Earth's surface.

This chapter presents a basic discussion of the thematic units and models that make up human geography, a major branch of geography. These thematic units include population and migration, agriculture and rural use, land use and natural resources, renewable energy, industrialization and economic development, and political and cultural geography. It is important to point out that these themes and models work in an ideal world but rarely work to perfection in the real world; they are presented in this manner to assist in the explanation of spatial patterns.

Population Geography

Because population is the basis for understanding a wide variety of human issues, population geography is a designated division or thematic unit of human geography. Specifically, it is the study of the ways in which spatial variations in the distribution, composition, migration, and growth of populations are related to the nature of places. The key issues in population geography evolve around questions related to where the world's population is distributed; where the world's population has increased/decreased; why population increases at different rates in different countries; and why the world might face an overpopulation problem. Population geography involves demography in a geographical perspective. It focuses on the characteristics of population distributions that change in a spatial context. Examples can be shown through population density maps. A few types of maps that show the spatial layout of population are choropleth (i.e., areas are shaded or patterned in proportion to the measurement), isoline (i.e., made up of contour lines), and dot maps.

Population geography (i.e., study of population) owes its beginning to the contributions of Thomas Malthus, an English clergyman, who in 1798 published his *Essay on the Principle of Population.* Malthus introduced the concept that at some point in time an expanding population must exceed supply of prerequisite natural resources—the "Struggle for Existence Concept." Malthus's theories profoundly influenced Charles Darwin in his 1859 *On the Origin of Species*—for example, the "Survival of the Fittest Concept."

THE 411 ON POPULATION GEOGRAPHY

Let's begin with the basics:

Population—Defined by the Word Masters

Webster's Third New International Dictionary defines population as follows:

- "The total number or amount of things especially within a given area."
- "The organisms inhabiting a particular area or biotype."
- "A group of interbreeding biotypes that represents the level of organization at which speciation begins."

Population—Defined by an Ecologist (Abedon 2007)

- A population in an ecological sense is a group of organisms, of the same species, which roughly occupy the same geographical area at the same time.
- Individual members of the same population can either interact directly or may interact with the dispersing progeny of the other members of the same population (e.g., pollen).
- Population members interact with a similar environment and experience similar environmental limitations.

Population—Defined by the Geographer

It refers to the inhabitants of a place (http://population.com).

Population system: Population system or life-system (population system is definitely better, however) is a population with its effective environment (Berryman 1981; Clark et al. 1967; Sharov 1992).

MAJOR COMPONENTS OF A POPULATION SYSTEM

1. *Population itself*: organisms in the population can be subdivided into groups according to their age, stage, sex, and other characteristics.

2. *Resources*: food, shelters, nesting places, space, and so forth.
3. *Enemies*: predators, parasites, pathogens, and so forth.
4. *Environment*: air (water, soil) temperature, composition, and variability of these characteristics in time and space (Sharov 1997).

POPULATION ECOLOGY (SHAROV 1996)

Population ecology is the branch of ecology that studies the structure and dynamics of populations. The term "population" is interpreted differently in various sciences, for example, in human demography a population is a set of humans in a given area. In genetics, a population is a group of interbreeding individuals of the same species, which is isolated from other groups. In population ecology a population is a group of individuals of the same species inhabiting the same area.

Main Axiom of Population Ecology—Organisms in a Population Are Ecologically Equivalent

Ecological equivalency means the following:

1. Organisms undergo the same life cycle.
2. Organisms in a particular stage of the life cycle are involved in the same set of ecological processes.
3. The rates of these processes (or the probabilities of ecological events) are basically the same if organisms are put into the same environment (however, some individual variation may be allowed). (Sharov 1996)

Did You Know?

An area where humans can live is called an *ecumene* (habitable land).

PROPERTIES OF POPULATIONS (ABEDON 2007)

1. *Population size*—depends on how the population is defined.
2. *Population density*—the number of individual humans (and/or organisms) per unit area. Population density is typically described as arithmetic or physiologic. *Arithmetic density* (often called *population density*) divides the entire population of a nation by the total land area (area of land measured in square kilometers or square meters) to come up with a population density for the nation as a whole. *Physiologic density* is a more meaningful and accurate way to measure a country's population density; it only takes into account the amount of arable land that is being used by

humans. Density can also be described as *agricultural density*, that is, the total rural population (farmers) per amount of agricultural (arable) land. In geography, the term *rural density* is also used to describe the number of people inhabiting a rural area per the total area of rural land.

3. *Patterns of dispersion (distribution)*—individual members of populations may be distributed over a geographical area in a number of different ways including clumped, uniform, and random distribution. There are five main areas of population density in the world. Sometimes these areas are called the population distributions of the world. The five areas or regions display significant differences in pattern of occupancy of the land. Despite these differences, we can see some significant similarities. Most inhabitants live near a river with easy access to an ocean or they live near an ocean. The five population clusters occupy mostly low-lying areas, with temperate climate and fertile soil.

   East Asia—the largest cluster of inhabitants, one-fifth of the world's population (>1.5 billion people), live here. This region contains the countries of China, South and North Korea, and Japan.

   South Asia—the second-largest concentration of people live in this region. This region includes the countries of India, Pakistan, and Bangladesh.

   Southeast Asia—the fourth-largest concentration of people live in this region. This region includes Vietnam, Indonesia, and Thailand.

   Western and Central Europe—the third-largest concentration of people live in this region. This region includes Western Europe, Eastern Europe, and the European portion of Russia.

   Northeastern United States and Canada—the fifth-largest concentration of people live in this region. This region includes the northeastern United States (Interstate 95 corridor) and the southeastern section of Canada, including the cities of Toronto, Ottawa, and Montreal.

4. *Demographics*—a population's vital statistics, including the following:

   - Education
   - Parental status
   - Work environment
   - Geographic location
   - Religious beliefs
   - Marital status
   - Income

   Additionally, demographics also includes the following:

   - Sex
   - Race
   - Gender
   - Ethnicity

- Age
- Sexual orientation
- Physical ability

5. *Population growth*—simply, population growth occurs when many more people are born than die (i.e., when there are no limitations on growth within the environment). When this occurs, two situations are present: (1) the population displays its intrinsic rate of increase (i.e., the rate of growth of a population when that population is growing under ideal conditions and without limits); and (2) the population experiences exponential growth (i.e., exponential growth means that a population's size at a given time is equal to the population's size at an earlier time, times some greater-than-one number) (Abedon 2007).
6. *Limits on population growth*—exponential growth cannot go on forever; sooner or later, any population will run into limits in its environment.

Note that all of these properties are not those of individual organisms but instead are properties that exist only if one considers more than one organism at any given time or over a period of time.

POPULATION GROWTH

The size of human populations is constantly changing due to natality, mortality, emigration, and immigration. As mentioned, the population size will increase if the natality and immigration rates are high. On the other hand, it will decrease if the mortality and emigration rates are high (*Malthusian Law* says that when birth and death rates are constant, a population will grow [or decline] at an exponential rate).

Each population has an upper limit on size, often called the carrying capacity. *Carrying capacity* can be defined as being the "optimum number of species' individuals that can survive in a specific area over time" (Enger et al. 1989). Stated differently, the carrying capacity is the maximum number of species that can be supported in a bioregion. Carrying capacity is based on the quantity of food supplies, the physical space available, the degree of predation, availability of medical treatment/medicine, and several other environmental factors.

DEMOGRAPHIC TRANSITION

A model known as the Demographic Transition Model is often used in demography and geography to explain the transition from high birth rates and high death rates to low birth rates and low death rates as a country's economy develops from preindustrial to industrialized. It is based on three primary factors: the birth rate, the death rate, and the total population. Most Demographic Transition Models have four stages. Each country must go through these stages, and barring major catastrophe, it is irreversible (it does not go backward). Geographers use the model as a tool for examining population development in countries and regions around the world.

Stage 1: Low Growth

This stage is common of preindustrial societies and is typified by high birth rates, which are balanced by high death rates.

Stage 2: High Growth

Death rates drop rapidly while birth rates remain high. Common of developing countries after 1750, during the Industrial Revolution. Increases in wealth cause the death rates to drop due to advances in sanitation and increased food supply, which improves health levels and health care, and increases life spans. Because birth rates remain high while death rates decline, a country in this stage experiences a high increase in population growth. During this stage, the S-curve, reflecting the total population number in the Demographic Transition Model, begins to take shape (see figure 12.1).

Stage 3: Moderate Growth

The population begins to head toward equilibrium in this stage because birth rates begin to decline. This is the case because of increasing urbanization, increasing wages, access to birth control measures, a reduction in child labor, increases in the quantity and quality of food, greater equality between men and women including better education of women, and additional social changes.

Stage 4: Low Growth

This stage is characterized by the shift from an industrialized to service-based economy. More people are involved with selling and maintaining products than in produc-

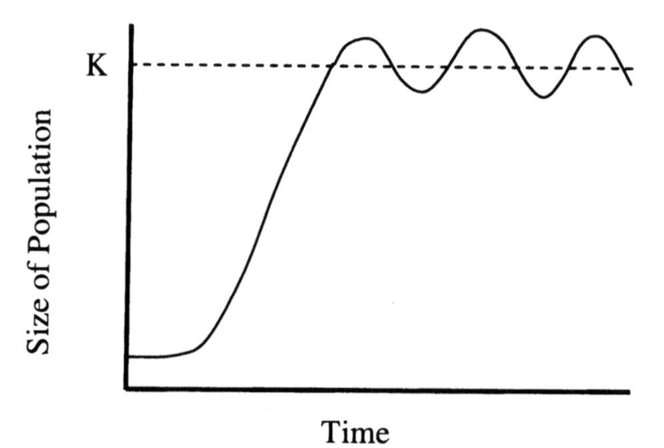

Figure 12.1. S-shaped (sigmoidal) growth curve showing world population growth. As you can see from the figure, total population begins to increase dramatically, eventually stabilizing in stage 4, thus creating an S-Shaped curve.

tion. The crude birth rate declines to the point where it equals the crude death rate, a condition known as *zero population growth.*

POPULATION PYRAMIDS

Geography (demography in particular) is concerned with not only population growth but also the characteristics of the population itself. In answering questions on the ages of people or the gender breakdown of a particular population, a tool—a bar graph—called a population pyramid can be used. A *population pyramid,* also called *age-sex pyramid* or *age structure diagram,* displays a country's population by age and gender groups. The population pyramid can be analyzed in terms of the Demographic Transition Model to determine in which stage a society is located. For example, four stages of the model are typically constructed showing a stage with a stable pyramid with unchanging patterns; a stage where stationary pyramids show low rates; a stage where an expansive pyramid shows a broad base, indicating a rapid rate of population growth and a low proportion of older people; and a final stage showing a constrictive pyramid, indicating lower numbers of younger people—in this case the country will have a graying population.

POPULATION: A GLOBAL PERSPECTIVE

In the recent past, countries with an increasing population were viewed in a positive light. A larger population means an increased military capability, more workers for either industrial or farming operations, and a larger tax base. Only in recent times have countries attempted to reverse the trend on expanding their populations. In viewing table 12.1 it is obvious that Asia contains the greatest number of high-population countries.

Population increases and decreases for a variety of reasons. Four primary factors lead to an increase in population:

1. Medical advances
2. Quantity and quality of food
3. Ethnic and religious issues
4. Economic issues

The three primary factors that lead to population decline are as follows:

1. Natural disasters
2. War or political turmoil
3. Economic issues

SUMMARY OF KEY POPULATION CONCEPTS

- Carrying capacity is the maximum number of people that can be supported by the resources and technology available.

Table 12.1. Twenty-Five Most Populated Countries

Country	Population in 2009
1. China	1,325.1 million
2. India	1,129.6 million
3. United States of America	301.9 million
4. Indonesia	224.5 million
5. Brazil	186.8 million
6. Pakistan	167.8 million
7. Nigeria	162.1 million
8. Russia	143.4 million
9. Bangladesh	137.5 million
10. Japan	128.6 million
11. Mexico	106.5 million
12. Philippines	87.2 million
13. Vietnam	85.1 million
14. Germany	82.5 million
15. Turkey	75.9 million
16. Ethiopia	73.9 million
17. Egypt	72.5 million
18. Iran	70.4 million
19. Thailand	67.3 million
20. France	61.4 million
21. United Kingdom	60.4 million
22. Congo	60.3 million
23. Italy	59.5 million
24. Myanmar	54.9 million
25. South Korea	51.3 million

- The systematic study of human population is called demography.
- The first stage of the demographic transition is marked by high birth rates and high but fluctuating death rates.
- Zero population growth (ZPG) refers to an exact equation of births and deaths.
- The theory of the demographic transition holds that both birth and death rates decrease with urbanization.
- One check to rapid population increase formerly available to European countries but not a possibility in today's developing societies was mass emigration.
- When the average fertility rate of a population drops to the replacement level, population continues to grow for a generation or more.
- The continent with the highest total fertility rates overall is Africa.
- The Malthusian theory is based on the assumption that population tends to increase more rapidly than do the food supplies to support that population.
- The portion of the Earth's surface permanently inhabited by humans makes up the ecumene.
- Common characteristics of world population distribution are that people congregate in lowland areas; people congregate along continental margins; and the majority of the world's population is rural.

- A broad-based population pyramid suggests that a country is in the second stage of the demographic transition.
- The crude death rate for wealthy Western Europe is much higher than that for Central America because Western Europe has a higher proportion of old people.
- Infant mortality ratio is not evident from the population pyramid.

Migration

Geography is about the use of space for different purposes around the globe. Therefore, migration (a form of mobility) of humans is an area of geography that has been (and continues be) intensely studied by geographers. *Migration* is a specific type of relocation diffusion—a permanent move to a new location. *Emigration* (E for exit) is migration from a location; *immigration* is migration to a location (I for into). The difference between the number of immigrants and the number of emigrants is the *net migration.*

Did You Know?

The short-term, repetitive, or cyclical movements that recur on a regular basis, such as daily, monthly, or annually, are called *recirculation.*

Most people migrate in search of three objects: economic opportunity (e.g., mass exodus of people from economically depressed areas such as Michigan to areas where employment opportunities are more favorable), cultural freedom (mass exodus because of political instability such as from Bosnia and Darfur to locations of relative safety), and environmental comfort (from the colder climates to warmer regions such as from North Dakota to Florida or Arizona).

Again, migration may indicate that people move across town or across the country or world, again for a variety of reasons. It is important to point out that once a person moves from one location to another their movement on the Earth does not affect the world's population, only a city, country, state, or country's population.

TYPES OF MIGRATIONS

Migrations can be of various types.

- *Cyclical movement*—commuting, seasonal movement, or nomadism.
- *Periodic movement*—consists of migrant labor, military service, or pastoral farming—seasonal movement of people with their livestock over relatively short distances, typically to higher pastures in summer and to lower valleys in winter.

- *Migratory movement*—moving from one part of the country to another (e.g., from the East Coast to the West Coast of the United States).
- *Rural exodus*—movement from rural areas to cities (e.g., from the farm to the city) in the hope of finding jobs.

LAWS OF MIGRATION

Ernst Georg Ravenstein (1834–1913), a German-English geographer, proposed certain laws to describe human migration. The laws are as follows:

- Every migration flow generates a return or countermigration.
- The majority of migrants move a short distance.
- Migrants who move longer distance tend to choose big-city destinations.
- Urban residents are less migratory than inhabitants of rural areas.
- Families are less likely to make international moves than young adults.

PUSH AND PULL FACTORS

Push and pull factors (like north and south poles on a magnet) are those factors that either forcefully push people into migration or attract them. A *push factor* is a negative perception about a location that induces a person to move away from that location. A *pull factor* is a positive perception about a location that induces a person to move there.

Push Factors

- Harsh or primitive conditions
- Natural disasters
- Loss of wealth
- Lack of opportunities
- Few jobs
- Few opportunities
- Poor medical care
- Death threats
- Harassment/bullying
- Slavery
- Landlords
- Political persecution
- Pollution
- Few opportunities for courtship
- Poor housing
- Lack of community services
- Drought or famine

Pull Factors

- Entertainment/enjoyment
- Job opportunities
- Climate
- Education
- Better medical care
- Better living conditions
- Political/religious freedom
- Security
- Lower taxes
- Family ties
- Better courtship opportunities
- Industry
- Cleaner environment
- Recreation
- Lower cost of living

OBSTACLES TO MIGRATION

Obstacles can adversely affect migrate between areas. Some factor, physical or cultural, can halt migration. The principal obstacle faced by migrants is distance, that is, the long, arduous, and expensive trek from one place to another. Another common obstacle to migration is quota laws that limit people from entering various countries. Cultural factors such as language differences can also hinder migration.

ECONOMIC RESULTS OF MIGRATION

The economic impact of migration is one of the most important dimensions of migration. The flow of skilled laborers or technicians from one country to another is termed *brain drain.* On the other hand, a gain of skilled laborers and technicians is known as *brain gain.* Migration can also cause a flow of money and skill into one country and out of the other. Another form of migration has to do with outsourcing. Big companies find that they are economically better off—no labor unions or OSHA regulators to put up with—by moving their entire operations overseas to places like Mexico, China, and India.

Agriculture

Agriculture refers to the raising of animals or the growing of crops to obtain food for primary consumption by the farm family or for sale off the farm. Agriculture was the key development that led to the rise of civilization.

Historically, the origins of agriculture cannot be documented with certainty because it began before recorded history. Historians use logic and fragments to reconstruct a sequence of likely developments. However, we can state with some certainty that agriculture developed in stages.

Before the invention of agriculture, the first way humans obtained food was by hunting, fishing, and gathering. Early people, living in small groups (hunting and gathering practices do not support large groups), relied on nomadic practices such as traveling frequently and depending on migratory animals and the existence of wild berries and fruits for sustenance. However, during severe weather events such as droughts and natural disasters, such as forest fires, for example, the supply of berries and fruits was limited; also, it was difficult if not impossible to store foods, causing starvation on a large scale, limiting population growth.

In describing the advance of humans and agriculture, scientists, historians, lecturers, geographers, and others describe the sequence of advancements in agriculture by referring to stages of development or three specific agricultural revolutions. The word revolution is used with some license here because we normally associate the word with radical, sudden changes. Though we can say the first two agricultural revolutions were indeed radical, they were in no way sudden events.

In each of the three agricultural revolutions, climate and technology played major roles, though the part climate played is difficult to describe and to quantify (i.e., we can describe and quantify the results of climate but cannot predict it, explain its causes, or determine its exact timing to any great extent). Technology, on the other hand, is rather easy to cite because the process of creating farming implements or instruments is well documented.

FIRST AGRICULTURAL REVOLUTION

The first agricultural revolution—the end of the last ice age and the beginning of domestication of plants—is often characterized as being the dawn of civilization. This period is characterized by development of seed agriculture and the use of the plow. These are linked to specific sorts of social change: village and town systems, the development of hereditary kingships, and so forth. This revolution allows for greater population growth. Although this revolution began about 10,000 BC, there are still some present-day societies making this transition.

SECOND AGRICULTURAL REVOLUTION

The second revolution took place beginning about 500 or 600 AD and is centered upon the medieval period. This period is associated with the dramatic improvements in outputs in crop and livestock yields. The advance of this period rests upon a favorable climate and upon two special features—an improved plow and the horse. In addition, during this period, advancements were made in applying better fertilizer and designing improved field drainage systems. As time passed, this agricultural revolution and the industrial revolution were linked. In fact, it can be accurately stated that the agricultural revolution made the industrial revolution possible. Advancements in agriculture made population growth possible, creating an industrial army. During this period, dozens of inventions increased the productivity of agriculture and helped centralize capital. All of these advances occurred along with changes in organization, that is, communal lands were disappearing due to the enclosure movement (i.e., land was deeded, fenced, and privately owned). The second revolution had a huge effect on farmers—more of them left the farm for the city because less work was being done manually on the farm.

THIRD AGRICULTURAL REVOLUTION

The third agricultural revolution (sometimes called the Green Revolution) is a product of the last century (1960s in the United States) and, based on crop yield data, is the most abrupt transformation of the three. The third agricultural revolution basically consists of three phases, including mechanization, the development of chemical farming (new chemical fertilizers, herbicides, fungicides, pesticides, etc.), and food manufacturing. Moreover, this is the period whereby the farm moved from being the centerpiece of agricultural production to become a part of an integrated string of vertically organized industrial processes including production, storage, processing, distribution, marketing, and retailing. Also, this period is characterized by the change in agriculture and labor roles, development of inputs-hybrid seeds (biotechnology/genetic engineering), and development of industrial substitutes for agricultural products.

AGRICULTURAL HEARTHS

In geography the term *hearth* generally designates the region from which innovative ideas originate—the spreading of ideas from one area to another via diffusion. Carl Sauer, a preeminent geographer in the history of the profession, mapped out the agricultural regions of both vegetative planting and seed agriculture. Sauer suggested that there are two distinct types (or hearths; locus of domestication) of agriculture: vegetative planting hearths and seed agricultural hearths.

Vegetative Planting Hearths

Vegetative planting refers to removing part of a plant and putting it in the ground to grow a new plant. Vegetative planting preceded agriculture and probably began in Southeast Asia with the taro, yam, banana, and palm plants. Early vegetative planting also took place in West Africa with oil palm and yam and in Andean South America with manioc, sweet potato, arrowroot, and potatoes.

Seed Agriculture

Seed agriculture means taking seeds from existing plants and planting them to produce new plants. Early seed agricultural practices began in the foothills overlooking the

Indus River Valley and Mesopotamia with the planting of wheat and barley; in Northern China with the planting of millet seed; in Ethiopia with the planting of millet and sorghum seed; and in Mesoamerica with maize and squash in Mexico and bean, cotton, and squash in Peru. Animal domestication coincided with seed agriculture.

KINDS OF FARMING

There are two predominant types of farming in the world today: subsistence and commercial farming. *Subsistence farmers* produce the food that they need to survive on a daily basis. They depend on the crops that they grow and the animal products they raise for their daily sustenance. Subsistence farming consists of shifting cultivation (moving from one farm field to another) and crop rotation (planting different crops each year). Pastoral nomadism is another type of subsistence farming involving the moving of animals on a seasonal basis to areas that have the necessary resources to meet the needs of the herd.

Commercial farming, usually conducted in developed countries, is the farming of products for sale off the farm. Commercialized farming consists of Mediterranean agriculture (growing of grapes, olives, and dates); dairy farming; mixed livestock with crop production; livestock ranching; truck farms; grain farms; and others.

VON THUNEN'S MODEL

In 1826 in northern Germany, Johann Heinrich von Thunen developed an agricultural, land use, quantitative-geographical model shown in figure 12.2. Von Thunen's model examines relationships between land values, distance to market, and agricultural land use. Generally, the model suggests that certain crops are grown in direct relation to their distance from market.

SUMMARY OF KEY AGRICULTURAL CONCEPTS

- Agriculture can be defined as the growing of crops and tending livestock, for sale or subsistence.
- Roughly 10 percent of the total land on Earth is for crop farming.
- There is a declining trend in agriculture employment in developing countries.
- In subsistence agriculture, there is nearly total self-sufficiency on the part of its members, that is, no exchange (minimal, if any)—food is for themselves only.
- Subsistence agriculture consists of two types:

Extensive: large areas of land and minimal labor input per unit area. Production and population density are low.

Intensive: cultivation of small landholdings through the expenditure of great amounts of labor per unit area. Production and population density are both high.

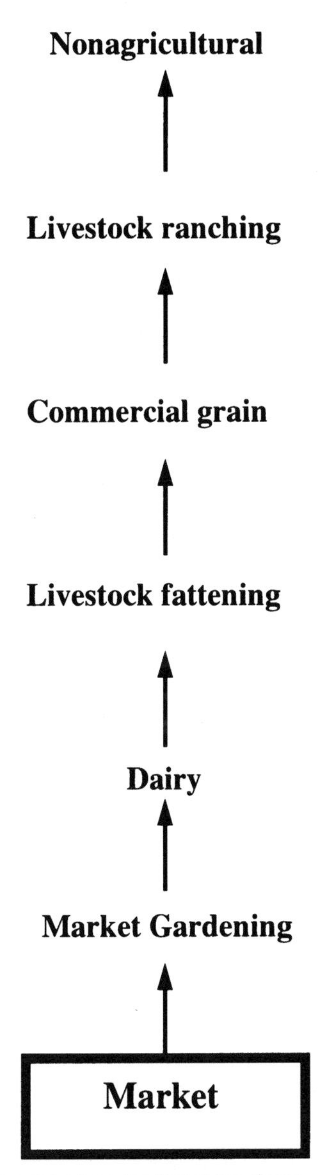

Figure 12.2. Von Thunen model of agricultural land use.

Land Use

Geographers are concerned with land use. Land use questions are a big part of the bigger questions that geographers always ask: Where? Why there? Why do we care? In regard to geography and geographers' questions, the preceding discussion, agriculture, is the perfect segue to the next four sections on land use, resources, industry, and economic development. Nothing uses more land in the world today than agriculture.

With the increasing demand for and dwindling supplies of fuel, this trend may become exponential in the near future. More and more arable (and some not so arable) land is being looked at to grow more corn, soybean, and other crops to produce ethanol, biofuel, and biomass ingredients (some from native perennial grasses, etc.) for creating synthetic or human-made fuel substitutes for hydrocarbons. Moreover, the present concern about global climate change and its ramifications is increasing the push for more substitute products (including agricultural products) and techniques or processes to fuel our needs in the future.

The pressing need to find substitutes for hydrocarbon-based fuels (for economical and environmental reasons) by using biofuel (and other sources) is not without some controversy, however. It is not only the difficulty of finding suitable agricultural land for use in growing the corn and soybeans or native grasses needed but also the availability of the land—there is only so much arable land on Earth. In addition, recent scientific research points out that there may be an environmental impact price to pay for using so much land to grow the biofuel ingredients or products. The Union of Concerned Scientists (UCS 2008), for example, points out that recent studies of the impact that land conversion has on the global warming pollution created by crop-based biofuels are changing the science of measuring biofuel risks and rewards. When deforestation or other damaging land conversion actions are necessary to grow more crop-based biofuels, the net result could be more pollution.

Land use dedicated to growing crops for biofuels can have another indirect impact when the corn and soy are taken out of the market for food and animal feed. Timothy Searchinger et al. (2008) used models and historical data on land conversion to estimate where new crops will be planted, what land will be converted, and what emission will result. Based on their research, not only will crop prices increase for consumers, but also the expanded use of corn ethanol will produce almost double the amount of global warming pollution as gasoline.

LAND USE MODELS

Land use models help geographers determine the best use of natural resources for industrial applications and how best to use the land. Generally, four different models, each having its advantages and disadvantages, can be used to help make decisions regarding issues related to economics, sustainability, environmental issues, and preservationist issues.

- Economic land use model—used to develop and build on the landscape for profit. For example, an automotive dealership or shopping mall developer would use this model to analyze the construction of both facilities. This model is sometimes called the *topocide* model because it kills off certain landscapes for commercial development.
- Sustainability land use model—taking something from the land and replacing it with something else. A good example of this model is the current U.S. forestry

industry practice of logging and then replanting with new seedlings. Crop rotation by farmers is another example of the sustainability land use model.

- Environmental land use model—based on the U.S. National Park system, suggests that we use land but must leave it in its natural state.
- Preservationist land use model—suggests that the landscape is sacred and therefore should be left alone; it should be untouched by humans.

Resources

As long as capitalism drives most modern economies, people will desire material things—precipitating a high level of consumption.[1] For better or for worse, the human desire to lead the "good life" (which Americans may interpret as a life enriched by material possessions) is a fact of life. Arguing against someone who wants to purchase a new, modern home with all the amenities or who wants to purchase the latest, greatest automobile is difficult. Arguing against the person wanting to make a better life for his or her children by making sure they have all they need and want to succeed in their chosen pursuit is even harder. How do you argue against such goals with someone who earns his or her own way and spends his or her hard-earned money at will? Look at the tradeoffs, though. The tradeoff often affects the environment. That new house purchased with hard-earned money may sit in a field of radon-rich soil or on formerly undeveloped land. That new SUV may get only eight miles to the gallon. The boat they use on weekends gets even worse mileage and exudes wastes into the local lake, river, or stream. Their weekend retreat on the five wooded acres is part of the watershed of the local community and disturbs breeding and migration habitat for several species.

The environmental tradeoffs never enter the average person's mind. Most people don't commonly think about it. In fact, most of us don't think much about the environment until we damage it, until it becomes unsightly, until it is so fouled that it offends us. People can put up with a lot of environmental abuse, especially with our surroundings—until the surroundings no longer please us. We treat our resources the same way. How often do we think about the air we breathe, the water we drink, or the soil our agribusiness conglomerates plant our vegetables in? Not often enough.

The typical attitude toward natural resources is often deliberate ignorance. Only when someone must wait in line for hours to fill the car gas tank does gasoline become a concern. Only when people can see—and smell—the air they breathe and cough when they inhale does air become a visible resource. Water, the universal solvent, causes no concern (and very little thought) until shortages occur, or until it is so foul that nothing can live in it or drink it. Only when we lack water or the quality is poor do we think of water as a resource to "worry" about. Is soil a resource or is it "dirt"? Unless you farm, or plant a garden, soil is only "dirt." Whether you pay any heed to the soil/dirt debate depends on what you use soil for—and on how hungry you are.

Resource utilization and environmental degradation are tied together. While people depend on resources and must use them, this use impacts the environment. A *resource* is usually defined as anything obtained from the physical environment that is of use to humans. Some resources, such as edible growing plants, water (in many places), and fresh air, are directly available to humans. But most resources, like coal, iron, oil, groundwater, game animals, and fish are not. They become resources only when humans use science and technology to find them, extract them, process them, and convert them, at a reasonable cost, into usable and acceptable forms. Natural gas, found deep below the Earth's surface, was not a resource until the technology for drilling a well and installing pipes to bring it to the surface became available. For centuries, humans stumbled across stinky, messy pools of petroleum and had no idea of its potential uses or benefits. When its potential was realized, humans exploited petroleum by learning how to extract it and convert (refine) it into heating oil, gasoline, sulfur extract, road tar, and other products.

Earth's natural resources and processes that sustain other species and us are known as *Earth's natural capital.* This includes air, water, soil, forests, grasslands, wildlife, minerals, and natural cycles. Societies are the primary engines of resource use, converting materials and energy into wealth, delivering goods and services, and creating waste or pollution. This provision of necessities and luxuries is often conducted in ways that systematically degrade the Earth's natural capital—the ecosystems that support all life.

Excluding *perpetual resources* (solar energy, tides, wind, and flowing water), two different classes (types) of resources are available to us: renewable and nonrenewable (see figure 12.3). *Renewable resources* (fresh air, fresh water, fertile soil, fish, grasslands,

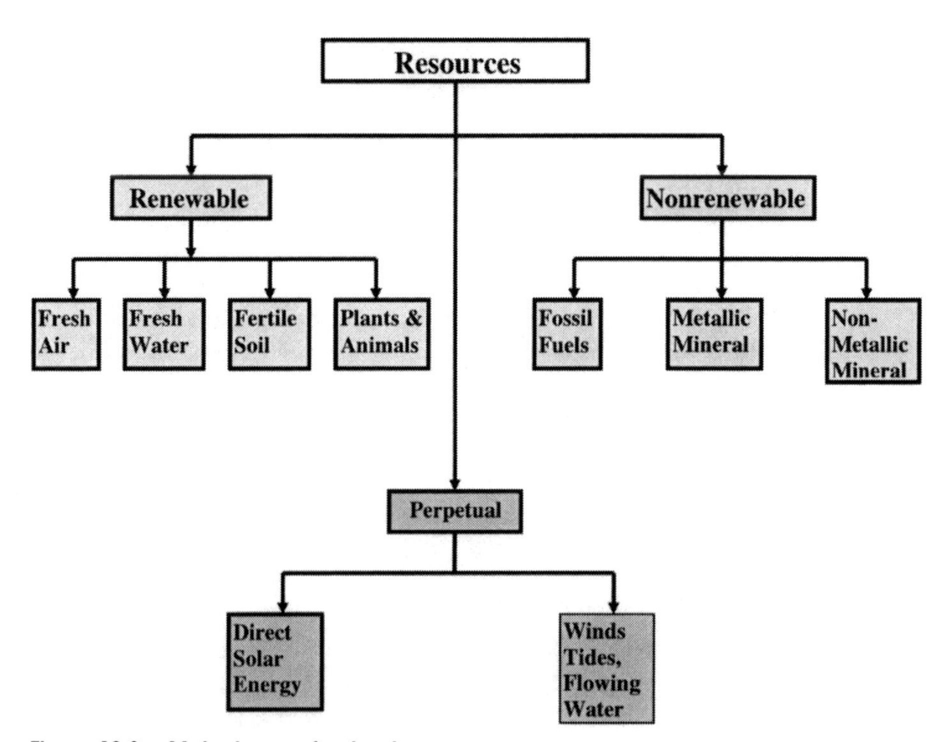

Figure 12.3. Major types of natural resources.

forests, and animals via genetic diversity) can be depleted in the short run if used or contaminated too rapidly but normally will be replaced through natural processes in the long run. Because renewable resources are relatively plentiful, we often ignore, overlook, destroy, contaminate, or mismanage them.

Mismanaged? Yes. Classifying anything as "renewable" is a double-edged sword. Renewable resources are renewable only to a point. Timber or grass used for grazing must be managed for *maximum sustainable yield* (the highest rate at which a renewable resource can be used without impairing or damaging its ability to be fully renewed). If timber or grass yield exceed this rate, the system gives ever-diminishing returns. Recovery is complicated by the time factor, which is life-cycle dependent. Grass can renew itself in a season or two. Timber takes decades. Any length of time is problematic when people get impatient.

Exceeding maximum sustainable yield is only the tip of the iceberg—other environmental, social, and economic problems may develop. Let's look at *overgrazing* (depleting) grass on livestock lands. The initial problem occurs when the grass and other grazing cover is depleted. But secondary problems kick in fast. Without grass, the soil erodes quickly. In very little time, so much soil is gone that the land is no longer capable of growing grass—or anything else. Productive land converted to nonproductive deserts (*desertification*) is a process of *environmental degradation*—and it impacts social and economic factors. Those who depend on the grasslands must move on, and moving on costs time, energy, and money—and puts more land at risk. Should the same level of poor stewardship of land resources continue on more acreage?

Nonrenewable resources (copper, coal, tin, and oil, among many others) have built up or evolved in a geological time span. They can't be replaced at will—only over a similar time scale. In this age of advanced technology, we often hear that, for example, when high-grade tin ore runs out (when 80 percent of its total estimated supply has been removed and used), low-grade tin ore (the other 20 percent) will become economically workable. This erroneous view neglects the facts of energy resource depletion and increasing pollution with lower-grade burdens. In short, to find, extract, and process the remaining 20 percent generally costs more than the result is worth. Even with unlimited supplies of energy (impossible according to the Laws of Thermodynamics), what if we could extract that last 20 percent? When it is gone, nothing is going to bring it back except time measured in centuries and millennia, paired with the elements that produce the resource.

Advances in technology have allowed us to make great strides in creating "the good life." These same technological advances have increased environmental degradation. But not all the news is bad. Technological advances have also let us (via recycling and reuse) conserve finite resources—aluminum, copper, iron, plastics, and glass, for example. *Recycling* involves collecting household waste items (aluminum beverage cans, for example) and reprocessing usable portions. *Reuse* involves using a resource over and over in the same form (refillable beverage bottles or water).

We discussed the so-called good life earlier—modern homes, luxury cars and boats, and the second home in the woods. With the continuing depletion of natural resources, prices must be forced upward until, economically, attaining the good life, or even gaining a foothold toward it, becomes difficult or impossible—and maintaining it becomes precarious. Ruthless exploitation of natural resources and the environment—overfishing a diminishing species (look at countless marine species populations, for example), intense exploitation of energy and mineral resources, cultivation of marginal land without proper conservation practices, degradation of habitat by unbalanced populations or introduced species, and the problems posed by further technological advances—will result in environmental degradation that will turn the good life into something we don't want to even think about.

So—what's the answer? What are we to do? What should we do? Can we do anything?

Some would have us all "return to nature." Those people suggest returning to Thoreau's Walden Pond on a large scale, to give up the "good life" to which we have become accustomed. They think that giving up the cars, boats, fancy homes, bulldozers that make construction and farming easier, pesticides that protect our crops, and medicines that improve our health and save our lives—the myriad material improvements that make our lives comfortable and productive—will solve the problem. Is this approach the answer—or even realistic? To a small (vocal and impractical) minority, it is—although for those who realize how urban Walden Pond was, the idea is amusing.

To the rest of us? Get real! This is a pipe dream, founded in romance, not logic. It cannot, should not, and will not happen. We can't abandon ship—we must prevent the need for abandoning our society from ever happening. Technological development is a boon to civilization and will continue to be. Technological development isn't the problem—improper use of technology is. But we must continue to make advances in technology, we must find further uses for technology, and we must learn to use technology for the benefit of humankind and the environment. Technology and the environment must work hand in hand, not stand opposed. We must also foster respect for, and care for, what we have left.

Just how bad are the problems of technology's influence on the environment?

Major advances in technology have provided us with enormous transformation and pollution of the environment. While transformation is generally glaringly obvious (damming a river system, for example), "polluting" or "pollution" is not always as clear. What do we mean by pollution? To *pollute* means to impair the purity of some substance or environment. *Air pollution* and *water pollution* refer to alteration of the normal compositions of air and water (their environmental quality) by the addition of foreign matter (gasoline, sewage, etc.). Ways technology has contributed to environmental transformation and pollution include the following:

- extraction, production, and processing of raw natural resources, such as minerals, with accompanying environmental disruption;
- manufacturing enormous quantities of industrial products that consume huge amounts of natural resources and produce large quantities of hazardous waste and water/air pollutants;
- agricultural practices resulting in intensive cultivation of land, irrigation of arid lands, drainage of wetlands, and application of chemicals;
- energy production and use accompanied by disruption and contamination of soil

by strip mining, emission of air pollutants, and pollution of water by release of contaminants from petroleum production, and the effects of acid rain;
- transportation practices (particularly reliance on the airplane) that cause scarring of land surfaces from airport construction, emission of air pollutants, and greatly increased demands for fuel (energy) resources; and
- transportation practices (particularly reliance on automobiles) that cause loss of land by road and storage construction, emission of air pollutants, and increased demand for fuel (energy) resources.

FOSSIL FUELS

The U.S. Department of Energy (DOE; 2009) reports that fossils fuels—oil, coal, and natural gas—currently provide more than 85 percent of all the energy consumed in the United States, nearly two-thirds of our electricity, and virtually all of our transportation fuels. The United States imports approximately 60 percent of its liquid fuel supply, totaling twelve million barrels per day (forty-two gallons per barrel).

Oil

Oil is the lifeline of most countries and is fundamental to most modern industrialized societies. The majority of the oil in the United States comes from the Gulf of Mexico, Alaska, and California. Many products are refined from crude oil, including gasoline, methane, propane, butane, kerosene, diesel fuels, and jet fuel.

Most of the easily recoverable oil has already been discovered. Because we now have to go to great technological lengths and expense to not only find the remaining oil but also recover it, oil has become more expensive, adding a significant cost in our daily lives—this trend is not expected to change. The proven reserves of petroleum are about 1.3 trillion barrels, about 175 trillion cubic meters of natural gas, and about 1 quadrillion metric tons of coal.

As shown in table 12.2, the United States is the third-largest oil producer, with over 7.5 million barrels per day; Saudi Arabia is the largest producer with just under 10.2 million barrels per day; and Russia is second with 9.9 million barrels of oil produced per day.

The United States uses approximately twenty-one million barrels of oil every day. Because the United States only produces 7.5 million barrels a day, it must import a large amount to make up for the imbalance.

Coal

Because of its abundant reserves of coal, the United States burns more coal than any other energy source in the production of electricity. Worldwide, China is the leader in coal production, with the United States in second place. Other large produces of coal include Australia, Russia, and India. The top twelve countries with proved recoverable coal reserves are listed in table 12.3.

Table 12.2. Top Twelve Oil Producers (2008)

Country	Production (10^6 bbl/d)
Saudi Arabia	10.2
Russia	9.9
United States	7.5
Iran	4.0
Canada	3.3
United Arab Emirates	2.9
Venezuela	2.7
Kuwait	2.6
Nigeria	2.4
Iraq	2.1
Libya	1.7
Kazakhstan	1.4

Source: PennWell Corporation, "Estimated Reserves by Country," *Oil and Gas Journal* 105, no. 48 (December 24, 2007).

Natural Gas

Natural gas, another important fossil fuel, was created deep within the Earth by the decomposition of organic matter and animals, which in turn formed methane gas, the major constituent of natural gas. Large volumes of methane are trapped in the subsurface of the Earth in Russia (33 percent), Iran (16 percent), and the United States (10 percent); these three countries possess the largest reservoirs of natural gas. The majority of homes in the United States are heated with natural gas.

METALS

After fuel, one of the major natural resources of great importance is the metals. Countries rich in metal natural resources have a distinct economic advantage and the poten-

Table 12.3. Top Twelve Proved Recoverable Coal Reserves (2006)

Country	Percent of world's recoverable coal
United States	27.1
China	12.6
India	10.2
Australia	8.6
South Africa	5.4
Ukraine	3.8
Kazakhstan	3.4
Poland	1.5
Brazil	1.1
Germany	0.7
Colombia	0.7
Canada	0.7

Source: U.S. Department of Energy (DOE), World Steam Coal Flows, 2009, www.eia.doe.gov/oiaf/aeo/supplement/pdf/suptab_114.pdf (accessed November 18, 2009).

tial for industrial dominance. Countries that possess large stores of readily available, easy-to-access fuels and metallic ores have huge advantages over other counties. The primary metals of interest and particular significance to economic health include gold and iron ore.

FORESTRY

Forestry is defined as the art and science of managing forests, tree plantations, and related natural resources. Countries possessing large areas of forested land generally value this natural resource as one of their most important treasures. Forestry is a huge industry in North America. American and Canadian sustainability practices have ensured a steady supply of this valuable renewable resource in North America for many years to come.

FISHING INDUSTRY

Fishing entails the capture of fish from the wild. The fishing industry also includes processing, preserving, storing, transporting, marketing, and selling fish or fish products. Commercial fishing is one of the most dangerous professions but remains viable and profitable because fish and fish products remain a high-demand item for consumers. Obviously, fishing is a renewable resource, but there are concerns, especially among environmentalists, that fish are being overharvested.

Alternative and Renewable Energy

In regard to the current status of the worldwide use of fossil fuels, politics and other persistent forces are pushing for substitute, alternate, and renewable fuel sources.[2] This is the case, of course, because of the current and future economic problems that $4-per-gallon gasoline has generated (especially in the United States) and because of the perceived crisis developing with high carbon dioxide emissions, the major contributing factor of global climate change.

Before proceeding with an introductory discussion of alternative and renewable energy sources, it is important to make a clear distinction between the two terms, the current buzzwords, alternative and renewable energy. *Alternative energy* is an umbrella term that refers to any source of usable energy intended to replace fuel sources without the undesired consequences of the replaced fuels. The use of the term "alternative" presupposes an undesirable connotation (for many people, fossil fuels has joined that endless list of four-letter words)—that is, fueled energy that does not use up natural resources or harm the environment—against which alternative energies are opposed. Examples of alternate fuels include petroleum as an alternative to whale oil; coal as an alternate to wood; alcohol as an alternate to fossil fuels; and coal gasification as alternative to petroleum. These alternate fuels need not be renewable.

Renewable energy is energy generated from natural resources—such as sunlight, wind, water (hydro), ocean thermal, wave and tide action, biomass, and geothermal heat—which are naturally replenished (and thus renewable). Renewable energy resources are virtually inexhaustible—they are replenished at the same rate as they are used—but limited in the amount of energy that is available per unit time. If we have not come full circle in our cycling from renewable to nonrenewable, we are getting close to that. Consider, for example, that in 1850, about 90 percent of the energy consumed in the United States was from renewable energy resources (hydropower, wind, burning wood, etc.). Now, however, the United States is heavily reliant on the nonrenewable fossil fuels of natural gas, oil, and coal. In 2009, about 7 percent of all energy consumed (see table 12.4) and about 8.5 percent of total electricity production is from renewable energy resources.

Most of the renewable energy is used for electricity generation, heat in industrial processes, heating and cooling buildings, and transportation fuels. Electricity producers (utilities, independent producers, and combined heat and power plants) consumed 51 percent of total U.S. renewable energy in 2007 for producing electricity. Most of the rest of the remaining 49 percent of renewable energy was biomass consumed for industrial applications (principally paper making) by plants producing only heat and steam. Biomass is also used for transportation fuels (ethanol) and to provide residential and commercial space heating. The largest share of the renewable-generated electricity comes from hydroelectric energy (71 percent), followed by biomass (16 percent), wind (9 percent), geothermal (4 percent), and solar (0.2 percent). Wind-generated electricity increased by almost 21 percent in 2007 over 2006, more than any other energy source. Its growth rate was followed closely by solar, which increased by over 19 percent in 2007 over 2006.

From table 12.4 it is obvious that currently there are five primary forms of renewable energy: solar, wind, biomass, geothermal, and hydroelectric energy sources. Each of theses holds promise and poses challenges regarding future development.

Table 12.4. U.S. Energy Consumption by Energy Source, 2007 (Quadrillion Btu)

Energy Source	*2007*
Total	**101.605**
Renewable	6.830
Biomass (biofuels, waste, wood, and wood derived)	3.615
Biofuels	1.018
Waste	0.431
Wood-derived fuels	2.165
Geothermal	0.353
Hydroelectric conventional	2.463
Solar/PV	0.080
Wind	0.319

Source: U.S. Energy Information Administration (EIA), "U.S. Energy Consumption by Energy Source," 2007, www.eia.doe.gov/cneaf/alternate/page/renew_energy_consump/table1.html (accessed June 12, 2009).

SOLAR ENERGY

It is fitting to begin our discussion of the various kinds of renewable energy with the Sun—the star that symbolizes life, power, strength, force, clarity, and, yes, energy. The Sun nourishes our planet. When we consider the Sun and solar energy first, we quickly realize that there is nothing new about renewable energy. The Sun was the first energy source; it has been around for 4.5 billion years, as long as anything else we are familiar with. On Earth without the Sun there is nothing—absolutely nothing. The Sun provided light and heat to the first humans. During daylight, the people searched for food. They hunted and gathered and probably stayed together for safety. When nightfall arrived, and in the dark, we can only imagine that they huddled together for warmth in the light of the stars and Moon, waiting for the Sun and its live-giving and sustaining light to return.

Solar energy (a term used interchangeably with solar power) uses the power of the Sun, using various technologies, "directly" to produce energy. Solar energy is one of the best renewable energy sources available because it is one of the cleanest sources of energy. Direct solar radiation absorbed in solar collectors can provide space heating and hot water. Passive solar can be used to enhance the solar energy use in buildings for space heating and lighting requirements. Solar energy can also be used to produce electricity, and this is the renewable energy area that is the focus of attention in this section.

According to DOE (2009), the two solar electric technologies with the greatest potential are photovoltaics (PV) and concentrating solar power (CSP).

Photovoltaics (PV)

Photovoltaic (Gr. *photo*, light, and *volt*, electricity pioneer Alessandro Volta) technology makes use of the abundant energy in the Sun, and it has little impact on our environment. *Photovoltaics* is the direct conversion of light into electricity at the atomic level. Some materials exhibit a property known as the *photoelectric effect* (discovered and described by Alexandre-Edmond Becquerel in 1839) that causes them to absorb photons of light and release electrons. When these free electrons are captured, an electric current results (i.e., electricity is the flow of free electrons) that can be used as electricity. The first photovoltaic module (billed as a solar battery) was built by Bell laboratories in 1954. In the 1960s, the space program began to make the first serious use of the technology to provide power aboard spacecraft. Space program use helped this technology make giant advancements in reliability and helped to lower cost. However, it was the oil embargo of the 1970s (the so-called energy crisis) that propelled photovoltaic technology to the forefront of recognition for use other than space applications only. Photovoltaics can be used in a wide range of products, from small consumer items to large, commercial, solar electric systems.

Figure 12.4 illustrates the operation of a basic *photovoltaic cell*, also called a *solar cell*. Solar cells are made of silicon, and other semiconductor materials such as germanium, gallium arsenide, and silicon carbide are used in the microelectronics industry. For solar cells, a thin semiconductor wafer is specially treated to form an electric field,

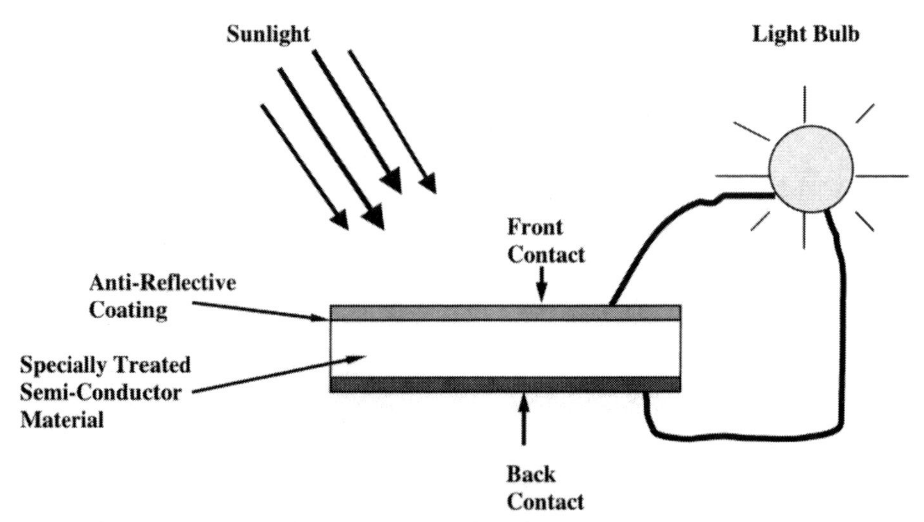

Figure 12.4. Operation of basic photovoltaic cell.

positive on one side and negative on the other. When light energy strikes the solar cell, electrons are jarred loose from the atoms in the semiconductor material. If electrical conductors are attached to the positive and negative sides, forming an electrical circuit, the electrons can be captured in the form of an electrical current—again, electron flow is electricity. This electricity can then be used to power a load, such as a light, tool, toaster, and other electrical appliance or apparatus.

A number of solar cells electrically connected to each other and mounted on a support panel or frame is called a photovoltaic module (see figure 12.5). Modules are designed to supply electricity at a certain voltage, such as a common twelve-volt system. The current produced is directly dependent on how much light strikes the module.

Multiple modules can be wired together to form an array. In general, as the size of the module or array increases, the amount of electricity produced also increases. Photovoltaic modules and arrays produce direct-current (DC) electricity. They can be connected in both series and parallel electrical arrangements to produce any required voltage and current combination.

Concentrating Solar Power (CSP)

CSP offers a utility-scale, firm, dispatchable, renewable energy option that can help meet a nation's demand for electricity. CSP plants produce power by first using mirrors to focus sunlight to heat a working fluid. Ultimately, this high-temperature fluid is used to spin a turbine or power an engine that drives a generator that produces electricity.

CSP systems can be classified by how they collect solar energy, by using linear concentrator, dish/engine, or power tower systems (NREL 2009).

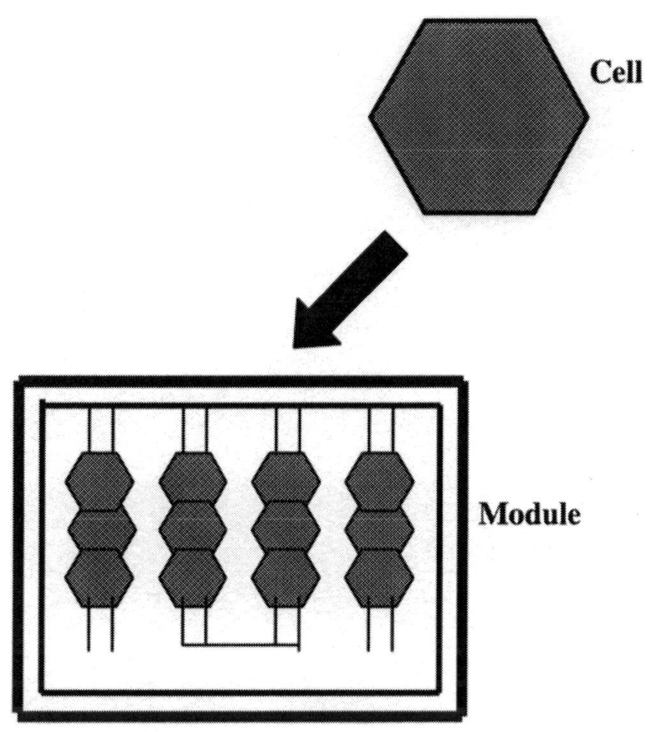

Figure 12.5. Single photovoltaic cell and module of cells.

- Linear concentrator—collects the Sun's energy using long, rectangular, curved (U-shaped) mirrors. The mirrors are tilted toward the Sun, focusing sunlight on tubes (or receivers) that run the length of the mirrors. The reflected sunlight heats a fluid flowing through the tubes. The hot fluid is then used to boil water in a conventional, steam-turbine generator to produce electricity. There are two major types of linear concentrator systems: parabolic trough systems, where receiver tubes are positioned along the focal line of each parabolic mirror; and linear Fresnel reflector systems, where one receiver tube is positioned above several mirrors to allow the mirrors greater mobility in tracking the Sun.
- Dish/engine system—uses a mirrored dish similar to a very large satellite dish. The dish-shaped surface directs and concentrates sunlight onto a thermal receiver, which absorbs and collects the heat and transfers it to the engine generator. The most common type of heat engine used today in dish/engine systems is the Stirling engine (conceived in 1816). This system uses the fluid heated by the receiver to move pistons and create mechanical power. The mechanical power is then used to run a generator or alternator to produce electricity.
- Power tower system—uses a large field of flat, Sun-tracking mirrors known as heliostats to focus and concentrate sunlight onto a receiver on the top of a tower. A heat-transfer fluid heated in the receiver is used to generate steam, which, in turn, is used in a conventional turbine generator to produce electricity. Some power towers use

water/steam as the heat-transfer fluid. Other advanced designs are experimenting with molten nitrate salt because of its superior heat-transfer and energy-storage capabilities. The energy-storage capability, or thermal storage, allows the system to continue to dispatch electricity during cloudy weather or at night.

Smaller CSP systems can be located directly where the power is needed. For example, a single, dish/engine system can produce three to twenty-five kilowatts of power and is well suited for such distributed applications. Larger, utility-scale CSP applications provide hundreds of megawatts of electricity for the power grid. Both linear concentrator and power tower systems can be easily integrated with thermal storage, helping to generate electricity during cloudy periods or at night. Alternatively, these systems can be combined with natural gas, and the resulting hybrid power plants can provide high-value, dispatchable power throughout the day.

Solar energy has some obvious advantages in that the source is free; however, the initial investment in operating equipment is not. Solar energy is also environmentally friendly, requires almost no maintenance, and reduces our dependence on foreign energy supplies. Probably the greatest downside of solar energy use is that, in areas without direct sunlight during certain times of the year, solar panels cannot capture enough energy to provide heat for home or office. Geographically speaking, the higher latitudes do not receive as much direct sunlight as tropical areas. Because of the position of the Sun in the sky, solar panels must be placed in Sun-friendly locations such as the U.S. Desert Southwest and the Sahara region of northern Africa.

WIND ENERGY

Wind energy is the movement of wind to create power.[3] Since early recorded history, people have been harnessing the energy of the wind for milling grain, pumping water, and other mechanical power applications. Wind energy propelled boats along the Nile River as early as 5000 BC. By 200 BC, simple windmills in China were pumping water, while vertical-axis windmills with woven reed sails were grinding grain in Persia and the Middle East.

The use of wind energy spread around the world, and by the eleventh century, people in the Middle East were using windmills extensively for food production; returning merchants and crusaders carried this idea back to Europe. The Dutch refined the windmill and adapted it for draining lakes and marshes in the Rhine River Delta. When settlers took this technology to the New World in the later nineteenth century, they began using windmills to pump water for farms and ranches and, later, to generate electricity for homes and industry. Today, there are several hundred thousand windmills in operation around the world, many of which are used for water pumping. But it is the use of wind energy as a pollution-free means of generating electricity on a significant scale that is attracting most current interest in the subject. As a matter of fact, with the present and pending shortage and high cost of fossil fuels to generate electricity and the green movement toward the use of cleaner fuels, wind energy is the world's fastest-growing energy source and will power industry, businesses, and homes with clean, renewable electricity for many years to come. In the United States since 1970, wind-based electricity-generating capacity has increased markedly although (at present) it remains a small fraction of total electric capacity. But this trend is beginning to change—with the advent of $4-per-gallon of gasoline, high heating and cooling costs, subsequent increases in the cost of electricity, and worldwide political unrest or uncertainty in oil-supplying countries, one only needs to travel the "wind corridors" of the United States (encompassing parts of Arizona, New Mexico, Texas, Missouri, and north through the Great Plains to the Pembina Escarpment and Turtle Mountains of North Dakota and elsewhere) to witness the considerable activity and the seemingly exponential increase in wind energy development and wind turbine installations; these machines are being installed to produce and provide electricity to the grid.

When you get right down to it, we can classify wind energy as a form of solar energy. As described in chapter 9, winds are caused by uneven heating of the atmosphere by the Sun, irregularities of the Earth's surface, and the rotation of the Earth. As a result, winds are strongly influenced and modified by local terrain, bodies of water, weather patterns, vegetative cover, and other factors. The wind flow, or motion of energy when harvested by wind turbines, can be used to generate electricity.

As with any other source of energy, nonrenewable or renewable, there are advantages and disadvantages associated with their use. On the positive side, it should be noted that wind energy is a free, renewable resource, so no matter how much is used today, there will still be the same supply in the future. Wind energy is also a source of clean, nonpolluting electricity. One huge advantage of wind energy is that it is a domestic source of energy, produced in the United States or another country where wind is abundant.

Wind turbines can be installed on farms or ranches, thus benefiting the economy in rural areas, where most of the best wind sites are found. Moreover, farmers and ranchers can continue to work the land because the wind turbines use only a fraction of the land.

On the other side of the coin, wind energy does have a few negatives. Wind power must compete with conventional generation sources on a cost basis. Even though the cost of wind power has decreased dramatically in the past ten years, the technology requires a higher initial investment than fossil-fueled generators. The challenge to using wind as a source of power is that the wind is intermittent and does not always blow when electricity is needed. Wind energy cannot be stored (unless batteries are being used), and not all winds can be harnessed to meet the timing of electricity demands. Another problem is that good sites are often located in remote locations, far from cities where the electricity is needed. Moreover, wind resource development may compete with other uses for the land, and those alternative uses may be more highly valued than electricity generation. Finally, in regard to the environment, wind power plants have relatively little impact on the environment compared to other conventional power plants; however, there is some concern over the noise produced by the rotor blades, aesthetic (visual) impacts, and sometimes birds are killed by flying into the rotors. Most of these problems have been resolved or greatly reduced through technological development or by properly siting wind plants.

In regard to wind energy and its future, one thing is certain: it continues to be one of the fastest-growing energy technologies, and it looks set to become a major generator of electricity throughout the world.

HYDROPOWER

When we look ar rushing waterfalls and rivers, we may not immediately think of electricity. But hydroelectric (water-powered) power plants are responsible for lighting many of our homes and neighborhoods. *Hydropower* is the harnessing of water to perform work. The power of falling water has been used in industry for thousands of years (see table 12.5). The Greeks used water wheels for grinding wheat into flour more than two thousand years ago. Besides grinding flour, the power of water was used to saw wood and power textile mills and manufacturing plants.

The technology for using falling water to create hydroelectricity has existed for more than a century. The evolution of the modern hydropower turbine began in the mid-1700s when a French hydraulic and military engineer, Bernard Forest de Belidor wrote a four-volume work describing using a vertical-axis versus a horizontal-axis machine.

Water turbine development continued during the 1700s and 1800s. In 1880, a brush-arc light dynamo driven by a water turbine was used to provide theater and storefront lighting in Grand Rapids, Michigan; and in 1881, a brush dynamo connected to a turbine in a flour mill provided street lighting at Niagara Falls, New York. These two projects used direct-current (DC) technology.

Alternating current (AC) is used today. That breakthrough came when the electric generator was coupled to the turbine, which resulted in the world's, and the United States', first hydroelectric plant located on the Fox River in Appleton, Wisconsin, in 1882. The U.S. Library of Congress (2009) lists the Appleton hydroelectric power plant as one of the major accomplishments of the Gilded Age (1878–1889). Soon, people across the United States were enjoying electricity in homes, schools, and offices, reading by electric lamp instead of candlelight or kerosene. Today, we take electricity for granted, not able to imagine life without it.

Ranging in size from small systems (one hundred kilowatts to thirty megawatts) for a home or village to large projects (capacity greater than thirty megawatts) producing electricity for utilities, hydropower plants are of three types: impoundment, diversion, and pumped storage. Some hydropower plants use dams and some do not. Many dams were built for other purposes and hydropower was added later. In the United States, there are about 80,000 dams of which only 2,400 produce power. The other dams are for recreation, stock/farm ponds, flood control, water supply, and irrigation. The types of hydropower plants are described below.

Impoundment

The most common type of hydroelectric power plant is an impoundment facility. An impoundment facility, typically a large hydropower system, uses a dam to store river

Table 12.5. History of Hydropower

Date	*Hydropower Event*
BC	Hydropower used by the Greeks to turn water wheels for grinding wheat into flour, more than two thousand years ago.
Mid-1770s	French hydraulic and military engineer Bernard Forest de Belidor wrote a four-volume work describing vertical- and horizontal-axis machines.
1775	U.S. Army Corps of Engineers founded, with establishment of Chief Engineer for the Continental Army.
1880	Michigan's Grand Rapids electric Light and Power Company, generating electricity by dynamo belted to a water turbine at the Wolverine Chair Factory, lit up sixteen brush-arc lamps.
1881	Niagara Falls city street lamps powered by hydropower.
1882	World's first hydroelectric power plant began operation on the Fox River in Appleton, Wisconsin.
1886	About forty-five water-powered electric plants in the United States and Canada.
1887	San Bernardino, Calif., opens first hydroelectric plant in the West.
1889	Two hundred electric plants in the United States use water power for some or all generation.
1901	First *Federal Water Power Act.*
1902	Bureau of Reclamation established.
1907	Hydropower provided 15 percent of U.S. electrical generation.
1920	Hydropower provided 25 percent of U.S. electrical generation. *Federal Power Act* establishes Federal Power Commission authority to issue licenses for hydro development on public lands.
1933	Tennessee Valley Authority established.
1935	Federal Power Commission authority extended to all hydroelectric projects built by utilities engaged in interstate commerce.
1937	Bonneville Dam, first federal dam, begins operation on the Columbia River; Bonneville Power Administration established.
1940	Hydropower provided 40 percent of electrical generation. Conventional capacity tripled in United States since 1920.
1980	Conventional capacity nearly tripled in United States since 1940.
2003	About 10 percent of U.S. electricity comes from hydropower. Today, there is about eighty thousand megawatts of conventional capacity and eighteen thousand megawatts of pumped storage.

Source: U.S. Department of Energy (DOE), Office of Energy Efficiency and Renewable Energy (EERE), "History of Hydropower," 2008, www.eere.energy.gov/windandhydro/printable_versions/hydro_history.html (accessed June 15, 2009).

water in a reservoir. Water released from the reservoir flows through a turbine, spinning it, which in turn activates a generator to produce electricity. The water may be released either to meet changing electricity needs or to maintain a constant reservoir level.

Diversion

A diversion (sometimes called run-of-river) facility channels all or a portion of the flow of a river from its natural course through a canal or penstock. It may not require the use of a dam.

Pumped Storage

When the demand for electricity is low, a pumped storage facility stores energy by pumping water from a lower reservoir to an upper reservoir. During periods of high electrical demand, the water is released back to the lower reservoir to generate electricity.

Hydropower offers advantages over the other energy sources but faces unique environmental challenges. The advantages of using hydropower begin with the fact that hydropower does not pollute the air like power plants that burn fossil fuels, such as coal and natural gas. Moreover, hydropower does not have to be imported into the United States like foreign oil does; it is produced in the United States. Because hydropower relies on the water cycle, driven by the Sun, it's a renewable resource that will be around for at least as long as humans. Hydropower is controllable, that is, engineers can control the flow of water through the turbines to produce electricity on demand. Finally, hydropower impoundment dams create huge lake areas for recreation, irrigation of farm lands, reliable supplies of potable water, and flood control.

Hydropower also has some disadvantages. For example, fish populations can be impacted if fish cannot migrate upstream past impoundment dams to spawning grounds or if they cannot migrate downstream to the ocean. Many dams have installed fish ladders or elevators to aid upstream fish passage. Downstream fish passage is aided by diverting fish from turbine intakes using screens or racks or even underwater lights and sounds, and by maintaining a minimum spill flow past the turbine. Hydropower can also impact water quality and flow. Hydropower plants can cause low dissolved oxygen (DO) levels in the water, a problem that is harmful to riparian (riverbank) habitats and is addressed using various aeration techniques, which oxygenate the water. Maintaining minimum flows of water downstream of a hydropower installation is also critical for the survival of riparian habitats. Hydropower is also susceptible to drought. When water is not available, the hydropower plants can't produce electricity. Finally, construction of new hydropower facilities impact investors and others by competing with other uses of the land. Preserving local flora and fauna and historical or cultural sites is often more highly valued than electricity generation.

BIOMASS

Biomass (all the Earth's living matter) or *bioenergy* (the energy from plants and plant-derived materials; stored energy from the Sun) has been used since people began burning wood to cook food and keep warm. Wood is still the largest biomass energy resource today, but other sources of biomass can also be used. These include food crops, grassy and woody plants, residues from agriculture or forestry, and the organic component of municipal and industrial wastes. Even the fumes from landfills (which are methane, a natural gas) can be used as a biomass energy source. It excludes organic material that has been transformed by geological processes into substances such as coal or petroleum. The biomass industry is one of the fastest-growing industries in the United States.

A variety of biomass feedstocks can be used to produce transportation fuels, bio-based products, and power. Currently, a majority of the ethanol produced in the United States is made from corn or other starch-based crops. However, the current trend in research is to develop biomass fuels from nonfoodstocks. For example, the focus is on the development of cellulosic feedstocks—nongrain, non-food-based feedstocks such as switchgrass, corn stover, and woody material—and on technologies to convert cellulosic material into transportation fuels and other products. Using cellulosic feedstocks not only can alleviate the potential concern of diverting food crops to produce fuel but also has a variety of environmental benefits.

Environmental benefits include the use of biomass energy to greatly reduce greenhouse gas emissions. Burning biomass releases about the same amount of carbon dioxide as burning fossil fuels. However, fossil fuels release carbon dioxide captured by photosynthesis millions of years ago—an essentially "new" greenhouse gas. Biomass, on the other hand, releases carbon dioxide that is largely balanced by the carbon dioxide captured in its own growth (depending on how much energy was used to grow, harvest, and process the fuel).

Another benefit of biomass use for fuel is that it can reduce dependence on foreign oil because biofuels are the only renewable liquid transportation fuels available.

Finally, biomass energy supports U.S. agricultural and forest-product industries. The main biomass feedstocks for power are paper mill residue, lumber mill scrap, and municipal waste. For biomass fuels, the feedstocks are corn (for ethanol) and soybeans (for biodiesel), both surplus crops. In the near future—and with developed technology—agricultural residues such as corn stover (the stalks, leaves, and husks of the plant) and wheat straw will also be used. Long-term plans include growing and using dedicated energy crops, such as fast-growing trees and grasses that can grow sustainably on land that will not support intensive food crops.

GEOTHERMAL ENERGY

Approximately 4,000 miles below the Earth's surface is the Earth's core where temperatures can reach 9000°F.[4] This heat—geothermal energy (*geo*, meaning earth, and *thermos*, meaning heat)—flows outward from the core, heating the surrounding area, which can form underground reservoirs of hot water and steam. These reservoirs can be tapped for a variety of uses, such as to generate electricity or heat buildings.

The geothermal energy potential in the uppermost six miles of the Earth's crust amounts to fifty thousand times the energy of all oil and gas resources in the world.

In the United States, most geothermal reservoirs are located in the western states, Alaska, and Hawaii. However, geothermal heat pumps (GHPs), which take advantage of the shallow ground's stable temperature for heating and cooling buildings, can be used almost anywhere.

Again, it is important to point out that there is nothing new about renewable energy. From solar power to burning biomass (wood) in caves and elsewhere, humans have taken advantage of renewable resources from time immemorial. For example, hot springs have been used for bathing since Paleolithic times or earlier (DOE 2009). The early Romans used hot springs to feed public baths and provide under-floor heating. The world's oldest geothermal district heating system, in France, has been operating since the fourteenth century (Lund 2007). The history of geothermal energy use in the United States is interesting and lengthy. In the following, a brief chronology of major geothermal events in the United States is provided (DOE 2006).

Geothermal Timeline

8,000 BC (and Earlier)

Paleo-Indians used hot springs for cooking and for refuge and respite. Hot springs were neutral zones where members of warring nations would bathe together in peace. Native Americans have a history with every major hot spring in the United States.

1807

As European settlers moved westward across the continent, they gravitated toward these springs of warmth and vitality. In 1807, the first European to visit the Yellowstone area, John Colter ([c. 1774–c. 1813] widely considered to be the first mountain man), probably encountered hot springs, leading to the designation "Colter's Hell." Also in 1807, settlers found the city of Hot Springs, Arkansas, where, in 1830, Asa Thompson charged one dollar each for the use of three spring-fed baths in a wooden tub, and the first-known commercial use of geothermal energy occurred.

1847

William Bell Elliot, a member of John C. Fremont's survey party, stumbles upon a steaming valley just north of what is now San Francisco, California. Elliot calls the area the Geysers—a misnomer—and thinks he has found the gates of hell.

1852

The Geysers is developed into a spa called the Geysers Resort Hotel. Guests include J. Pierpont Morgan, Ulysses S. Grant, Theodore Roosevelt, and Mark Twain.

1862

At springs located southeast of the Geysers, businessman Sam Brannan pours an estimated half million dollars into an extravagant development dubbed "Calistoga," replete with hotel, bathhouse, skating pavilion, and racetrack. Brannan's was one of many spas reminiscent of those of Europe.

1864

Homes and dwellings have been built near springs through the millennia to take advantage of the natural heat of these geothermal springs, but the construction of the Hot Lake Hotel near La Grande, Oregon, marks the first time that the energy from hot springs is used on a large scale.

1892

Boise, Idaho, provides the world's first district heating system as water is piped from hot springs to town buildings. Within a few years, the system is serving two hundred homes and forty downtown businesses. Today, there are four district heating systems in Boise that provide heat to over five million square feet of residential, business, and governmental space. There are now seventeen district heating systems in the United States and dozens more around the world.

1900

Hot springs water is piped to homes in Klamath Falls, Oregon.

1921

John D. Grant drills a well at the Geysers with the intention of generating electricity. This effort is unsuccessful, but one year later Grant meets with success across the valley at another site, and the United States' first geothermal power plant goes into operation. Grant uses steam from the first well to build a second well, and, several wells later, the operation is producing 250 kilowatts, enough electricity to light the buildings and streets at the resort. The plant, however, is not competitive with other sources of power, and it soon falls into disuse.

1927

Pioneer Development Company drills the first exploratory wells at Imperial Valley, California.

1930

The first commercial greenhouse use of geothermal energy is undertaken in Boise, Idaho. The operation uses a one-thousand-foot well drilled in 1926. In Klamath Falls, Charlie Lieb develops the first downhole heat exchanger (DHE) to heat his house. Today, more than five hundred DHEs are in use around the country.

1940

The first residential space heating in Nevada begins in the Moan area in Reno.

1948

Geothermal technology moves east when Carl Nielsen develops the first ground-source heat pump, for use at his residence. J. D. Krocker, an engineer in Portland, Oregon, pioneers the first commercial building use of a groundwater heat pump.

1960

The country's first large-scale geothermal electricity-generating plant begins operation. Pacific Gas and Electric operates the plant located at the Geysers. The first turbine produces eleven megawatts of net power and operates successfully for more than thirty years. Today, sixty-nine generating facilities are in operation at eighteen resource sites around the country.

1978

Geothermal Food Processors, Inc., opens the first geothermal food-processing (crop-drying) plant in Brady Hot Springs, Nevada. The Loan Guaranty Program provides $3.5 million for the facility.

1979

The first electrical development of a water-dominated geothermal resource occurs, at the east Mesa field in the Imperial Valley in California. The plant is named for B. C. McCabe, the geothermal pioneer who, with his Magma Power Company, did field development work at several sites, including the Geysers.

1980

TAD's Enterprises of Nevada pioneers the use of geothermal energy for the cooking, distilling, and drying processes associated with alcohol fuels production. UNOCAL builds the country's first flash plant, generating ten megawatts at Brawley, California.

1982

Economical electrical generation begins at California's Salton Sea geothermal field through the use of crystallizer-clarifier technology. The technology resulted from a government/industry effort to manage the high-salinity brines at the site.

1984

A twenty-megawatt plant begins generating power at Utah's Roosevelt Hot Springs. Nevada's first geothermal electricity, generated with a 1.3-megawatt binary power plant, begins operation.

1987

Geothermal fluids are used in the first geothermal-enhanced heap leaching project for gold recovery, near Round Mountain, Nevada.

1989

The world's first hybrid (organic Rankine/gas engine) geopressure-geothermal power plant begins operation at Pleasant Bayou, Texas, using both the heat and the methane of a geopressured resource.

1992

Electrical generation begins at the twenty-five-megawatt geothermal plant in the Puna field of Hawaii.

1993

A twenty-three-megawatt binary power plant is completed at Steamboat Springs, Nevada.

1995

Integrated Ingredients dedicates a food-dehydration facility that processes fifteen million pounds of dried onions and garlic per year at Empire, Nevada. A DOE low-temperature resource assessment of ten western states identifies nearly nine thousand thermal wells and springs and 271 communities collocated with a geothermal resource greater than 50°C.

2002

Organized by GeoPowering the West, geothermal development working groups are active in five states—Nevada, Idaho, New Mexico, Oregon, and Washington. Group members represent all stakeholder organizations. The working groups are identifying barriers to geothermal development in their states and bringing together all interested parties to arrive at mutually beneficial solutions.

2003

The Utah Geothermal Working Group is formed.

Geothermal energy can be and already is accessed by drilling water or steam wells in a process similar to drilling for oil. Geothermal energy is an enormous, underused heat and power resource that is clean (emits little or no greenhouse gases), reliable (average system availability of 95 percent), and homegrown (making us less dependent on foreign oil).

Geothermal resources range from shallow ground to hot water and rock several miles below the Earth's surface and even farther down to the extremely hot molten rock called magma. Mile-or-more-deep wells can be drilled into underground reservoirs to tap steam and very hot water that can be brought to the surface for use in a variety of applications. In the United States, most geothermal reservoirs are located in the western states, Alaska, and Hawaii.

OCEAN ENERGY

The ocean can produce two types of energy: *thermal energy* from the Sun's heat and *mechanical energy* from the tides and waves.

Open thermal energy can be used for many applications, including electricity generation. Electricity conversion systems use either the warm surface water or boil the seawater to turn a turbine, which activates a generator.

The electricity conversion of both tidal and wave energy usually involves mechanical devices. It is important to distinguish tidal energy from hydropower. Recall that hydropower is derived from the hydrological climate cycle, powered by solar energy, which is usually harnessed via hydroelectric dams. In contrast, tidal energy is the result of the interaction of the gravitational pull of the Moon and, to a lesser extent, the Sun, on the seas. Processes that use tidal energy rely on the twice-daily tides, the resultant upstream flows and downstream ebbs in estuaries and the lower reaches of some rivers, as well as, in some cases, tidal movement out at sea. A dam is typically used to convert tidal energy into electricity by forcing the water through turbines, activating a generator. Meanwhile, wave energy, a very large potential resource to be tapped, uses mechanical power to directly activate a generator, to transfer to a working fluid, water, or air, which then drives a turbine/generator.

HYDROGEN

Containing only one electron and one proton, hydrogen, chemical symbol H, is the simplest element on Earth.[5] Hydrogen is a diatomic molecule—each molecule has two atoms of hydrogen (which is why pure hydrogen is commonly expressed as H_2). Although abundant on Earth as an element, hydrogen combines readily with other elements and is almost always found as part of another substance, such as water, hydrocarbons, or alcohols. Hydrogen is also found in biomass, which includes all plants and animals.

- Hydrogen is an energy carrier, not an energy source. Hydrogen can store and deliver usable energy, but it doesn't typically exist by itself in nature; it must be produced from compounds that contain it.
- Hydrogen can be produced using diverse, domestic resources including nuclear; natural gas and coal; and biomass and other renewables including solar, wind, hydroelectric, or geothermal energy. This diversity of domestic energy sources makes hydrogen a promising energy carrier and important to our nation's energy security.

It is expected and desirable for hydrogen to be produced using a variety of resources and process technologies (or pathways).

- DOE focuses on hydrogen-production technologies that result in near zero net greenhouse gas emissions and use renewable energy sources, nuclear energy, and coal (when combined with carbon sequestration). To ensure sufficient clean energy for our overall energy needs, energy efficiency is also important.
- Hydrogen can be produced via various process technologies, including thermal (natural gas reforming, renewable liquid and bio-oil processing, and biomass and coal gasification), electrolytic (water splitting using a variety of energy resources), and photolytic (splitting water using sunlight via biological and electrochemical materials).
- Hydrogen can be produced in large, central facilities (fifty to three hundred miles from point of use) or smaller, semicentral facilities (located within twenty-five to one hundred miles of use) and distributed (near or at point of use).
- In order for hydrogen to be successful in the marketplace, it must be cost competitive with the available alternatives. In the light-duty vehicle transportation market, this competitive requirement means that hydrogen needs to be available untaxed at two to three dollars per gge (gasoline gallon equivalent). This price would result in hydrogen fuel cell vehicles having the same cost to the consumer on a cost-per-mile-driven basis as a comparable conventional internal-combustion engine or hybrid vehicle.
- DOE is engaged in research and development of a variety of hydrogen production technologies. Some are further along in development than others—some can be cost competitive for the transition period (beginning in 2015), and others are considered long-term technologies (cost competitive after 2030).

Infrastructure is required to move hydrogen from the location where it's produced to the dispenser at a refueling station or stationary power site. Infrastructure includes the pipelines, trucks, railcars, ships, and barges that deliver fuel, as well as the facilities and equipment needed to load and unload them.

Delivery technology for hydrogen infrastructure is currently available commercially, and several U.S. companies deliver bulk hydrogen today. Some of the infrastructure is already in place because hydrogen has long been used in industrial applications, but it's not sufficient to support widespread consumer use of hydrogen as an energy carrier. Because hydrogen has a relatively low volumetric energy density, its transportation, storage, and final delivery to the point of use comprise a significant cost and result in some of the energy inefficiencies associated with using it as an energy carrier.

Options and tradeoffs for hydrogen delivery from central, semicentral, and distributed production facilities to the point of use are complex. The choice of a hydrogen production strategy greatly affects the cost and method of delivery.

For example, larger, centralized facilities can produce hydrogen at relatively low costs due to economies of scale, but the delivery costs for centrally produced hydrogen are higher than the delivery costs for semicentral or distributed production options (because the point of use is farther away). In comparison, distributed production facilities have relatively low delivery costs, but the hydrogen production costs are likely to be higher—lower volume production means higher equipment costs on a per-unit-of-hydrogen basis.

Key challenges to hydrogen delivery include reducing delivery cost, increasing energy efficiency, maintaining hydrogen purity, and minimizing hydrogen leakage. Further research is needed to analyze the tradeoffs between the hydrogen production options and the hydrogen delivery options taken together as a system. Building a national hydrogen delivery infrastructure is a big challenge. It will take time to develop and will likely include combinations of various technologies. Delivery infrastructure needs and resources will vary by region and type of market (e.g., urban, interstate, or rural). Infrastructure options will also evolve as the demand for hydrogen grows and as delivery technologies develop and improve.

Hydrogen Storage

Storing enough hydrogen on board a vehicle to achieve a driving range of greater than three hundred miles is a significant challenge. On a weight basis, hydrogen has nearly three times the energy content of gasoline (120 megajoule (MJ)/kg for hydrogen versus 44 MJ/kg for gasoline). However, on a volume basis the situation is reversed (8 MJ/liter for liquid hydrogen versus 32 MJ/liter for gasoline). On-board hydrogen storage in the range of 5–13 kg H_2 is required to encompass the full platform of light-duty vehicles.

Hydrogen can be stored in a variety of ways, but for hydrogen to be a competitive fuel for vehicles, the hydrogen vehicle must be able to travel a comparable distance to conventional hydrocarbon-fueled vehicles.

Hydrogen can be physically stored as either a gas or a liquid. Storage as a gas typically requires high-pressure tanks (five to ten thousand pounds per square inch [psi] tank pressure). Storage of hydrogen as a liquid requires cryogenic temperatures because the boiling point of hydrogen at one atmosphere pressure is $-252.8°C$.

Hydrogen can also be stored on the surfaces of solids (by adsorption) or within solids (by absorption). In adsorption, hydrogen is attached to the surface of material either as hydrogen molecules or as hydrogen atoms. In absorption, hydrogen is dissociated into H-atoms, and then the hydrogen atoms are incorporated into the solid, lattice framework.

Hydrogen storage in solids may make it possible to store large quantities of hydrogen in smaller volumes at low pressures and at temperatures close to room temperature. It is also possible to achieve volumetric storage densities greater than liquid hydrogen because the hydrogen molecule is dissociated into atomic hydrogen within the metal hydride lattice structure.

Finally, hydrogen can be stored through the reaction of hydrogen-containing materials with water (or other compounds such as alcohols). In this case, the hydrogen is effectively stored in both the material and in the water. The term "chemical hydrogen storage" or chemical hydrides is used to describe this form of hydrogen storage. It is also possible to store hydrogen in the chemical structures of liquids and solids.

Hydrogen Fuel Cell

The fuel cell uses the chemical energy of hydrogen to cleanly and efficiently produce electricity with water and heat as byproducts. Fuel cells are unique in terms of variety of their potential applications; they can provide energy for systems as large as a utility power station and as small as a laptop computer.

Fuel cells have several benefits over conventional combustion-based technologies currently used in many power plants and passenger vehicles. They produce much smaller quantities of greenhouse gases and none of the air pollutants that create smog and cause health problems. If pure hydrogen is used as a fuel, fuel cells emit only heat and water as byproducts.

Did You Know?

Hydrogen fuel cell vehicles (FCVs) emit approximately the same amount of water per mile as vehicles using gasoline-powered internal combustion engines (ICEs).

A *fuel cell* is a device that uses hydrogen (or hydrogen-rich fuel) and oxygen to create electricity by an electrochemical process. A single fuel cell consists of an electrolyte and two catalyst-coated electrodes (a porous anode and cathode). While there are different fuel cell types, all fuel cells work similarly:

- Hydrogen, or a hydrogen-rich fuel, is fed to the anode where a catalyst separates hydrogen's negatively charged electrons from positively charged ions (protons).
- At the cathode, oxygen combines with electrons and, in some cases, with species such as protons or water, resulting in water or hydroxide ions, respectively.
- For polymer electrolyte membrane and phosphoric acid fuel cells, protons move through the electrolyte to the cathode to combine with oxygen and electrons, producing water and heat.
- For alkaline, molten carbonate, and solid oxide fuel cells, negative ions travel through the electrolyte to the anode where they combine with hydrogen to generate water and electrons.
- The electrons from the anode cannot pass through the electrolyte to the positively charged cathode; they must travel around it via an electrical circuit to reach the other side of the cell. This movement of electrons is an electrical current.

Pollution

When asked to define pollution, most people have little trouble doing so, having witnessed some form of it firsthand. They usually come up with an answer that is a description of its obvious effects. But pollution is complicated, and it cannot be easily defined because what pollution is and isn't is a judgment call. In nature, even the most minute elements are intimately connected with every other element, and so, too, are pollution's effects.

When needing a definition for any environmental term, the first place to look is in pertinent Environmental Protection Agency (EPA) publications. With the term pollution, however, EPA's definition was neither particularly helpful nor complete. The EPA (1989) defines pollution as "generally . . . the presence of matter or energy whose nature, location or quantity produced undesired environmental effects . . . impurities producing an undesirable change in an ecosystem." Under the *Clean Air Act* (CAA), for example, the term is defined as "the man-made or man-induced alteration of the physical, biological, and radioactive integrity of water" (USEPA/89–12). Though their definition is not inaccurate, it leaves out too much to suit our needs. The EPA does, however, provide an adequate definition of the term *pollutant* as "any substance introduced into the environment that adversely affects the usefulness of a resource." Pollution is often classed as point source or nonpoint source pollution. However, the EPA's definition of *pollution* seems so general as to be useless, perhaps because it fails to add material on what such a broadly inclusive term may cover. Definitions from other sources presented similar problems. Anyone who seriously studies pollution quickly realizes that there are five major categories of pollution, each with its own accompanying subsets (types); these are shown in table 12.6.

The categories and types of pollution listed in table 12.6 can also be typed or classified as to whether they are *biodegradable* (subject to decay by microorganisms) or *nonbiodegradable* (cannot be decomposed by microorganisms). Moreover, nonbiodegradable pollutants can also be classified as *primary pollutants* (emitted directly into the environment) or *secondary pollutants* (result of some action of a primary pollutant).

To understand the basic concepts of environmental pollution, you'll need to learn the core vocabulary. Here are some of the key terms that were used in this section.

Scientists gather information and draw conclusions about the workings of the environment by applying the *scientific method*, a way of gathering and evaluating information. It involves observation, speculation (hypothesis formation), and reasoning.

The science of pollution may be divided among the study of air pollution (atmosphere), water pollution (hydrosphere), soil pollution (geosphere), and life (biosphere). Again, the emphasis in this text is on the first three, air, water, and soil, because without any of these, life as we know it is impossible.

The *atmosphere* is the envelope of thin air around the Earth. The role of the atmosphere is multifaceted: (1) it serves as a reservoir of gases; (2) it moderates the Earth's temperature; (3) it absorbs energy and damaging ultraviolet (UV) radiation from the Sun; (4) it transports energy away from equatorial regions; and (5) it serves as a pathway for vapor-phase movement of water in the hydrologic cycle. *Air*, the mixture of gases that constitutes the Earth's atmosphere, is by volume, at sea level, 78.0 percent nitrogen, 21.0 percent oxygen, 0.93 percent argon, and 0.03 percent carbon dioxide, together with very small amounts of numerous other constituents.

Table 12.6. Categories and Types of Pollution

Pollution Categories	*Type of Pollution*
Air pollution	Acid rain Chlorofluorocarbon Global warming Global dimming Global distillation Particulate Smog Ozone depletion
Water pollution	Eutrophication Hypoxia Marine pollution Marine debris Ocean acidification Oil spill Ship pollution Surface runoff Thermal pollution Wastewater Waterborne diseases Water quality Water stagnation
Soil contamination	Bioremediation Electrical resistance heating Herbicide Pesticide Soil guideline values (SGVs)
Radioactive contamination	Actinides in the environment Environmental radioactivity Fission product Nuclear fallout Plutonium in the environment Radiation poisoning Radium in the environment Uranium in the environment
Others	Invasive species Light pollution Noise pollution Radio spectrum pollution Visual pollution

The *hydrosphere* is the water component of the Earth, encompassing the oceans, seas, rivers, streams, swamps, lakes, groundwater, and atmospheric water vapor. *Water* (H_2O) is a liquid that when pure is without color, taste, or odor. It covers 70 percent of the Earth's surface and occurs as standing (oceans and lakes) and running (rivers and streams) water, rain, and vapor. It supports all forms of Earth's life.

The *geosphere* consists of the solid Earth, including *soil*—the *lithosphere*, the topmost layer of decomposed rock and organic matter that usually contains air, moisture, and nutrients and can therefore support life.

The *biosphere* is the region of the Earth and its atmosphere in which life exists, an envelope extending from up to six thousand meters above to ten thousand meters below sea level. Living organisms and the aspects of the environment pertaining directly to them are called *biotic* (biota), and other portions, nonliving parts, of the physical environment are *abiotic*.

A series of biological, chemical, and geological processes by which materials cycle through ecosystems are called *biogeochemical cycles*. We are concerned with two types, the *gaseous* and the *sedimentary*. Gaseous cycles include the carbon and nitrogen cycles. The main *sink* (the main receiving area for material, for example, plants are sinks for carbon dioxide) of nutrients in the gaseous cycle is the atmosphere and the ocean. The sedimentary cycles include sulfur and phosphorous cycles. The main sink for sedimentary cycles is soil and rocks of the Earth's crust.

Formerly known as natural science, and now more commonly known as ecology, *ecology* is critical to the study of geography and environmental science, as the study of the structure, function, and behavior of the natural systems that comprise the biosphere. As mentioned earlier, the terms "ecology" and "interrelationship" are interchangeable; they mean the same thing. In fact, ecology is the scientific study of the interrelationships among organisms and between organisms, and all aspects, living and nonliving, of their environment.

Ecology is normally approached from two viewpoints: (1) the environment and the demands it places on the organisms in it or (2) organisms and how they adapt to their environmental conditions. An *ecosystem*, a cyclic mechanism, describes the interdependence of species in the living world (the biome or community) with one another and with their nonliving (abiotic) environment. An ecosystem has physical, chemical, and biological components, as well as energy sources and pathways.

An ecosystem can be analyzed from a functional viewpoint in terms of several factors. The factors important in this discussion include *biogeochemical cycles*, *energy*, and *food chains*. Each ecosystem is bound together by biogeochemical cycles through which living organisms use energy from the Sun to obtain or "fix" nonliving, inorganic elements such as carbon, oxygen, and hydrogen from the environment and transform them into vital food, which is then used and recycled. The environment in which a particular organism lives is a *habitat*. The role of an organism in a habitat is its *niche*.

Additional key terms and definitions include the following:

- *Acid rain*—is any form of precipitation made more acidic from falling though air pollutants (primarily sulfur dioxide) and dissolving them.
- *Actinides in the environment*—refers to the sources, environmental behavior, and effects of radioactive actinides in the environment.
- *Air quality index*—is a standardized indicator of the air quality in a given location.
- *Atmospheric dispersion modeling*—is the mathematical simulation of how air pollutants disperse in the ambient atmosphere.

- *Bioremediation*—is any process that uses microorganisms, fungi, green plants, or their enzymes to return the natural environment altered by contaminants to its original condition.
- *Chlorofluorocarbons (CFCs)*—are synthetic chemicals that are odorless, nontoxic, nonflammable, and chemically inert.
- *Electrical resistance heating remediation*—is an in situ environmental remediation method that uses the flow of alternating current electricity to heat soil and groundwater and evaporate contaminants.
- *Emerging pollution (contaminants); PPCPs*—are any synthetic or naturally occurring chemical or any microorganism that is not commonly monitored in the environment but has the potential to enter the environment and cause known or suspected adverse ecological or human health effects. PPCPs (pharmaceuticals and personal care products) comprise a very broad, diverse collection of thousands of chemical substances, including prescription and over-the-counter therapeutic drugs, fragrances, cosmetics, sunscreen agents, diagnostic agents, nutrapharmaceuticals, biopharmaceuticals, and many others.
- *Environmental radioactivity*—is the study of radioactive material in the human environment.
- *Eutrophication*—is a natural process in which lakes receive inputs of plant nutrients as a result of natural erosion and runoff from the surrounding land basin.
- *Fission product*—are the atomic fragments left after a large nucleus fission.
- *Global dimming*—is the gradual reduction in the amount of global direct irradiance at the Earth's surface.
- *Global distillation (or grasshopper effect)*—is the geochemical process by which certain chemicals, most notably persistent organic pollutants (POPs), are transported from warmer to colder regions of the Earth.
- *Global warming*—is the long-term, average rise of temperature of the Earth.
- *Herbicide*—is used to kill unwanted plants.
- *Hypoxia*—is a phenomenon that occurs in aquatic environments as dissolved oxygen (DO) becomes reduced in concentration to a point detrimental to aquatic organisms living in the system.
- *Indoor air quality*—is a term referring to the air quality within and around buildings and structures, especially as it relates to the health and comfort of building occupants.
- *Invasive species*—are nonindigenous species (e.g., plants or animals) that adversely affect the habitats they invade economically, environmental, or ecologically.
- *Light pollution*—is excessive or obtrusive artificial light (photo-pollution or luminous pollution).
- *Marine debris*—is human-created waste that has deliberately or accidentally become afloat in a waterway, lake, ocean, or sea.
- *Marine pollution*—occurs when harmful effects, or potentially harmful effects, can result from the entry into the ocean of chemicals, particles, industrial, agricultural, and residential waste, or the spread of invasive organisms.
- *Noise pollution*—is unwanted sound that disrupts the activity or balance of human or animal life.

- *Nuclear fallout*—is the residual radiation hazard from a nuclear explosion, so named because it "falls out" of the atmosphere into which it is spread during the explosion.
- *Ocean acidification*—is the ongoing decrease in the pH of the Earth's oceans, caused by their uptake of anthropogenic carbon dioxide from the atmosphere (Caldeira and Wickett 2003).

Fish Poop to the Rescue

Catherine Brahic (2009) reports that an unlikely ally to buffering the carbon dioxide that acidifies seawater is fish poop. There are two billion tons of fish in the world's oceans. Fish poop seem to play a key role in maintaining the ocean's delicate pH balance.

- *Oil spill*—is the release of a liquid petroleum hydrocarbon into the environment due to human activity and is a form of pollution.
- *Ozone depletion*—while ozone concentrations vary naturally with sunspots, the seasons, and latitude, these processes are well understood and predictable. Scientists have established records spanning several decades that detail normal ozone levels during these natural cycles. Each natural reduction in ozone levels has been followed by a recovery. Recently, however, convincing scientific evidence has shown that the ozone shield is being depleted well beyond changes due to natural processes (EPA 2009).
- *Particulate*—normally refers to fine dust and fume particles; travels easily through air.
- *Pesticide*—is a substance or mixture of substances used to kill pests.
- *Plutonium in the environment*—is an article (part) of the actinides series in the environment.
- *Radiation poisoning*—is a form of damage to organ tissue due to excessive exposure to ionizing radiation.
- *Radio spectrum pollution*—is the straying of waves in the radio and electromagnetic spectrums outside their allocations that cause problems for some activities.
- *Radium and radon*—radium is highly radioactive, and its decay product, radon gas, is also radioactive.
- *Smog*—term used to describe visible air pollution; a dense, discolored haze containing large quantities of soot, ash, and gaseous pollutants such as sulfur dioxide and carbon dioxide.
- *Soil guideline values (SGVs)*—are a series of measurements and values used to measure contamination of the soil.
- *Surface runoff*—is the water flow that occurs when soil is infiltrated to full capacity and excess water, from rain, snowmelt, or other sources flows over the land.
- *Thermal pollution*—is the increase in water temperature with harmful ecological effects on aquatic ecosystems.

- *Uranium*—is a naturally occurring element found in low levels within all rock, soil, and water.
- *Visual pollution*—is the unattractive or unnatural (human-made) visual elements of a vista, a landscape, or any other thing at which a person might not want to look.
- *Wastewater*—is the liquid wastestream primarily produced by the five major sources: human and animal waste, household wastes, industrial waste, stormwater runoff, and groundwater infiltration.
- *Waterborne diseases*—are caused by pathogenic microorganisms that are directly transmitted when contaminated drinking water is consumed.
- *Water quality*—is the physical, chemical, and biological characteristics of water.
- *Water stagnation*—is water at rest; it allows for the growth of pathogenic microorganisms to take place.

According to Edward Keller (1988), pollution is "a substance that is in the wrong place in the environment, in the wrong concentrations, or at the wrong time, such that it is damaging to living organisms or disrupts the normal functioning of the environment" (496). Again, this definition seems incomplete, though it makes the important point that often pollutants are or were useful—in the right place, in the right concentrations, at the right time.

Let's take a look at some of the definitions for pollution that have been used over the years.

1. Pollution is the impairment of the quality of some portion of the environment by the addition of harmful impurities.
2. Pollution is something people produce in large enough quantities that it interferes with our health or well-being.
3. Pollution is any change in the physical, chemical, or biological characteristics of the air, water, or soil that can affect the health, survival, or activities of human beings or other forms of life in an undesirable way. Pollution does not have to produce physical harm; pollutants such as noise and heat may cause injury but more often cause psychological distress, and aesthetic pollution such as foul odors and unpleasant sights affects the senses.

Pollution that initially affects one medium frequently migrates into the other media; air pollution falls to Earth, contaminating soil and water; soil pollutants migrate into groundwater; acid precipitation, carried by air, falls to Earth as rain or snow, altering the delicate ecological balance in surface waters.

In our quest for the definitive definition, the source of last resort was consulted: the common dictionary. According to one dictionary, pollution is a synonym for contamination. A contaminant is a pollutant—a substance present in greater than natural concentrations as a result of human activity and having a net detrimental effect upon its environment or upon something of value in the environment. Every pollutant originates from a source. A receptor is anything that is affected by a pollutant. A sink is a longtime repository of a pollutant. What is actually gained from the dictionary definition is that, since pollution is a synonym for contamination, contaminants are things that contaminate the three environmental mediums (air, water, and soil) in some manner. The bottom line is that we have come full circle; the impact and the exactness of what we stated in the beginning of this text: "Pollution is a judgment call."

Why a judgment call? Because people's opinions differ in what they consider to be a pollutant on the basis of their assessment of benefits and risks to their health and economic well-being. For example, visible and invisible chemicals spewed into the air or water by an industrial facility might be harmful to people and other forms of life living nearby. However, if the facility is required to install expensive pollution controls, forcing the industrial facility to shut down or move away, workers who would lose their jobs and merchants who would lose their livelihoods might feel that the risks from polluted air and water are minor weighed against the benefits of profitable employment. The same level of pollution can also affect two people quite differently. Some forms of air pollution, for example, might cause a slight irritation to a healthy person but cause life-threatening problems to someone with chronic obstructive pulmonary disease (COPD) like emphysema. Differing priorities lead to differing perceptions of pollution (concern at the level of pesticides in foodstuffs generating the need for wholesale banning of insecticides is unlikely to help the starving). No one wants to hear that cleaning up the environment is going to have a negative impact on them. Public perception lags behind reality because the reality is unbearable.

Although pollution is difficult to define, its adverse effects are often relatively easy to see. For example, some rivers are visibly polluted, have an unpleasant odor, or apparent biotic population problems (fish kill, for example). The infamous Cuyahoga River in Ohio became so polluted it twice caught on fire from oil floating on its surface. Air pollution from automobiles and unregulated industrial facilities is obvious. In industrial cities, soot often drifts onto buildings and clothing and into homes. Air pollution episodes can increase hospital admissions and kill people sensitive to the toxins. Fish and birds are killed by unregulated pesticide use. Trash is discarded in open dumps and burned, releasing impurities into the air. Traffic fumes in city traffic plague commuters daily. Ozone levels irritate the eyes and lungs. Sulfate hazards obscure the view.

Even if you are not in a position to see pollution, you are still made aware of it through the media. How about the 1984 Bhopal Incident, the 1986 Chernobyl nuclear plant disaster, the 1991 pesticide spill into the Sacramento River, the Exxon Valdez, or the 1994 oil spill in Russia's Far North? Most of us do remember some of them, even though most of us did not directly witness any of these tragedies. Events, whether man-made (Bhopal) or natural disasters (Mt. St. Helens erupting), sometimes impact us directly, but if not directly, they still get our attention. Worldwide, we see constant reminders of the less dramatic, more insidious, continued, and increasing pollution of our environment. We see or hear reports of dead fish in streambeds, or litter in national parks; or decaying buildings and bridges, leaking landfills, and dying lakes and forests. On the local scale, air quality alerts may have already begun in your community.

Some people experience pollution more directly, firsthand—what we call "in your face," "in your nose," "in your mouth," or "in your skin" type pollution. Consider the train and truck accidents that result in the release of toxic pollutants that force us to evacuate our homes (see case study 12.1). We become ill after drinking contaminated water, breathing contaminated air, or eating contaminated (salmonella-laced) peanut butter products. We can no longer swim at favorite swimming holes because of sewage contamination. We restrict fish, shellfish, and meat consumption because of the presence of harmful chemicals, cancer-causing substances, and hormone residues. We are exposed to nuclear contaminants released to the air and water from uranium-processing plants.

Case Study 12.1: Toxic Sulfuric Acid

At 6:30 p.m., Monday, October 5, 1998, sixteen railroad cars derailed on the Buffalo and Pittsburgh Railroad, at the edge of the Allegheny National Forest near the Clarion River, not far from Erie, Pennsylvania. One of the derailed cars spilled its load of toxic sulfuric acid.

Emergency workers contained the spill about eight hours after the accident occurred, and the leaking tank car was sealed about three hours later. Once the tank was sealed, the acid, which hung in the air in a light mist, dissipated.

No injuries were reported, though one hundred people were evacuated from their homes in nearby Portland Mills overnight. Route 949 was closed while workers from a remediation company finished cleaning up the spill.

Emergency workers had been concerned about acid contamination of the Clarion River, but the spill's flow had been contained in a ditch between the tracks and the road. None of the sulfuric acid reached the river (Associated Press 1998).

In this particular hazardous materials emergency, proper planning and emergency procedures prevented both human health and environmental damage.

Because of the awareness of the potential for the occurrence of hazardous materials incidents, as the one described in case study 12.1, proper hazardous materials emergency preplanning and responder training, using a well-thought-out emergency response procedure, can lessen the impact of chemical spills on the environment—gross environmental pollution can be averted.

Most of our effort to prevent occurrences of environmental pollution has focused on preplanning and dry-run practice exercises; the results of such efforts are clearly demonstrated in events related in case study 12.1. It is important to point out, however, that we are more reactive than proactive in preventing or mitigating such events. Simply, we are not always so proactive in our pollution control planning techniques. Instead, it is quite common (too common) to have reactive (after the fact) responses to such incidents. Of course, as clearly demonstrated in case study 12.1, not all pollution events can be prevented or prepared for. Consider, for example, the tragic events of 9/11. Because we can't get into the minds of terrorists (and other cold-blooded murderers or anyone else for that matter), we have difficulty imagining the deliberate crashing of perfectly good airplanes full of fuel and passengers into buildings and a farm field. We all recognize that this tragic event occurred. However, there might be some people out there wondering what those tragic incidents have to do with environmental pollution.

If you were not present in New York City or the Pentagon or in that Pennsylvania farm field and not up close and personal with any of these events, then you might not be aware of the catastrophic unleashing of various contaminants into the environment because of the crashes. Or maybe you did not have access to television coverage where it clearly showed the massive cloud of dust, smoke, and other ground-level debris engulfing New York City. Maybe you have not had a chance to speak with any of the emergency response personnel who climbed through the contaminated wreckage looking for survivors—responders who were exposed to chemicals and various hazardous materials, many of which we are still uncertain about their exact nature. Days later, when rescue turned to recovery, you may not have noticed personnel garbed in moon suits (level A hazmat response suits) with instruments attempting to sample and monitor the area for harmful contaminants. If you had not witnessed or known about any of the reactions after the 9/11 event, then it might be reasonable to assume you are not aware these were indeed pollution-emitting events.

In addition to terrorism, vandalism, and other deliberate acts, we pollute our environment with apparent abandon. Many of us who teach various environmental science and health subjects to undergraduate and graduate students often hear students complain that the human race must have a death wish. Students quickly adopt this view based on their research and intern work with various environmental-based service entities. During their exposure to all facets of pollution—air, water, and soil contamination—they come to understand that everything we do on Earth contributes pollution of some sort or another to one or all three environmental mediums.

Science and technology notwithstanding, we damage the environment through use, misuse, and abuse of technology. Frequently, we use technological advances before we fully understand their long-term effects on the environment. We weigh the advantages a technological advance can give us against the environment and discount the importance of the environment, through greed, hubris, lack of knowledge, or stupidity. We often only examine short-term plans without fully developing how problems may be handled years later. We assume that, when the situation becomes critical, technology will be there to fix it. The scientists will figure it out, we believe; thus, we ignore the immediate consequences of our technological abuse.

Consider this: while technological advances have provided us with nuclear power, the light bulb and its energy source, plastics, the internal combustion engine, air conditioning, and refrigeration (and scores of other advances that make our modern lives pleasant and comfortable), these advances have affected the Earth's environment in ways we did not expect, in ways we deplore, and in ways we may not be able to live with. In this text, the argument is made that this same science and technology that created or exacerbated pollution events can, in turn, be used to mitigate the misuse of science and technology (Spellman 2009).

POLLUTION: GLOBAL CONCERNS

For millions of years, human beings inhabited the planet Earth without having any perceptible impact on the quality of the global environment (note that, unless I otherwise stipulate, "global environment" refers to the three main media—soil, air, and water—the focus of this text) that accounts for the planet's uniquely habitable conditions. The situation started to change several millennia ago when anthropocentric—human-generated—agricultural practices transformed land cover over large areas. These small-scale practices had perceptible impact primarily on regional (local) environments. More substantial changes to the global environment have come about as a result of the greatly expanded human use of soil, water, and air as the convenient sinks for a myriad of waste products and materials (pollutants). The advent of the Industrial Revolution in the eighteenth century led to great increases in the use of natural resources such as fossil fuels. The human assault on the global environment has quickened ever more dramatically since World War II. An exploding world population, an exponential rise in industrial activity, the massive spread of agriculture, and the wholesale clearing of forests have dramatically altered our planet's condition—its environmental health.

Prior to the twentieth century, significant environmental pollution was almost exclusively a localized problem in the vicinity of the emission sources, in particular the large cities and industrial zones of North America, the British Isles, and Western Europe. Even today, pollution is often still a localized problem. However, its global implications are now not only recognized—but they are also being felt. In this chapter (and throughout this text), I discuss pollution in the context of its global impact.

Evidence of the growing trend toward global interdependence, international commitment and awareness of the global impact of pollution can be found in the notes generated from the economic summit meeting in Paris, July 1989 (UNEP 1989). At this meeting, the leaders of the seven major democracies released a communiqué covering virtually every environmental issue and calling for immediate awareness and a plan of action, one designed to remediate, monitor, measure, and correct anthropogenic (human-caused) emissions.

We are in the midst of what can be called a global awakening; the long-term consequences of this increasing interdependence are still unclear but will certainly include history-making changes in the structures of societies and governments, in levels of multilateral commitment and involvement, in patterns and directions of economic activity, and in the lifestyles, rights, and responsibilities of individuals. Obviously, this is good news for the environment. However, the road to reducing the environmental impact of pollution worldwide is strewn with many boulders.

Addressing global pollution issues in the world forum has obvious advantages. Shifting the focus from pollution at the local level is a major step in the international progress toward solving worldwide pollution problems. (Note that, for our purposes, local refers to a town, a neighborhood, and extending to smaller scales at home and work.)

Shifting the focus from local to global scale is a hard sell. Local and regional pollution issues are familiar to most of us—we encounter many of them daily. They include smog; toxic effects of locally generated air pollutants such as eye irritants, organic vapors, particulate matter, and persistent environmental toxins; indoor air pollution, including radon, formaldehyde, tobacco smoke, biogenic pollutants, and indoor water pollution; and acid rain and acid fog. These local and regional pollution issues are real—and obvious. On the global scale, however, the threat of worldwide pollution and its terrible consequences is not so clear.

How do local pollution concerns shift to the larger problems (the worldwide impact potential) faced at the global scale? In reality, the immediate human impacts of pollution reach a global level in a patchwork and cumulative fashion. Harold Brookfield (1989) points out that changes that are local in domain but widely replicated in sum constitute change in the whole human environment. Brookfield's point is that individual, localized pollution events pose issues in less than global domains, but they add up to a global effect. More significantly, they also connect with much longer chains of environmental consequences (explained below), some of them reaching global environmental systems—climate, for example.

Case Study 12.2: Persistent Organic Pollutants

Pesticide Action Network
North America
Updates Service
July 17, 1998
www.panna.org/

GLOBAL MEETING ON PERSISTENT ORGANIC POLLUTANTS (POPS)

Global representatives from ninety-two countries concluded their first round of talks on how to reduce and eliminate worldwide use and emissions of POPs, highly toxic chemicals such as DDT, and dioxins that remain in the environment for years. The meeting, held in Montreal, June 29 to July 3, 1998, focused on a list of twelve persistent chemicals, including nine pesticides. Eight of these nine pesticides are on Pesticide Action Network's Dirty Dozen list: aldrin, chlordane, DDT, dieldrin, endrin, heptachlor, hexachlorobenzene, and toxaphen. The remaining chemicals on the list are dioxins, furans, mirex, and PCBs.

"These substances travel readily across international borders to even the most remote region, making this a global problem that requires a global solution," said Klaus Toepfer, executive director of the United Nations Environment Program (UNEP), which sponsored the meetings. A growing body of scientific evidence indicates that exposure to very low doses of certain POPs can lead to cancer, damage to the central and peripheral nervous systems, diseases of the immune system, reproductive disorders, and interference with normal infant and child development. POPs can travel through the atmosphere thousands of miles from their source. In addition, these substances concentrate in living organisms and are found in people and animals worldwide.

GLOBAL POLLUTION: CAUSAL FACTORS

Before beginning a brief overview of global pollution problems, let us point out that many nations (in regard to environmental degradation attributed to pollution) lack the ability to measure environmental change, to monitor change brought about by pollution, and to anticipate the impact of interaction among factors such as population size, availability of natural resources (energy), supplies of food and water, and environmental quality. Let us also point out that the United States and most other nations fail to link the results of any existing projections to current decision making.

It is also important to point out that when discussing or explaining causal factors, there are two levels of causation: proximate and ultimate causation. *Proximate causation* (closest cause) answers the question "How?" (How did the local stream become polluted with raw sewage?). This factor is causal at the immediate, direct level. But underlying that, there is a second, deeper cause. This level is referred to as the *ultimate causation* (distal cause), which answers the question "Why?" (Why was the stream vulnerable to sewage pollution?) (Spellman 2009).

Case Study 12.3: Frontier Mentality

It was one of the most famous trails ever. It was the longest overland trail in North America. In 1843, Americans were encouraged by the U.S. government to travel the two thousand miles from Independence, Missouri, to their final destination in the Oregon Territory to homestead the land. By settling the land with more Americans than British, Oregon would belong to the United States.

The two-thousand-mile trek to the Oregon Territory was no walk in the woods; it was a great migration of hordes of people traveling in covered wagons over mostly uncharted lands and trails (or so this was the case early on). The seemingly endless lines of prairie schooners snaked their way along the trail for more than six months before they arrived at their destinations.

The pioneers took what they could carry in their wagons. For example, it was not unusual for each wagon to be loaded with food that included yeast for baking, crackers, cornmeal, bacon, eggs, dried meat, potatoes, rice, beans, and a big barrel of water. They would also take a cow if they had one. Pioneers made their own clothing, so they brought cloth to sew, needles, thread, pins, scissors, and leather to fix worn-out shoes. They had to make their own repairs, so they brought saws, hammers, axes, nails, string, and knives.

Occasionally, and against the advice of the wagon master, pioneers would also pack away personal treasures; heirlooms such as pianos, family trunks and furniture, mirrors, assorted chinaware, silver, paintings, and other decorative household goods of the day. These items not only took up a lot of space within the wagon but also were clumsy to handle and added extra weight to the wagon.

It usually did not take too many miles before many of those who had packed too much started to drop off or discard various personal treasures along the trail. So many personal property items were discarded along the trail that by the late 1840s it was no longer necessary to follow the tracks of the wagons that had preceded them to follow and stay on the trail. All one needed to do was to follow the strewn and rotting personal treasures (and assorted gravesites) that marked the trail. The parting with their personal treasures must have been heartbreaking to many of the pioneers as they made the difficult trek along the Oregon Trail to their new homesteads in the West.

For decades, many have asked what was it that drove or enticed these people to undertake such a perilous adventure into lands unknown. The main draw, of course, was the promise of free land. For others, it was the freedom from the squalor of eastern city life or the drudgery of farm life in trying to eke out a living on worn-out soils as tenant farmers that drove them to undertake the arduous western adventure. For others it was the wide-open untamed spaces and the quest for adventure that was the drawing card.

This burning desire to conquer new lands, exploit it, and become rich was like a powerful magnet attracting and aligning the metal filings of their mentalities. This outlook, mode, or way of thought that drove the pioneers to trek across barren, rugged, unforgiving wilderness is commonly called a frontier mentality, that is, we live as though we can't effectively harm the natural world in a significant way because it is so big and we are so little, and if we damage one place, there will always be a new frontier to move to. The reality is that this mode of thought, this frontier mentality, can be summed up today by simply stating that in the United States there has always been a western frontier, a place to go and start over where there are riches just for the taking for a hardy spirit and determined worker.

Let's get back to the pioneers' trek along the Oregon Trail where they have marked the trail with ruts carved by their wagon wheels and various other signposts provided by their discarded personal goods. Again, after years of traversing the Oregon Trail, wagon train after wagon train discarded goods that dotted the prairie areas for miles. Humans have this tendency, that is, when some object they own is no longer needed or no longer pleases them or has outlived its usefulness, they simply discard it—out of sight, out of mind. The pioneers' frontier mentality about the West being an expanse of wide-open, untamed space gave them no qualms about leaving their personal goods to rot along the trail—no qualms about polluting the landscape.

This same frontier mentality—"I no longer want or need it, so I'll throw it away"—did not stop when the pioneers reached their destinations. The mindset alters a bit, but did it stop? No. It began with settlements that were almost exclusively built along rivers. The rivers provided a convenient means of disposing of the unwanted. Beginning with simple discards such as coffee grounds to more complex items such as white goods (washers, dryers, refrigerators, etc.), to even more complex and persistent chemical compounds and mixtures, the river was the repository of choice for all. The thinking was, of course, that no matter what you threw into the river the running water would purify itself every ten miles or so. But, when there are several settlements with hundreds or thousands of people up and down the length of the river, the purification capacity of the running water is exhausted. However, even though we are running out of pristine areas to pollute, our throw-away society continues to be strongly influenced by our consumerism and excessive disposal of short-lived items.

POPULATION GROWTH

A contributing factor to degradation of the world's environment is a result of overpopulation. How serious is overpopulation's impact on Earth's environment? With our population increasing rapidly, pollution is a problem of increasing proportions. To gain perspective on the overpopulation problem, let's take a look at the record. From about two billion in 1960 and around three billion in 1995, the world's population has rapidly grown to its current level of approximately 6.7 billion—and counting. The world's population increases by more than eighty to ninety million each year, and is projected to reach ten billion by the year 2050, unless significant increases in the use of birth control occur worldwide.

The opinion that population is the fundamental cause of environmental pollution is arguable. There are those, however, who hold that this opinion has merit. Garrett Hardin (1968), for example, writing in *Tragedy of the Commons*, argues that there wouldn't be large problems with air pollution and land degradation if there weren't so many people. In illustrating his point, he uses the example of the "commons": a resource (like air) owned by no one but utilized by many (a pasture, in the case of his example). He argues that, in a commons, there is no incentive to be conserving of resources, for he who is conserving "loses," and he extends the analogy to people's right to have as many children as they wish (Hardin 1986).

In contrast to Hardin's view, Barry Commoner holds that rather than too many people being blamed for environmental pollution, the real root cause of our environmental pollution problems is the inappropriate use of technology (Goldfarb 1989). He suggests that, if we used resources more efficiently and cleanly, there wouldn't be problems, even though there are some five billion (currently over six billion) of us.

Whenever we discuss differing opinions on any topic, it is not too long before we run across an even more extreme position than those previously discussed. In this discussion about population being a possible contributor to environmental pollution, one such extreme opinion is expressed by noted economist Julian Simon. Simon says that there are not too many people and that the quality of life will only improve as the human population increases. This is because he believes our supply of resources is virtually infinite, by virtue of an infinite capacity to substitute one resource for another. Simon said that it was only human ingenuity that limited our use of available resources as substitutes for exhausted resources. Hence, he argued that our condition will improve as population increases because there will be more clever people who will be able to arrive at innovative solutions (Simon 1980).

Along with the rapidly increasing population growth is a trend associated with population pressure: *population concentrations.* These population concentrations tend to exacerbate the pollution problem by increasing the level of pollution and the accompanying environmental degradation—increased deforestation, desertification, and soil erosion. Not only do population concentrations increase pollution, but they also change the nature of the pollutants, by producing pollutants that become hazardous because of the sheer quantities involved, as is the case with all types of municipal wastes.

In recent decades, population concentrations within different regions of the world have demonstrated a certain dynamism—a trend toward increasingly rapid change. This problem is most apparent in developing countries, those that have not only rapid increases in overall population (compared to developed countries) but also large populations that have been lured by economic opportunities to more congested areas. These congested areas typically are responsible for consuming more natural resources than rural areas and produce mountains of waste products. Not only is more waste produced per capita than in rural areas, but in urban areas, this waste is also more hazardous. It places strains on the city's infrastructures, increasing demand on the city's ability to absorb and handle wastes. Pressure is also increased on the nonurban residents (the agrarian sector) to produce more food on less land, straining soil productivity and aggravating the agricultural pollution problem. The level of urbanization in developing countries increased at a rate of four to one as compared to the level of urbanization in developed countries from 1970 to 1995.

Another pollution-related problem associated with overpopulation and the unprecedented rush toward urbanization by developed and undeveloped countries can be seen in the difference in views on the pollution problem itself. Many industrialized countries have experienced industrialization for more than a century and, more importantly, have also felt the effects of pollution for a longer period of time. These developed countries have been modifying their polluting activities for several years, working to clean up existing pollution. Developing countries, where population is on the increase and economic resources limited, are inadvertently escalating activities that lead to pollution. Pollution and its effects take a back seat to more pressing concerns, namely, the daily struggle to survive. However, the problem is more than attitudinal—it is not just a free-will expression of the view that survival is number 1, so pollution is not a concern. Historically, pollution problems are generally exacerbated by lack of financial resources and other economic problems.

One measure of the impact of global population is the fraction of the basic energy supply of all terrestrial animals directly consumed, co-opted, or eliminated by human activity. According to Peter Vitousek et al. (1986), this figure has reached 40 percent. This level of exploitation could double as population growth meets expectations.

So, what is the bottom line on the impact of a growing population on environmental pollution (and the environment in general)? Well, using science and mathematics we can resort to a well-known equation, $I = PAT$, to help us answer this question. The impact (I) of any population can be expressed as a product of three characteristics: the population's size (P), its affluence or per-capita consumption (A), and the environmental damage (T) inflicted by the technologies used to extract resources and supply each unit of consumption (Daily and Ehrlich 1992).

Thus, human impact on the environment is a function of the population size, the level of resource use, and the environmental impacts associated with obtaining and using those resources. This formulation suggests that no one factor alone is responsible for our environmental problems. The relative importance of the various factors will differ depending on the particular problem (Daily and Ehrlich 1992).

DEVELOPMENT

Increasing worldwide population without a corresponding growth in development is virtually impossible—though the development is not always as we would envision it. As a case in point, consider the following comment by Walter Reid et al. (1988): "The Third World is littered with the rusting good intentions of projects that did not achieve social and economic success; environmental problems are now building even more impressive monuments to failure in the form of sediment-choked reservoirs and desertified landscapes" (1).

One thing is certain: economic development cannot proceed without natural resources. The environmental degradation problem develops when natural resources are mismanaged, misused, wasted, and then exhausted. When this occurs, development leads to degradation—not only of the economic well-being of the inhabitants but of the environment as well. Soil, fresh water, and air are all, in one way or another, degraded.

Case Study 12.4: Transnational Corporations and Environmental Pollution

Transnational corporations drive the global economy and marketplace. These massive corporations, while providing the monetary base for industry, mining, distribution, technical knowledge, agriculture, and trade, are also heavily involved in most of the world's serious environmental crises. Considering the record of corporate-generated pollution problems in the United States, and what environmentalists have had to do to prevent contamination, pollution, and ecological abuse, this single issue, with its potential of thousands of incarnations, may be the biggest environmental problem of the new century.

Globalization expands environmentally hazardous activities around the world. Corporations commonly shift industrial practices that draw heavy environmental focus in the United States to third-world countries, dumping into these environments the wastes they are prohibited from releasing into our own. International trade and investment undermines environmentalists' attempts to curb abuse; cases supported in the United States can be overturned by global organizations for financial reasons.

Global economics and industry have helped turn environmental pollution into a global problem—one that must necessarily be solved globally as well. Environmentalists will not be able to fight this battle alone. A global community, with members as diverse as scientists, artists, unions, elected officials, lawyers, consumer and environmental activists, as well as the everyday people in the affected countries, must work to solve this issue across borders and divisions, both physical and social ones (Karliner 1998).

DEVELOPMENT AND SOIL DEGRADATION

Soil degradation takes on many forms. Serious soil erosion, which commonly occurs in most of the world's important agricultural regions (i.e., from overgrazing of animals; planting of a monoculture; row cropping; tilling or plowing; crop removal; and land use conversion), is one form of degradation. Another form, just as serious as erosion, develops from our waste disposal practices. In the past, throw-away societies were able to dispose of unwanted materials and wastes with little impact on the environment. Most of the waste products of the past were biodegradable, and Earth's natural systems were able to self-purify environmental mediums, including soil to an extent. The problem today is that we are introducing thousands of substances into our environment that are not biodegradable; instead, they are quite persistent and hang around for a very long time.

Did You Know?

According to Barry Commoner (1971), everything must go somewhere. There is no "waste" in nature, and there is no "away" to which things can be thrown.

DEVELOPMENT AND FRESHWATER DEGRADATION

Poor management, lack of adequate conservation, pollution, and rapid local increases in demand create localized shortages of potable water worldwide. In developing nations, the problem is even worse; only about half of the people have access to safe drinking water. In the Western world we have come to believe that the waterborne intestinal diseases that killed so many in the past are a problem of the past. This is not the case, however. In developing nations, an estimated ten million deaths each year result from waterborne intestinal diseases. Even in the United States, recent localized outbreaks of illness and death from *Cryptosporidium*- and *E. Coli*-contaminated water supplies have forced communities to recognize that we all are at risk at times.

These problems are not limited to developing nations. Surface and underground water supplies in industrial nations are being polluted by industrial and municipal wastes and by surface runoff from urban and agricultural areas. Heavy demands for water by industry, agriculture, and municipalities are rapidly depleting groundwater supplies.

DEVELOPMENT AND ATMOSPHERIC AIR DEGRADATION

We have all read news accounts on El Niño, El Niña, global climate change due to global warming, the greenhouse effect, acid rain, and damage to the ozone layer. The terms have become part of our common vocabulary. However, no consensus on what these issues will really mean to us has yet been reached. Note the two typical responses that we usually hear to arguments related to atmospheric degradation: One side argues that "the sky is falling" or "the world as we know it is doomed by pollution" or some other scare tactic to grab our attention and to provide a lead-in to some political statement. The other side disputes the doom-and-gloom reports and states that all such statements are simply hyperbole and worse—that the real truth of the matter is that our environment is doing just fine, thank you very much.

What is the truth? Are we destroying our environment—or not?

The truth lies somewhere in the middle, of course. We are affecting our environment—of this there is and can be little doubt. Doomed? Not exactly. Every problem has a solution. What we need is the motivation to find solutions to environmental degradation problems and then apply them on a global scale.

Sounds simple enough, doesn't it?

Just a piece of cake. It's not rocket science, we simply need to do it—to do whatever it is that is needed to prevent pollution and protect the global environment. Again, simply, we need to do something.

However, in response to the alarmists—the producers of needless warnings—we warn about shooting from the hip. Easy, feel-good science (versus real science) and bandage solutions are not the answer. We should be aware of not only the heavily inflated pronouncements of "gloom and doom" or of the "nothing's wrong" but also any quick fixes we may be tempted to implement to "take care of" the problem. Too often quick fixes simply slap a coat of paint over the layers of rust—allowing us to walk away with the warm, fuzzy glow of accomplishment without correcting the underlying problems. All remediation and mitigation should be accomplished based on the sound principles of science and an ounce of common sense.

Extremists, who in the past have sounded the warning calls of environmental degradation so stridently (even to the point of spiking trees, monkey wrenching equipment, or burning the forest to the ground), may inadvertently create problems as well as serve the useful purpose of drawing national attention to environmental problems. Knowing practitioners of science learn to quiet the strident and radical alarmist by simply hearing their views, then ignoring them completely, quelling conflict or confrontation. These perceptive people of science have found the magic off switch: when you take away or ignore the argument, there is no argument.

The alarmists are correct when they state that humankind and their practices are adversely affecting our environment. You don't need an Einstein mentality to see the proverbial writing on the wall. Every day, many toxic gases and fine particles enter the air we breathe and pose hazards to our health; some of these air pollutants cause cancer, genetic defects, and respiratory disease, as well as exacerbating existing medical problems such as heart and lung disease.

Gaseous chemical substances such as nitrogen and sulfur oxides, ozone, and other air pollutants from fossil fuels inflict damage in many countries throughout the world. Ozone and acids of nitrogen and sulfur are damaging forests, crops, soils, lakes, streams, coastal waters, and man-made structures.

The Earth's protective ozone layer is being depleted by CFCs and other pollutants. In some regions, this depletion is beginning to increase the amount of harmful UV radiation that reaches the Earth's surface, which could cause skin cancers and cataracts, damage immune systems, disrupt marine food chains, reduce crop yields, and cause significant climate change.

The amount of carbon dioxide entering the Earth's atmosphere is increasing. Combustion of fossil fuels is the primary culprit, but other heat-absorbing atmospheric gases are raising global temperatures. This, in turn, could alter weather patterns, worsen storms, disrupt agriculture, and destroy natural systems. Another problem with global warming is sea-level rise. Glacier and ice cap melting may release large quantities of water that could lead to the flooding of many low-lying regions.

These changes are now becoming so widespread, and the individual pieces of the problem so complex and interwoven, that the only way to approach the problem is on a global scale. Some sort of accord is needed to uniformly deal with these issues in both industrial and developing nations; otherwise, solving the problem is only postponed, increasing the risk to all.

So, when it comes to pollution, what is the answer?

Awareness is the answer—or at least a good place to start. One thing is certain: we cannot live without soil, water, and air. Likewise, we cannot live if these media are polluted beyond use. We must take steps to ensure that we have a future—one we can look forward to.

Consider Carol Christensen's (1982, 2) quote on the subject of the future: "When it comes to the future, there are three kinds of people: those who let it happen, those who make it happen, and those who wonder what happened." When it comes to pollution and its impact on the future and on people, this text is designed to provide the information needed to equip the reader with the information he or she needs to be one of those who make it happen.

Political Geography

Political geography, sometimes called geopolitics, is an important subset of human geography because they both study human systems, which are important in understanding the constant state of spatial organizing and reorganizing of the land to fit the needs of humans. Generally, the spatial organizing and reorganizing referred to and briefly discussed here has to do with distribution by not only gender and ethnicity (i.e., the cultural aspects of a group of people) but also resources. Although politics does not create resources that, in turn, generate employment, prosperity, and wealth, politics distributes (sometimes fairly but more often unfairly) those resources. The primary concerns of the subset can be summarized as the interrelationships between people, state, and territory. Or, as Mark Blacksell (2005) points out, political geography is about the forces that go to shape the world we inhabit and how they play themselves out in the landscape across the globe. In particular, modern political geography considers the following:

- how and why states are formally and informally organized into regional groupings;
- the relationship between states and former colonies;
- the relationships between states including international trades and treaties;
- the relationship between a government and its people;
- the functions, demarcations, and governance of boundaries along with imagined geographies and the resulting political implications;
- the influence of political power on geographical space; and
- electoral geography, that is, the study of election results.

Cultural Geography

Cultural geography is another subset of or within human geography. Cultural geography focuses on describing and analyzing the ways language, religion, economy, government, and other cultural phenomena vary or remain constant, from one place to another, and on explaining how humans function spatially (Jordan-Bychkov, Mona, and Rowntree 2006). David Sibley et al. (2005) explain that cultural geography is actually a trinity of space, knowledge, and power situated right at the heart of the ways that contemporary cultural geographers make sense of society. Cultural geography looks at the differences, or forms of difference, and the material cultures of groups but also at the ideas that hold them together, that make them coherent. The ingredients of coherency can and often do include styles of clothing, ornaments, and lifestyles, and these are blended with different "worldviews," different priorities, different belief systems, and different ways of making sense of the world with its daily, weekly, monthly, and ongoing happenings (Crang 1998). In particular, modern cultural geography considers the following:

- the effect globalization has had (is having) on cultural convergence;
- processes such as westernization, Americanization, Islamization, and others;
- theories of cultural assimilation or cultural hegemony via cultural imperialism;
- studies in differences in way of life, attitudes, languages, practices, institutions, and structures of power;
- study of cultural landscapes;
- study of colonialism, postcolonialism, internationalism immigration, emigration, and ecotourism; and

- modern areas of study including children's geography, behavioral geography, sexuality and space, feminist geography, and tourism geography.

URBANIZATION

Urbanization is the process by which physical growth of urban areas from rural areas results as people immigrate into, live in, and work in the city—urbanization is attributed to growth of cities (UM 2006). This movement of people from rural to urban areas can result in concentrations of people in large, dense, diverse settlements. Some of these large, urban areas can and do develop into megacities—metropolitan areas with over ten million people. Similarly, urbanization can cause the development of megalopolis regions (extensive metropolitan areas) such as that commonly attributed to the U.S. East Coast area from Boston, Massachusetts, to New York. Another form or type of urbanization is known as conurbation, whereby a number of fairly close towns, villages, or cities are connected by transportation systems that link each separate entity into one region. A good example of conurbation is the Hampton Roads, Virginia, region. The Hampton Roads region is connected both by land and sea. The land area includes several cities, towns, and counties linked by a highway and tunnel/bridge system. Hampton Roads is also linked by water. This area contains one of the world's biggest natural (and ice-free) harbors.

People are drawn to urban areas for a variety of reasons. These reasons can be categorized into two factors: push and pull. In push factor urbanization, a rural area may have an overabundance of available (surplus) labor for a limited number of jobs. Moreover, the rural political conditions may be unbearable to some, pushing them to move into an urban area. Some people may also feel that a nearby rural area is being oppressed by an urban area. For example, when a city grows exponentially, it is in dire need of more space; thus, it will often look to annex rural regions, officially or unofficially. Probably one of the most salient push factors occurring at the present time is the price of gasoline and diesel fuel. In 2008, when the price of a gallon of gasoline exceeded four dollars, many people with long commutes from their rural setting to their urban workplace found themselves in a real financial bind. Some of these people switched from gas-guzzling, SUV-type automobiles to smaller, more efficient cars. Others chose to carpool to save money on gas. And then there were others who just pulled up stakes, so to speak, and left the rural area to move to urban areas to be closer to work, thus reducing the distance of the commute and increasing the ease of getting to work. Also, during 2008, and somewhat at present, the high cost of gasoline and diesel fuels was augmented with one of the worst recessions in U.S. history. Thus, the combination of high fuel prices and lack of jobs in rural areas pushed many people to move to the urban areas where they might have a better chance of obtaining employment and also a shorter distance to commute.

In the summer of 2009, I found it rather ironic and interesting that, through research and personal conversations with people in the Midwestern and West Coast regions of the United States, one of the most stable occupations appeared to be in rural areas; employment was steady in almost all aspects of agriculture. Not only was agriculture providing steady work, but it was also seeking additional seasonal workers to plant, care for, and harvest crops. In addition, during this same time frame, again the summer of 2009, every state in the Union was verging on bankruptcy or had serious budget shortfalls, except one: Nebraska. Nebraska is a conservative, well-managed farm state that produces food and food products. One thing is certain: even with high fuel costs and high unemployment, when it comes to purchasing the necessities of life—air, water, and food, even though air is free (at the present time), water and food are not—people will spend their money on what they need to sustain themselves. Thus, when the choice comes to new clothes, furniture, cars, or food, food will win out almost every time—especially if the purchaser and his or her family are hungry.

In my conversations with rural folks in the Midwest and other western locales, this ironic twist in the comparison between the flight of people from rural to urban settings and the steadfast stay-put folks who farm the rural lands is indicative of what many rural folks view as a possible future, that is, according to many of them, it might be a good idea to go back to the basics of providing for the necessities of life and to ignore the frivolous. Again, at least this is the view of many farmers and rural people I spoke to. One farm lady in South Dakota told me that, when she has a need for food, she simply goes to the back 160 acres and picks crops, to the chicken coop for eggs or fryers, and to the stable area to pick a swine or beef to slaughter. I asked about the harsh winter in the area and therefore the impossibility of growing crops during that season. She replied by showing me her pantry, which was stacked wall to wall and floor to ceiling with home-canned goods. While pointing to jam-packed shelves of canned goods she was humming that famous western; out loud she paraphrased a bit and sang: "And a country girl can survive."

On the pull side of the urbanization equation are several factors. For instance, industrialization has continued to be the big draw for those in rural areas to immigrate into urban areas. The draw, of course, is the possibility of obtaining a good job where hard work and imagination may lead to a fortune made and the possibility of social mobility. This type of thinking in the recent past certainly was a prime motive for people leaving rural areas for urban areas. At the current time, with the world in worldwide recession, with the high cost of fuel and the scarcity of high-paying positions, it is much more difficult to make a living in some urban areas—it is not uncommon for workers in the cities today to work two or three additional, part-time jobs just to scrape by. In addition, many of the urban areas are hurting financially because of the economic downturn and thus have had to raise taxes on just about everything. This increased cost-of-living expense just adds to the difficulty of finding a well-paying job.

Another urban draw is local government, which may be viewed positively by outsiders. Local religious customs and tolerance may also be draws. Many people also view the presence of financial institutions, marketplaces, and entertainment centers as incentives to leave their rural environment for the urban setting. According to the Associated Press (2008), the UN projected that half the world population would live in urban areas at the end of 2008.

SUMMARY OF KEY CULTURAL GEOGRAPHY ASPECTS

- The establishment of the state of Israel represented a return of a dispersed religion to its hearth region.
- Hindu and Urdu are essentially the same language written in different scripts.
- The acceptance of Christianity as the state religion of the Roman Empire initiated its hierarchical diffusion in Europe.
- Standard French is based on the dialect of the Paris region.
- Vernacular speech is the everyday native language or dialect of a locality.
- Language is rather more important than religion in shaping a culture's economic, social, and political institutions.
- The roots of modern English may be traced back to various southern European proto-Germanic dialects.
- The separation of the Roman Empire into western and eastern halves also served to divide the Christian religion.
- Ethnic and folk cultural traits tend to separate peoples; popular culture tends to reduce differences.
- Four major U.S. hearths of folk culture origin and dispersal are New England, southeastern Pennsylvania, Chesapeake Bay, and the Georgia-Carolina coast.
- Where popular cultures dominate, regional differences are reduced and the cultural landscape acquires increased uniformity.
- In modern America, folk culture is more an individual expression than a group characteristic.
- A particular interest of folk geography is the spatial association and co-variation of culture and environment.
- The East Coast North American hearths received their cultural concepts and artifacts through elevation diffusion.
- Vernacular housing is traditional in design but built without formal plans or blueprints.
- Log cabins were introduced into America by Finnish and German immigrants.
- Folk culture is controlled by tradition.
- Bluegrass music is a derivative of Scottish bagpipe and church congregational music.
- The smallest distinctive item of culture is a called culture trait.
- An assemblage of interrelated culture traits comprises a culture.
- Cultural convergence implies that world populations increasingly share a common technology.

Chapter Review Questions

12.1. Define population ecology.

12.2. What is the main axiom of population ecology?

12.3. When measuring populations, the level of ________ or ________ must be determined.

12.4. The arrival of new species to a population from other places is termed ________.
12.5. ________ studies the structure and dynamics of animal and plant communities.
12.6. ________ produce aggregation, the result of response by plants and animals to habitat differences.
12.7. ________ is the upper limit of population size.
12.8. The ________ is the actual maximum population density that a species maintains in an area.
12.9. ________ factors affect the size of populations.
12.10. ________ is the observed process of change in the species structure of an ecological community over time.

Notes

1. Much of the material in this section is from F. R. Spellman and N. Whiting, *Environmental Science and Technology*, 2nd ed. (Lanham, Md.: Government Institutes Press, 2006).
2. Much of the information and data in this section is from the U.S. Energy Information Administration (EIA)'s "Renewable Energy Trends," 2004, www.eia.doe.gov/cneaf/solar.renewables/page/trends/rentrends04.html (accessed June 12, 2009); and EIA's "How Much Renewable Energy Do We Use?" 2007, http://tonto.eia.doe.gov/energy_in_brief/renewable_energy.cfm (accessed June 12, 2009).
3. Much of the information in this section is from U.S. Department of Energy (DOE), Office of Energy Efficiency and Renewable Energy (EERE), "History of Wind Energy," 2005, www1.eere.energy.gov/windandhydro/printable_versions/wind_history.html (accessed June 14, 2009).
4. Based on information from U.S. Department of Energy (DOE), *Renewable Energy: An Overview* (Washington, D.C.: Author, 2001).
5. Information in this section is from U.S. Department of Energy (DOE), Office of Energy Efficiency and Renewable Energy (EERE), "Hydrogen, Fuel Cells and Infrastructure Technologies Program," 2008, www1.eere.energy.gov/hydrogenandfuelcells/production/basics.html (accessed November 18, 2009).

References and Recommended Reading

Abedon, S. T. 2007. Population ecology. www.mansfield.ohio-state.edu/~sabedon/campbl50.htm#population_ecology (accessed February 27, 2007).

Allee, W. C. 1932. *Animal aggregations: A study in general sociology*. Chicago: University of Chicago Press.

Associated Press. 1989. Town evacuated after acid spill. *Lancaster New Era*. Lancaster, PA, September 6.

———. 2008. UN says half the world's population will live in urban areas by end of 2008. www.iht.com/aritcles/ap/2008/02/26/news/UN-GEN-UB-Growing-Cities.php (accessed June 24, 2009).

Berryman, A. A. 1981. *Population systems: A general introduction.* New York: Plenum Press.
———. 1993. Food web connectance and feedback dominance, or does everything really depend on everything else? *Oikos* 68:13–185.
———. 1999. *Principles of population dynamics and their application.* Cheltenham, UK: Stanley Thornes.
———. 2002. *Population cycles: The case for trophic interactions.* New York: Oxford University.
———. 2003. On principles, laws and theory in population ecology. *Oikos* 103:695–701.
Blacksell, M. 2005. *Political geography.* New York: Routledge.
Bonner, J. T. 1965. *Size and cycle.* Princeton, N.J.: Princeton University Press.
Brahic, C. 2009. Fish "an ally" against climate change. *New Scientist.* January 16. www.newscientist.com/article/dn16432-fish-an-ally-against-clmate-change.html (accessed November 18, 2009).
Brookfield, H. C. 1989. Sensitivity to global change: A new task for old/new geographers. Norma Wilkinson Memorial Lecture, University of Reading, UK, May.
Brundtland, G. H. 1987. *Our common future.* New York: Oxford University Press.
Caldeira, K., and M. E. Wickett. 2003. Anthropogenic carbon and ocean pH. *Nature* 425 (6956): 365.
Calder, W. A. 1983. An allometric approach to population cycles of mammals. *Journal of Theoretical Biology* 100:275–82.
———. 1996. *Size, function and life history.* Mineola, N.Y.: Dover Publications.
Campbell, N. A., and J. B. Reece. 2004. *Biology*, 7th ed. San Francisco: Pearson/Benjamin Cummings.
Christensen, C. 1982. In *Making It Happen: A Positive Guide to the Future*, ed. J. M. Richardson. Washington, D.C.: U.S. Association for the Club of Rome.
Clark, L. R., P. W. Gerier, R. D. Hughes, and R. F. Harris. 1967. *The ecology of insect populations.* London: Methuen.
Cohen, J. E. 2003. Human population: The next half century. *Science* 302:1172–74.
Colyvan, M., and L. R. Ginzburg. 2003. Laws of nature and laws of ecology. *Oikos* 101:649–53.
Commoner, B. 1971. *The closing circle: Nature, man, and technology.* New York: Knopf.
Crang, M. 1998. *Cultural geography.* New York: Routledge.
Daily, G. C., and P. R. Ehrlich. 1992. Population, sustainability and Earth's carrying capacity. *BioScience* 42:761–71.
Damuth, J. 1981. Population density and body size in mammals. *Nature* 290:699–700.
———. 1987. Interspecific allometry of population density in mammals and other animals: The independence of body mass and population energy-use. *Biological Journal of the Linnean Society* 31:193–246.
———. 1991. Of size and abundance. *Nature* 351:268–69.
Debres, K. 2005. Burgers for Britain: A cultural geography of McDonald's UK. *Journal of Cultural Geography* 22 (2): 115–88.
Enger, E., J. R. Kormelink, B. F. Smith, and R. J. Smith. 1989. *Environmental science: The study of interrelationships.* Dubuque, Iowa: William C. Brown Publishers.
Fenchel, T. 1974. Intrinsic rate of natural increase: The relationship with body size. *Oecologia* 14:317–26.
Ginzburg, L. R. 1986. The theory of population dynamics: 1. Back to first principles. *Journal of Theoretical Biology* 122:385–99.
Ginzburg, L. R., and M. Colyvan. 2004. *Ecological orbits: How planets move and populations grow.* New York: Oxford University Press.
Ginzburg, L. R., and C. X. J. Jensen. 2004. Rules of thumb for judging ecological theories. *Trends in Ecology and Evolution* 19:121–26.
Goldfarb, T. D., ed. 1989. Is population control the key to preventing environmental deterio-

ration? In *Taking sides: Clashing views on controversial environmental issues*, issue 6. Guilford, Conn.: Dushkin Publishing.

Greenberg, N. 2005. Proximate and ultimate causation. *Deep Ethology*, 22 February.

Haemig, P. D. 2006. Laws of population ecology. *Ecology Info* 23.

Hardin, G. 1968. *Tragedy of the commons*. Washington, D.C.: American Association for the Advancement of Science.

———. 1986. Cultural carrying capacity: A biological approach to human problems. *BioScience* 36:599–606.

Hickman, C. P., L. S. Roberts, and F. M. Hickman. 1990. *Biology of animals*. St Louis: Time Mirror/Mosby College Publishing.

Holdren, J. P. 1991. Population and the energy problem. *Popul. Environm.* 12:231–55.

Hubbell, S. P., and L. K. Johnson. 1977. Competition and next spacing in a tropical stingless bee community. *Ecology* 58:949–63.

Jones, R. C. 2006. Cultural diversity in a "bi-cultural" city: Factors in the location of ancestry groups in San Antonio. *Journal of Cultural Geography* (March).

Jordan-Bychkov, T. G., D. Mona, L. Rowntree. 2006. *The human mosaic: A thematic introduction to cultural geography*, 11th ed. New York: HarperCollins College.

Karliner, J. 1998. Earth predators. *Dollars and Sense* (July/August).

Keller, E. A. 1988. *Environmental geology*. Columbus, Ohio: Merrill.

Krebs, R. E. 2001. Scientific laws, principles and theories. Westport, Conn.: Greenwood Press.

Kuhlken, R. 2002. Intensive agricultural landscapes of Oceania. *Journal of Cultural Geography* 19 (2): 161–95.

Liebig, J. 1840. *Chemistry and its application to agriculture and physiology*. London: Taylor and Walton.

Lotka, A. J. 1925. *Elements of physical biology*. Baltimore: Williams & Wilkens.

Lund, J. W. 2007. Characteristics, development and utilization of geothermal resources. *Institute of Technology* 28 (2): 1–9.

Malthus, T. R. 1798. *An essay on the principle of population*. London: J. Johnson.

Masters, G. M. 1991. *Introduction to environmental engineering and science*. Englewood Cliffs, N.J.: Prentice Hall.

Miller, G. T. 1988. *Environmental science: An introduction*. Belmont, Calif.: Wadsworth.

National Renewable Energy Laboratory (NREL). 2009. Concentrating solar power. www.nrel.gov/learning/re_csp.html (accessed June 13, 2009).

Odum, E. P. 1983. *Basic ecology*. Philadelphia: Saunders College Publishers.

Painter, J. 1995. *Politics, geography and political geography: A critical perspective*. London: Arnold.

PennWell Corporation. 2007. Estimated reserves by country. *Oil and Gas Journal* 105 (48) (December 24).

Pianka, E. R. 1988. *Evolutionary ecology*. New York: HarperCollins.

Population.com. 2009. World's population. www.population.com/ (accessed June 5, 2009).

Reid, W., et al. 1988. *Bankrolling successes*. Washington, D.C.: Environmental Policy Institute and National Wildlife Federation.

Searchinger, T., et al. 2008. Use of U.S. croplands for biofuels increases greenhouse gases through emissions from land-use change. *Science* 319 (5867): 1238–40.

Sharov, A. 1992. Life-system approach: A system paradigm in population ecology. *Oikos* 63: 485–94.

———. 1996. *What is population ecology*? Blacksburg, Va.: Department of Entomology, Virginia Tech. University.

———. 1997. Population ecology. www.gypsymoth.ento.vt.edu~Sharov/population/welcome (accessed February 28, 2007).

Sibley, D., et al., 2005. *Cultural geography: A critical dictionary of key ideas.* London: I. B. Tauris.

Simon, J. L. 1980. Resources, population, environment: An oversupply of false bad news. *Science* 208:1431–37.

Sinha, A. 2006. Cultural landscape of Pavagadh: The abode of mother goddess Kalika. *Journal of Cultural Geography* 23:89–103.

Smith, R. L. 1974. *Ecology and field biology.* New York: Harper and Row.

Spellman, F. R. 1996. *Stream ecology and self-purification.* Lancaster, Pa.: Technomic Publishing.

———. 2009. *The science of environmental pollution,* 2nd ed. Boca Ration, FL: CRC Press.

Spellman, F. R., and N. Whiting. 2006. *Environmental science and technology,* 2nd ed. Lanham, Md.: Government Institutes Press.

Tomera, A. N. 1990. *Understanding basic ecological concepts.* Portland, Maine: J. Weston Walch, Publisher.

Turchin, P. 2001. Does population ecology have general laws? *Oikos* 94:17–26.

———. 2003. *Complex population dynamics: A theoretical/empirical synthesis.* Princeton, N.J.: Princeton University Press.

Union of Concerned Scientists (UCS). 2008. Land use changes and biofuels. October 17. www.ucsusa.org / clean_vehicles / technologies_and_fuels / biofuels / Land-Use-Changes-and-Biofuels.html (accessed June 9, 2009).

United Nations Environment Program (UNEP). 1989. *UNEP North American News* 4 (4) (August).

University of Michigan (UM). 2006. Urbanization and global change. http://globalchange .umich.edu/globalchange2/current/lectures/urban_gc (accessed June 24, 2009).

U.S. Department of Energy (DOE). 2001. *Renewable energy: An overview.* Washington, D.C.: Author.

———. 2009. Fossil fuels. www.energy.gov/energysources/fossilfuels.htm (accessed June 10, 2009).

U.S. Department of Energy (DOE), Office of Energy Efficiency and Renewable Energy (EERE). History of wind energy. 2005. www1.eere.energy.gov/windandhydro/printable_ versions/wind_history.html (accessed June 14, 2009).

———. 2006. Geothermal technologies program: A history of geothermal energy in the United States. www1.eere.energy.gov/geothermal/printable_versions/history.html (accessed June 18, 2009).

———. 2008. History of hydropower. www.eere.energy.gov/windandhydro/printable_ver sions/hydro_history.html (accessed June 15, 2009).

———. 2009. Solar energy technologies program. www1.eere.Energy.gov/solar/printable_ versions/about.html (accessed June 13, 2009).

U.S. Energy Information Administration (EIA). 2004. Renewable energy trends. www.eia .doe.gov/cneaf/solar.renewables/page/trends/rentrends04.html (accessed June 12, 2009).

———. 2007. How much renewable energy do we use? http://tonto.eia.doe.gov/energy_ in_brief/renewable_energy.cfm (accessed June 12, 2009).

U.S. Environmental Protection Ageny. 1989. *Glossary of environmental terms and acronyms list.* Washington, DC: Environment Protection Agency.

———. (EPA). 2009. Ozone science: The facts behind the phase-out, ozone depletion. ww-w.epa.gov/ozone/science/sc_fact.html (accessed November 18, 2009).

U.S. Geological Survey (USGS). 1999. Hawaiian volcano observatory. http://hvo.wr.usgs.gov/ volcanowatch/1999 (accessed March 1, 2007).

U.S. Library of Congress. 2009. The world's first hydroelectric power plant began operation. www.americaslibrary.gov/cgi-bin/page.cgi/jb/gilded/hydro_1 (accessed June 15, 2009).

Verhulst, P. F. 1838. Notice sur la loi que la population suit dans son accrossement. *Corr. Math. Phys.* 10:113–21.

Vitousek, P. M., P. R. Ehrlich, A. H. Ehrlich, and P. A. Matson. 1986. Human appropriation of the products of photosynthesis. *BioScience* 36 (6): 368–73.

Volterra, V. 1926. Variazioni e fluttuazioni del numero d'indivudui in specie animali conviventi. *Mem. R. Accad. Naz. die Lincei Ser.* 6 (2).

Wilson, E. 1989. The value of biodiversity. *Sci. Am.* (September).

Wilson, E. O. 2001. *The future of life.* New York: Knopf.

Winstead, R. L. 2007. Population regulation. http://nsm1.nsm.iup.edu/rwinstea/popreg.shtm (accessed February 28, 2007).

World Commission on Environment and Development (WCED). 1987. *Our common future.* New York: Oxford University Press.

Zelinsky, W. 2004. Globalization reconsidered: The historical geography of modern Western male attire. *Journal of Cultural Geography* 22:83–134.

APPENDIX A

Worldwide Industry and Economic Development

> Standing at the Notch in the Badlands of South Dakota, on a clear day one can easily see the sacred Black Hills, more than 30 miles in the distance. There are those days, however, when it is impossible to see that striking black silhouette because the surrounding area is covered by a mask of pollution. And, when this is the case, I can't help myself from replaying that same refrain from deep within my bowels to the words formed at my lips: Salt, meet wound. Insult, greet injury. Pollution, say hello to the environment.
>
> —Frank R. Spellman (2009)

When describing an economic system we can say it is a system of production, distribution, and consumption of an economy. An economic system can also be described as the set of principles and techniques by which problems of economics are addressed, such as the economic problem of scarcity through allocation of finite productive resources (New Encyclopedia Britannica 2007).

In regard to industry, it is based on transportation and labor costs. At present, there are five main means of industrial transportation, each with its advantages and disadvantages: truck, train, plane, pipeline, and ship. Today, the importance of these means of transport to our economy and to basic industry as a whole has become more apparent to many more people in the real world. This new awareness came about in 2007, peaked in 2008, and remained pretty much the same since then because of the increase in gasoline, diesel, jet fuels, and heating oil prices. For example, long-haul truckers transporting goods from the West Coast to the East Coast in July of 2008 were paying more than five dollars per gallon for diesel fuel. Obviously, when the price of fuel goes up, everything else follows; thus, prices for any commodity that is shipped increases in cost. Consumers end up paying a premium for the goods they need to maintain their lifestyles. Thus, the increase in costs for just about everything, caused by increased fuel costs, puts pressure on consumers to make more money to cover increased costs. This in turn puts pressure on employers to increase prices to cover not only the increased price of fuel but also the increased cost of labor. In the end, the reality is that businesses will often downsize before raising the wages of its employees. This is the case because, for many businesses, there is no other alternative except bankruptcy, which is normally viewed by the business entity as not a viable option.

The bottom line on the future of worldwide industry and economics is clear. We must develop renewable energy sources to replace dwindling supplies of expensive hydrocarbon fuels. In addition, regardless of the fuel type we make as the alternative to hydrocarbon fuels, the fuel must not only be renewable but also as close to pollution free as possible. One not need be a Nobel laureate in economics to understand the basics of transportation and its importance to industry and economies. If the products that fuel our transportation systems can't be obtained and at a relatively low rate, then the stability of the economic systems as we know them today is in serious doubt. The question beginning to be asked is, have today's prosperous, developed countries reached and passed the apogee of their success; are they sliding down on that slope of regression to repeat the past—to remain in that lowered position forever?

Describing Development

All kinds of theoreticians, many of them well meaning, have spent years in attempting to list and describe those parameters that are indicative of a nation's successful development. Many presumptions can be made. If a nation sits on a veritable pile of uranium, gold, silver, or platinum, is it successfully developed? If a nation sits on a reservoir of crude oil, is it successfully developed? If a nation is literally the breadbasket of the world, is it successfully developed? If a nation sits on or is surrounded by easily accessible safe potable fresh water, is it successfully developed? If a nation has the best medical care system possible, whereby any and all patients receive quick and reliable treatment at no cost to them, is it successfully developed? If a nation is equipped with enough nuclear weapons to blow the world to smithereens many times over, is it successfully developed? If a nation has a high life expectancy, literacy, education, and standard of living, is it successfully developed?

In sorting through the list above, if you chose the last item as an indicator of a nation's achievement of successful development, then you are in agreement with the United Nations (UN). The UN uses what it calls the Human Development Index (HDI) to indicate whether a country has successfully achieved development. The UN feels there is a strong correlation between having a high HDI score (i.e., high life expectancy, literacy, education, and standard of living rate) and a prosperous economy. Moreover, the UN points out that HDI accounts for more than income or prosperity. The HDI takes into account how income is turned into education and health opportunities and therefore into higher levels of human development (UNDP 2008).

In data released December 18, 2008, compiled on the basis of data from 2006, the UN listed the countries that make up their HDI list. The listing is rated by highest to lowest HDI scores. In table A.1, the top thirty countries on the HDI are listed. This is followed by a brief overview of each country's economy.

Table A.1. United Nations Human Development Index (top thirty countries)

Rank	*Country*
1	Iceland
2	Norway
3	Canada
4	Australia
5	Ireland
6	Netherlands
7	Sweden
8	Japan
9	Luxembourg
10	Switzerland
11	France
12	Finland
13	Denmark
14	Austria
15	United States
16	Spain
17	Belgium
18	Greece
19	Italy
20	New Zealand
21	United Kingdom
22	Hong Kong (SAR)
23	Germany
24	Israel
25	South Korea
26	Slovenia
27	Brunei
28	Singapore
29	Kuwait
30	Cyprus

Economies of Top Thirty HDI Countries

1. ICELAND

Iceland's Scandinavian-type economy is basically capitalistic, yet with an extensive welfare system (including generous housing subsidies), low unemployment, and a remarkably even distribution of income.[1] In the absence of other natural resources (except for abundant geothermal power), the economy depends heavily on the fishing industry, which provides 79 percent of the export earnings and employs 6 percent of the workforce. The economy remains sensitive to declining fish stocks as well as to the fluctuations in world prices for its main exports: fish and fish products, aluminum, and ferrosilicon. Substantial foreign investment in the aluminum and hydropower sectors has boosted economic growth, which, nevertheless, has been volatile and characterized by recurring imbalances. Government policies include reducing the current account deficit, limiting foreign borrowing, containing inflation, revising agricultural and fishing policies, and diversifying the economy. The government remains opposed to European Union (EU) membership, primarily because of Icelanders' concern about losing control over their fishing resources. Iceland's economy has been diversifying into manufacturing and service industries in the last decade, and new developments in software production, biotechnology, and financial services are taking place. The tourism sector is also expanding, with the recent trends in ecotourism and whale watching. The 2006 closure of the U.S. military base at Keflavik had very little impact on the national economy; Iceland's low unemployment rate aided former base employees in finding alternate employment.

2. NORWAY

The Norwegian economy is a prosperous bastion of welfare capitalism, featuring a combination of free market activity and government intervention. The government controls key areas, such as the vital petroleum sector, through large-scale state enterprises. The country is richly endowed with natural resources—petroleum, hydropower, fish, forests, and minerals—and is highly dependent on its oil production and international oil prices, with oil and gas accounting for one-third of exports. Only Saudi Arabia and Russia export more oil than Norway. Norway opted to stay out of the EU during a referendum in November 1994; nonetheless, as a member of the European Economic Area, it contributes sizably to the EU budget. The government has moved ahead with privatization. Although Norwegian oil production peaked in 2000, natural gas production is still rising. Norwegians realize that once their gas production peaks they will eventually face declining oil and gas revenues; accordingly, Norway has been saving its oil-and-gas-boosted budget surpluses in a Government Petroleum Fund, which is invested abroad and now is valued at more than $250 billion. After lackluster growth of less than 1 percent in 2002–2003, gross domestic product (GDP) growth picked up to 3–5 percent in 2004–2007, partly due to higher oil prices. Norway's economy remains buoyant. Domestic economic activity is, and will continue to be, the main driver of growth, supported by high consumer confidence and strong investment spending in the offshore oil and gas sector. Norway's record high budget surplus and upswing in the labor market in 2007 highlighted the strength of its economic position going into 2008.

3. CANADA

As an affluent, high-tech industrial society in the trillion-dollar class, Canada resembles the United States in its market-oriented system, pattern of production, and affluent living standards. Since World War II, the impressive growth of the manufacturing, mining, and service sectors has transformed the nation from a largely rural economy into one primarily industrial and urban. The 1989 U.S.-Canada Free Trade Agreement (FTA) and the 1994 North American Free Trade Agreement (NAFTA, which includes Mexico) touched off a dramatic increase in trade and economic integration with the United States. Given its great natural resources, skilled labor force, and modern capital plant, Canada enjoys solid economic prospects. Top-notch fiscal management has produced consecutive balanced budgets since 1997, although public debate continues over the equitable distribution of federal funds to the Canadian provinces. Exports account for roughly a third of GDP. Canada enjoys a substantial trade surplus with its principal trading partner, the United States, which absorbs 80 percent of Canadian exports each year. Canada is the United States' largest supplier of energy, including oil, gas, uranium, and electric power. During 2007, Canada enjoyed good economic growth, moderate inflation, and the lowest unemployment rate in more than three decades.

4. AUSTRALIA

Australia has an enviable, strong economy with a per capita GDP on par with the four dominant West European economies. Robust business and consumer confidence and high export prices for raw materials and agricultural products are fueling the economy, particularly in mining states. Australia's emphasis on reform, low inflation, a housing market boom, and growing ties with China have been key factors behind the economy's sixteen solid years of expansion. Drought, robust import demand, and a strong currency have pushed the trade deficit up in recent years, while infrastructure bottlenecks and a tight labor market are constraining growth in export volumes and stoking inflation. Australia's budget has been in surplus since 2002 due to strong revenue growth.

5. IRELAND

Ireland is a small, modern, trade-dependent economy with growth averaging 6 percent in 1995–2007. Agriculture, once the most important sector, is now dwarfed by industry and services. Although the exports sector, dominated by foreign multinationals, remains a key component of Ireland's economy, construction has most recently fueled economic growth, along with strong consumer spending and business investment. Property prices have risen more rapidly in Ireland in the decade up to 2006 than in any other developed world economy. Per capita GDP is 40 percent above that of the four big European economies and the second highest in the EU behind Luxembourg and, in 2007, surpassed that of the United States. The Irish government has implemented a series of national economic programs designed to curb price and wage inflation, invest in infrastructure, increase labor force skills, and promote foreign investment. A slowdown in the property market, more intense global competition, and increased costs, however, have compelled government economists to lower Ireland's growth forecast slightly for 2008. Ireland joined in circulating the euro on January 1, 2002, along with eleven other EU nations.

6. NETHERLANDS

The Netherlands has a prosperous and open economy, which depends heavily on foreign trade. The economy is noted for stable industrial relations, moderate unemployment and inflation, a sizable current account surplus, and an important role as a European transportation hub. Industrial activity is predominantly in food processing, chemicals, petroleum refining, and electrical machinery. A highly mechanized agricultural sector employs no more than 3 percent of the labor force but provides large surpluses for exports. The Netherlands, along with eleven of its EU partners, began circulating the euro currency on January 1, 2002. The country continues to be one of the leading European nations for attracting foreign direct investment and is one of the five largest investors in the United States. The economy experienced a slowdown in 2005 but in 2006 recovered to the fastest pace in six years on the back of increased exports and strong investment. The pace of job growth reached ten-year highs in 2007.

7. SWEDEN

Aided by peace and neutrality for the whole of the twentieth century, Sweden has achieved an enviable standard of living under a mixed system of high-tech capitalism and extensive welfare benefits. It has a modern distribution system, excellent internal and external communications, and a skilled labor force. Timber, hydropower, and iron ore constitute the resource base of an economy heavily oriented toward foreign trade. Privately owned firms account for about 90 percent of industrial output, of which the engineering sector accounts for 50 percent of output and exports. Agriculture accounts for only 1 percent of GDP and 2 percent of employment. Sweden is in the midst of a sustained economic upswing, boosted by increased domestic demand and strong exports. This and robust finances have offered the center-right government considerable scope to implement its reform program aimed at increasing employment, reducing welfare dependence, and streamlining the state's role in the economy. The government plans to sell thirty-one billion dollars in state assets during the next three years to further stimulate growth and raise revenue to pay down the federal debt. In September 2003, Swedish voters turned down entry into the euro system due to concern about the impact on the economy and sovereignty.

8. JAPAN

Government-industry cooperation, a strong work ethic, mastery of high technology, and a comparatively small defense allocation (1 percent of GDP) helped Japan advance with extraordinary rapidity to the rank of the second-most technologically powerful economy in the world after the United States and the third-largest economy in the world after the United States and China, measured on a purchasing power parity (PPP) basis. One notable characteristic of the economy has been how manufacturers, suppliers, and distributors have worked together in closely-knit groups called keiretsu. A second basic feature has been the guarantee of lifetime employment of a substantial portion of the urban labor force. Both features have now eroded. Japan's industrial sector is heavily dependent on imported raw materials and fuels. The tiny agricultural sector is highly subsidized and protected, with crop yields among the highest in the world. Usually self-sufficient in rice, Japan must import about 55 percent of its food on a caloric basis. Japan maintains one of the world's largest fishing fleets and accounts for nearly 15 percent of the global catch. For three decades, overall real economic growth had been a spectacular 10 percent average in the 1960s, 5 percent average in the 1970s, and 4 percent average in the 1980s. Growth slowed markedly in the 1990s, averaging just 1.7 percent, largely because of the aftereffects of overinvestment and an asset price bubble during the late 1980s that required a protracted period of time for firms to reduce excess debt, capital, and labor. From 2000 to 2001, government efforts to revive economic growth proved short lived and were hampered by the slowing of the U.S., European, and Asian economies. In 2002–2007, growth improved and the lingering fear of deflation in prices and economic activity lessened, leading the central bank to raise interest rates to 0.25 percent in July 2006, up from the near 0 percent rate of the six years prior, and to 0.50 percent in February 2007. In addition, the ten-year privatization of Japan Post, which has functioned as not only the national postal delivery system but also, through its banking and insurance facilities, Japan's largest financial institution, was completed in October 2007, marking a major milestone in the process of structural reform. Nevertheless, Japan's huge government debt, which totals 182 percent of GDP, and the aging of the population are two major, long-run problems. Some fear that a rise in taxes could endanger the current economic recovery. Debate also continues on the role of and effects of reform in restructuring the economy, particularly with respect to increasing income disparities.

9. LUXEMBOURG

This stable, high-income economy—benefiting from its proximity to France, Belgium, and Germany—features solid growth, low inflation, and low unemployment. The industrial sector, initially dominated by steel, has become increasingly diversified to include chemicals, rubber, and other products. Growth in the financial sector, which now accounts for about 28 percent of GDP, has more than compensated for the decline in steel. Most banks are foreign owned and have extensive foreign dealings. Agriculture is based on small, family-owned farms. The economy depends on foreign and cross-border workers for about 60 percent of its labor force. Although Luxembourg, like all EU members, suffered from the global economic slump in the early part of this decade, the country continues to enjoy an extraordinarily high standard of living—GDP per capita ranks first in the world. After two years of strong economic growth in 2006–2007, turmoil in the world financial markets slowed Luxembourg's economy in 2008, but growth will remain above the European average.

10. SWITZERLAND

Switzerland is a peaceful, prosperous, and stable modern market economy with low unemployment, a highly skilled labor force, and per capita GDP larger than that of the big Western European economies. The Swiss in recent years have brought their economic practices largely into conformity with the EU's to enhance their international competitiveness. Switzerland remains a safe haven for investors because it has maintained a degree of bank secrecy and has kept up the franc's long-term external value. Reflecting the anemic economic conditions of Europe, GDP growth stagnated during the 2001–2003 period, improved during 2004–2005, and jumped to 2.9 percent in 2006 and 2.6 percent in 2007. Unemployment has remained at less than half the EU average.

11. FRANCE

France is in the midst of transition from a well-to-do modern economy that has featured extensive government ownership and intervention to one that relies more on market mechanisms. The government has partially or fully privatized many large companies, banks, and insurers and has ceded stakes in such leading firms as Air France, France Telecom, Renault, and Thales. It maintains a strong presence in some sectors, particularly power, public transport, and defense industries. The telecommunications sector is gradually being opened to competition. France's leaders remain committed to a capitalism in which they maintain social equity by means of laws, tax policies, and social spending that reduce income disparity and the impact of free markets on public health and welfare. Widespread opposition to labor reform has in recent years hampered the government's ability to revitalize the economy. In 2007, the government launched divisive labor reform efforts that continued into 2008. France's tax burden remains one of the highest in Europe (nearly 50 percent of GDP in 2005). France brought the budget deficit within the eurozone's 3 percent-of-GDP limit for the first time in 2007 and has reduced unemployment to roughly 8 percent. With at least seventy-five million foreign tourists per year, France is the most visited country in the world and maintains the third-largest income in the world from tourism.

12. FINLAND

Finland has a highly industrialized, largely free market economy with per capita output roughly that of the United Kingdom, France, Germany, and Italy. Its key economic sector is manufacturing—principally the wood, metals, engineering, telecommunications, and electronics industries. Trade is important; exports equal nearly two-fifth of GDP. Finland excels in high-tech exports, for example, mobile phones. Except for timber and several minerals, Finland depends on imports of raw materials, energy, and some components for manufactured goods. Because of the climate, agricultural development is limited to maintaining self-sufficiency in basic products. Forestry, an important export earner, provides a secondary occupation for the rural population. High unemployment remains a persistent problem. In 2007, Russia announced plans to impose high tariffs on raw timber exported to Finland. The Finnish pulp and paper industry will be threatened if these duties are put into place in 2008 and 2009, and the matter is now being handled by the EU.

13. DENMARK

The Danish economy has in recent years undergone strong expansion fueled primarily by private consumption growth but also supported by exports and investments. This thoroughly modern market economy features high-tech agriculture, up-to-date small-scale and corporate industry, extensive government welfare measures, comfortable living stands, a stable currency, and high dependency on foreign trade. Unemployment is low, and capacity constraints are limiting growth potential. Denmark is a net exporter of food and energy and enjoys a comfortable balance of payments surplus. Government objectives include streamlining the bureaucracy and further privatization of state assets. The government has been successful in meeting, and even exceeding, the economic convergence criteria for participating in the third phase (a common European currency) of the European Economic and Monetary Union (EMU), but so far Denmark has decided not to join fifteen other EU members in the euro. Nonetheless, the Danish krone remains pegged to the euro. Economic growth gained momentum in 2004, and the upturn continued through 2007. The controversy over caricatures of the Prophet Muhammad printed in a Danish newspaper in 2005 led to boycotts of some Danish exports to the Muslim world, especially exports of dairy products, but the boycotts did not have a significant impact on the overall Danish economy. Because of high GDP per capita, welfare benefits, a low Gini index (i.e., low measure of inequality of income or wealth), and political stability, the Danish living standards are among the highest in the world. A major long-term issue will be the sharp decline in the ratio of workers to retirees.

14. AUSTRIA

Austria, with its well-developed market economy and high standard of living, is closely tied to other EU economies, especially Germany's. The Austrian economy also benefits greatly from strong commercial relations, especially in the banking and insurance sectors, with central, eastern, and southeastern Europe. The economy features a large service sector, a sound industrial sector, and a small, but highly developed agricultural sector. Membership in the EU has drawn an influx of foreign investors attracted by

Austria's access to the single European market and proximity to the new EU economies. The outgoing government has successfully pursued a comprehensive economic reform program, aimed at streamlining government and creating a more competitive business environment, further strengthening Austria's attractiveness as an investment location. It has implemented effective pension reforms; however, lower taxes in 2005–2006 led to a small budget deficit in 2006 and 2007. Boosted by strong exports, growth nevertheless reached 3.3 percent in both 2006 and 2007, although the economy may have slowed in 2008 because of the strong euro, high oil prices, and problems in international financial markets. To meet increased competition—especially for new EU members and Central European countries—Austria will need to continue restructuring, emphasizing knowledge-based sectors of the economy, and encouraging greater labor flexibility and greater labor participation by its aging population.

15. UNITED STATES

The United States has the largest and most technologically powerful economy in the world, with per capita GDP of forty-six thousand dollars. In this market-oriented economy, private individuals and business firms make most of the decisions, and the federal and state governments buy needed goods and services predominately in the private marketplace. U.S. business firms enjoy greater flexibility than their counterparts in Western Europe and Japan in decisions to expand capital plant, lay off surplus workers, and develop new products. At the same time, they face higher barriers to enter their rivals' home markets than foreign firms face entering U.S. markets. U.S. firms are at or near the forefront in technological advances, especially in computers and in medical, aerospace, and military equipment; their advantage has narrowed since the end of World War II. The onrush of technology largely explains the gradual development of a "two-tier labor market" in which those at the bottom lack the education and the professional/technical skills of those at the top and, more and more, fail to get comparable pay raises, health insurance coverage, and other benefits.

Since 1975, practically all the gains in household income have gone to the top 20 percent of households. The response to the terrorist attacks of September 11, 2001, showed the remarkable resilience of the economy. The war in March–April 2003 between a U.S.-led coalition and Iraq, and the subsequent occupation of Iraq, required major shifts in national resources to the military. The rise in GDP in 2004–2007 was undergirded by substantial gains in labor productivity. Hurricane Katrina caused extensive damage to the Gulf Coast region in August 2005 but had a small impact on overall GDP growth for the year. Soaring oil prices in 2005–2007 threatened inflation and unemployment, yet the economy continued to grow through year-end 2007. Imported oil accounts for about two-thirds of U.S. consumption. Long-term problems include inadequate investment in economic infrastructure, rapidly rising medical and pension costs of an aging population, sizable trade and budget deficits, and stagnation of family income in the lower economic groups. The merchandise trade deficit reached a record $847 billion in 2007. Together, these problems caused a marked reduction in the value and status of the dollar worldwide in 2007.

16. SPAIN

The Spanish economy boomed from 1986 to 1990, averaging 5 percent annual growth. After a European-wide recession in the early 1990s, the Spanish economy resumed moderate growth starting in 1994. Spain's mixed capitalist economy supports a GDP that on a per capita basis is equal to that of the leading West European economies. The center-right government of former President Maria Aznar successfully worked to gain admission to the first group of countries launching the European single currency (the euro) on January 1, 1999. The Aznar administration continued to advocate liberalization, privatization, and deregulation of the economy and introduced some tax reforms to that end. Unemployment fell steadily under the Aznar administration but remains high at 7.6 percent. Growth averaging more than 3 percent annually during 2003–2007 was satisfactory given the background of a faltering European economy. The socialist president, Rodriguez Zapatero, has made mixed progress in carrying out key structure reforms, which need to be accelerated and deepened to sustain Spain's economic growth. Despite the economy's relatively solid footing, significant downside risks remain including Spain's continued loss of competitiveness, the potential for a housing market collapse, the country's changing demographic profile, and a decline in EU structural funds.

17. BELGIUM

This modern, private-enterprise economy has capitalized on its central geographic location, highly developed transport network, and diversified industrial and commercial base. Industry is concentrated mainly in the populous Flemish area in the north. With few natural resources, Belgium must import substantial quantities of raw materials and export a large volume of manufactures, making its economy unusually dependent on the state of world markets. Roughly three-quarters of its trade is with other EU countries. Public debt is more than 85 percent of GDP. On the positive side, the government has succeeded in balancing its budget, and income distribution is relatively equal. Belgium began circulating the euro currency in January 2002. Economic growth in 2001–2003 dropped sharply because of the global economic slowdown, with moderate recovery in 2004–2007. Economic growth and foreign direct investment were expected to slow down in 2008, due to credit tightening, falling consumer and business confidence, and above average inflation. However, with the successful negotiation of the 2008 budget and devolution of power within the government, political tensions seem to be easing and could lead to an improvement in the economic outlook for 2008.

18. GREECE

Greece has a capitalist economy with the public sector accounting for about 40 percent of GDP and with per capita GDP at least 75 percent of the leading eurozone economies. Tourism provides 15 percent of GDP. Immigrants make up nearly one-fifth of the workforce, mainly in agricultural and unskilled jobs. Greece is a major beneficiary of EU aid, equal to about 3.3 percent of annual GDP. The Greek economy grew by nearly 4.0 percent per year between 2003 and 2007, due partly to infrastructural spending related to the 2004 Athens Olympic Games, and in part to an increased availability of credit, which has sustained record levels of consumer spending. Greece violated the EU's Growth and Stability Pact budget deficit criteria of no more than 3 percent of GDP from 2001 to 2006 but finally met that criteria in 2007. Public debt, inflation, and unemployment are above the eurozone average but are falling. The Greek government continues to grapple with cutting government spending, reducing the size of the public sector, and reforming the labor and pension systems, in the face of often vocal opposition from the country's powerful labor unions and general public. The economy remains an important domestic political issue in Greece and, while the ruling New Democracy government has had some success in improving economic growth and reducing the budget deficit, Athens faces long-term challenges in its effort to continue its economic reforms, especially social security reform and privatization.

19. ITALY

Italy has a diversified industrial economy with roughly the same total and per capita output as France and the United Kingdom. This capitalistic economy remains divided into a developed industrial north, dominated by private companies, and a less-developed, welfare-dependent, agricultural south, with 20 percent unemployment. Most raw materials needed by industry and more than 75 percent of energy requirements are imported. Over the past decade, Italy has pursued a tight fiscal policy in order to meet the requirements of the Economic and Monetary Unions and has benefited from lower interest and inflation rates. The current government has enacted numerous short-term reforms aimed at improving competitiveness and long-term growth. Italy has moved slowly, however, on implementing needed structural reforms, such as lightening the high tax burden and overhauling Italy's rigid labor market and overgenerous pension systems, because of the current economic slowdown and opposition from labor unions. But the leadership faces a severe economic constraint: Italy's official debt remains above 100 percent of GDP, and the government has found it difficult to bring the budget deficit down to a level that would allow a rapid decrease in that debt. The economy continues to grow by less than the eurozone average, and growth was expected to decelerate from 1.9 percent in 2006 and 2007 to under 1.5 percent in 2008 as the eurozone and world economy slow.

20. NEW ZEALAND

Over the past twenty years, the government has transformed New Zealand from an agrarian economy dependent on concessionary British market access to a more industrialized, free market economy that can compete globally. This dynamic growth has boosted real incomes—but left behind many at the bottom of the ladder—and broadened and deepened the technological capabilities of the industrial sector. Per capita income has risen for eight consecutive years and reached $27,300 in 2007 in PPP terms. Consumer and government spending have driven growth in recent years. And exports picked up in 2006 after struggling for several years. Exports were equal to about 22 percent of GDP in 2007, down from 33 percent of GDP in 2001. Thus far the economy has been resilient, and the Labor government promises that expenditures on health, education, and pensions will increase proportionately to output. Inflationary pressures have built in recent years, and the central bank raised its key rate thirteen times since January 2004 to finish 2007 at 8.25 percent. A large balance of payments deficit poses another challenge in managing the economy.

21. UNITED KINGDOM

The United Kingdom, a leading trading power and financial center, is one of the quintet of trillion-dollar economies of Western Europe. Over the past two decades, the government has greatly reduced public ownership and contained the growth of social welfare programs. Agriculture is intensive, highly mechanized, and efficient by European standards, producing about 60 percent of food needs with less than 2 percent of the labor force. The United Kingdom has large coal, natural gas, and oil reserves; primary energy production accounts for 10 percent of GDP, one of the highest shares of any industrial nation. Services, particularly banking, insurance, and business services, account by far for the largest proportion of GDP, while industry continues to decline in importance. Since emerging from recession in 1992, Britain's economy has enjoyed the longest period of expansion on record; growth has remained in the 2–3 percent range since 2004, outpacing most of Europe. The economy's strength has complicated the Labor government's efforts to make a case for Britain to join the EMU. Critics point out that the economy is doing well outside of EMU, and public opinion polls show a majority of Britons are opposed to the euro. The Brown government has been speeding up the improvement of education, health services, and affordable housing at a cost in higher taxes and a widening public deficit.

22. HONG KONG (SAR)

Hong Kong has a free market economy highly dependent on international trade. In 2006, the total value of goods and services traded, including the sizable share of reexports, was equivalent to 400 percent of GDP. The territory has become increasingly integrated with mainland China over the past few years through trade, tourism, and financial links. The mainland has long been Hong Kong's largest trading partner, accounting for 46 percent of Hong Kong's total trade by value in 2006. As a result of China's easing travel restrictions, the number of mainland tourists to the territory has surged from 4.5 million in 2001 to 13.6 million in 2006, when they outnumbered visitors from all other countries combined. Hong Kong has also established itself as the premier stock market for Chinese firms seeking to list abroad. Bolstered by several successful initial public offerings in early 2007, by September 2007, mainland companies accounted for one-third of the firms listed on the Hong Kong Stock Exchange and more than half of the exchange's market capitalization. During the past decade, as Hong Kong's natural resources are limited, and food and raw materials must be imported, GDP growth averaged a strong 5 percent from 1989 to 2007, despite the economy suffering two recessions during the Asian financial crisis in 1997–1998 and the global downturn in 2001–2002. Hong Kong continues to link its currency closely to the U.S. dollar, maintaining an arrangement established in 1983.

23. GERMANY

Germany's affluent and technologically powerful economy—the fifth largest in the world in PPP terms—showed considerable improvement in 2007 with 2.6 percent growth. After a long period of stagnation with an average growth rate of 0.7 percent between 2001 and 2005 and chronically high unemployment, stronger growth led to a considerable fall in unemployment to about 8 percent near the end of 2007. Among the most important reasons for Germany's high unemployment during the past decade were macroeconomic stagnation, the decline level of investment in plant and equipment, company restructuring, flat domestic consumption, structural rigidities in the labor market, lack of competition in the service sector, and high interest rates. The modernization and integration of the eastern German economy continues to be a costly, long-term process, with annual transfers from west to east amounting to roughly eighty billion dollars. The former government of Chancellor Gerhard Schroeder launched a comprehensive set of reforms of labor market and welfare-related institutions. The current government of Chancellor Angela Merkel has initiated other reform measures, such as a gradual increase in the mandatory retirement age from sixty-five to sixty-seven and measures to increase female participation in the labor market. Germany's aging population, combined with high chronic unemployment, has pushed social security outlays to a level exceeding contributions, but higher government revenues from the cyclical upturn in 2006–2007 and a 3 percent rise in the value-added tax pushed Germany's budget deficit well below the EU's 3 percent debt limit. Corporate restructuring and growing capital markets are setting the foundations that could help Germany meet the long-term challenges of European economic integration and globalization, although some economists continue to argue the need for change in inflexible labor and services markets. Growth may fall below 2 percent in 2008 as the strong euro, high oil prices, tighter credit markets, and slowing growth abroad take their toll.

24. ISRAEL

Israel has a technologically advanced market economy with substantial, though diminishing, government participation. It depends on imports of crude oil, grains, raw materials, and military equipment. Despite limited natural resources, Israel has intensively developed its agricultural and industrial sectors over the past twenty years. Israel imports substantial quantities of grain but is largely self-sufficient in other agricultural products. Cut diamonds, high-technology equipment, and agricultural products (fruits and vegetables) are the leading exports. Israel usually posts sizable trade deficits, which are covered by foreign loans. Roughly half of the government's external debt is owed to the United States, its major source of economic and military aid. Israel's GDP, after contracting slightly in 2001 and 2002 due to the Palestinian conflict and troubles in the high-technology sector, has grown by about 5 percent per year since 2003. The economy grew an estimated 5.4 percent in 2007, the fastest pace since 2000. The government's prudent fiscal policy and structural reforms over the past few years have helped to induce strong foreign investment, tax revenues, and private consumption, setting the economy on a solid growth path.

25. SOUTH KOREA

Since the 1960s, South Korea has achieved an incredible record of growth and integration into the high-tech modern world economy. Four decades ago, GDP per capita was comparable with levels in the poorer countries of Africa and Asia. In 2004, South Korea joined the million-dollar club of world economies. Today its GDP per capita is roughly the same as that of Greece and Spain. This success was achieved by a system of close government/business ties including directed credit import restrictions, sponsorship of specific industries, and a strong labor effort. The government promoted the import of raw materials and technology at the expense of consumer goods and encouraged savings and investment over consumption. The Asian financial crisis of 1997–1998 exposed long-standing weaknesses in South Korea's development model, including high debt to equity ratios, massive foreign borrowing, and an undisciplined financial sector. GDP plunged by 6.9 percent in 1998, then recovered by 9.5 percent in 1999 and 8.5 percent in 2000. Growth fell back to 3.3 percent in 2001 because of the slowing global economy, falling exports, and the perception that much-needed corporate and financial reforms had stalled. Led by consumer spending and exports, growth in 2002 was an impressive 7 percent, despite anemic global growth. Between 2003 and 2007, growth moderated to about 4–5 percent annually. A downturn in consumer spending was offset by a rapid export growth. Moderate inflation, low unemployment, and an export surplus in 2007 characterize this solid economy, but inflation and unemployment are increasing in the face of rising oil prices.

26. SLOVENIA

Slovenia, which on January 1, 2007, became the first 2004 European Union entrant to adopt the euro, is a model of economic success and stability for the region. With the highest per capita GPD in Central Europe, Slovenia has an excellent infrastructure, a well-educated workforce, and a strategic location between the Balkans and Western Europe. Privatization has lagged since 2002, and the economy has one of highest levels of state control in the EU. Structural reforms to improve the business environment have allowed for somewhat greater foreign participation in Slovenia's economy and have helped to lower unemployment. In March 2004, Slovenia became the first country to graduate from borrower status country to donor partner at the World Bank. In December 2007, Slovenia was invited to begin the accession process for joining the Organization for Economic Cooperation and Development (OECD). Despite its economic success, foreign direct investment (FDI) in Slovenia has lagged behind the region average, and taxes remain relatively high. Furthermore, the labor market is often seen as inflexible, and legacy industries are losing sales to more competitive firms in China, India, and elsewhere.

27. BRUNEI

Brunei has a small, well-to-do economy that encompasses a mixture of foreign and domestic entrepreneurship, government regulations, welfare measures, and village tradition. Crude oil and natural gas production account for just over half of GDP and more than 90 percent of exports. Per capita GDP is among the highest in Asia, and substantial income from overseas investment supplements income from domestic production. The government provides for all medical services and free education through the university level and subsidizes rice and housing. Brunei's leaders are concerned that steadily increased integration in the world economy will undermine internal social cohesion. Plans for the future include upgrading the labor force, reducing unemployment, strengthening the banking and tourist sectors, and, in general, further widening the economic base beyond oil and gas.

28. SINGAPORE

Singapore has a highly developed and successful free market economy. It enjoys a remarkably open and corruption-free environment, stable prices, and a per capita GDP equal to that of the four largest West European countries. The economy depends heavily on exports, particularly in consumer electronics and information technology products. It was hard hit from 2001 to 2003 by the global recession, by the slump in the technology sector, and by an outbreak of severe acute respiratory syndrome (SARS) in 2003, which curbed tourism and consumer spending. Fiscal stimulus, low interest rates, a surge in exports, and internal flexibility led vigorous growth in 2004–2007 with real GDP growth averaging 7 percent annually. The government hopes to establish a new growth path that will be less vulnerable to the global demand cycle for information technology products—it has attracted major investments in pharmaceuticals and medical technology production—and will continue efforts to establish Singapore as Southeast Asia's financial and high-tech hub.

29. KUWAIT

Kuwait is a small, rich, relatively open economy with self-reported crude oil reserves of about 104 billion barrels—10 percent of world reserves. Petroleum accounts for nearly half of GDP, 95 percent of export revenues, and 80 percent of government income. High oil prices in recent years have helped build Kuwait's budget and trade surpluses and foreign reserves. As a result of the positive fiscal situation, the need for economic reforms is less urgent and the government has not earnestly pushed through new initiatives. Despite its vast oil reserves, Kuwait experienced power outages during the summer months in 2006 and 2007 because demand exceeded power-generating capacity. Power outages are likely to worsen, given its high population growth rates, unless the government can increase generating capacity. In May 2007 Kuwait changed its currency peg from the U.S. dollar to a basket of currencies in order to curb inflation and to reduce its vulnerability to external shocks.

30. CYPRUS

The area of the Republic of Cyprus under government control has a market economy dominated by the service sector, which accounts for 78 percent of GDP. Tourism, financial services, and real estate are the most important sectors. Erratic growth rates over the past decade reflect the economy's reliance on tourism, which often fluctuates with political instability in the region and economic conditions in Western Europe. Nevertheless, the economy in the area under government control grew by an average of 3.8 percent per year during the period of 2000–2006, well above the EU average. Cyprus joined the European Exchange Rate Mechanisms (ERM2) in May 2005 and adopted the euro as its national currency on January 1, 2008. An aggressive austerity program in the preceding years, aimed at paving the way for the euro, helped turn a soaring fiscal deficit (6.3 percent in 2003) into a surplus of 1.5 percent in 2007. As in the area administered by Turkish Cypriots, water shortages are a perennial problem; a few desalination plants are now online. After ten years of drought, the country received substantial rainfall from 2001 to 2004, alleviating immediate concerns. Rainfall in 2005 and 2006, however, was well below average, making water rationing a necessity in 2007.

Note

1. Data from CIA, "World Factbook," 2009, www.cia.gov/library/publications/the-world-factbook/index.html (accessed June 26, 2009).

References and Recommended Reading

New Encyclopedia Britannica. 2007. Economic systems, 357. vol. 4.

Spellman, F. R. 2009. *Geology for nongeologists*. Lanham, Md.: Government Institutes.

UN Development Programme (UNDP). 2008. New UN data (2006) shows progress in human development. December 18. http://hdr.undp.org/en/mediacentre/news/title,15493,en.html (accessed June 25, 2009).

APPENDIX B

Answers to Chapter Review Questions

Chapter 1

1.1. Location is important because every place on Earth can be located in either an absolute or relative way. Place is important because each place on the Earth has its own physical features and can be described in terms of its land, water, weather, soil, plant and animal life, and human features.
1.2. Regions make the study of geography more manageable.
1.3. Geographers use history to help them understand the way the places looked in the past. They use political science to help them see how people in different places are governed. They use sociology to help them understand societies in different places throughout the world. They use anthropology to help them gain insights into the culture of people in different places. They use economics to help them understand how the location of resources affects the ways people make, transport, and use natural resources and goods.
1.4. physical
1.5. urban
1.6. globes, maps
1.7. graticule
1.8. Rhumb line
1.9. direction indicator
1.10. Miller cylindrical

Chapter 2

2.1. 35
2.2. 7
2.3. Moho layer
2.4. flow
2.5. water cycle
2.6. lithosphere
2.7. misplaced soil
2.8. the part of Earth where life is found
2.9. huge, platelike sections of rock that rest, or float, on the heavier layers of the mantle

Chapter 3

3.1. rock type; structure; slope; climate; animals; time
3.2. physical weathering
3.3. chemical weathering
3.4. frost wedging
3.5. carbonic acid
3.6. lichens

Chapter 4

4.1. stream channel
4.2. evapotranspiration
4.3. infiltration capacity
4.4. thalweg
4.5. sinuosity
4.6. meander flow
4.7. balance
4.8. laminar
4.9. wash load; coarser materials
4.10. aquifers

Chapter 5

5.1. mountain
5.2. fjord
5.3. basal sliding
5.4. cirques
5.5. tarn
5.6. arêtes
5.7. drumlins
5.8. till
5.9. erratics
5.10. eskers

Chapter 6

6.1. magma
6.2. basaltic, andesitic, rhyolitic
6.3. gabbro
6.4. felsite

6.5. cider cone
6.6. lava spillways
6.7. intrusive rocks
6.8. sills
6.9. stocks or plutons
6.10. fumarole
6.11. plates
6.12. Pacific plate
6.13. convergent
6.14. divergent
6.15. transform

Chapter 7

7.1. transporter
7.2. saltation
7.3. deflation
7.4. ventifacts
7.5. eolian
7.6. barchan dunes
7.7. U
7.8. loess
7.9. mass wasting
7.10. slumps

Chapter 8

8.1. salinity
8.2. trench
8.3. guyots
8.4. tides
8.5. headlands

Chapter 9

9.1. climate
9.2. troposphere
9.3. stratosphere
9.4. desert air
9.5. cirrus
9.6. mixing ratio

9.7. relative humidity
9.8. hygroscopic nuclei
9.9. convectional precipitations
9.10. red wind
9.11. meteorology
9.12. Sun
9.13. inversion
9.14. albedo

Chapter 10

10.1. Bare rocks exposed to the elements; rocks become colonized by lichens; masses replace the lichens; grasses and flowering plants replace the mosses; woody shrubs begin replacing the grasses and flowering plants; a forest eventually grows where bare rocks once existed.
10.2. A medium for plant growth; regulate water supplies; recycle raw materials; provide habitat for organisms; an engineering medium; provide materials.
10.3. ground cover
10.4. air, water, mineral, and organic matter
10.5. structure

Chapter 11

11.1. ecology
11.2. Autecology is a major subdivision of ecology that studies the individual organism or a species.
11.3. Synecology is a major subdivision of ecology that studies groups of organisms associated together as a unit.
11.4. Pollution is an adverse alteration to the environment by a pollutant.
11.5. temperature; rainfall; light; mineral; wind; humidity; elevation; predominant land forms; tides; medium upon which the organisms exist (water, sand, mud, and rock).
11.6. The biogeochemical cycle describes the tendency of chemical elements, including all the essential elements of protoplast, to circulate in the biosphere in characteristic paths from environment to organism and back to the environment.
11.7. reservoirs
11.8. residence time
11.9. meteorological; geological; biological
11.10. sulfur

Chapter 12

12.1. Population ecology is the branch of ecology that studies the structure and dynamics of population.
12.2. Organisms in a population are ecologically equivalent.
12.3. species; density
12.4. immigration
12.5. community ecology
12.6. clumped distribution
12.7. carrying capacity
12.8. environmental carrying capacity
12.9. population controlling
12.10. ecological succession

Index

About the Author

Frank R. Spellman is assistant professor of Environmental Health at Old Dominion University in Norfolk, Virginia. He holds a BA in public administration, a BS in business management, an MBA, and an MS and a PhD in environmental engineering.

Spellman consults on homeland security vulnerability assessments (VAs) for critical infrastructure, including water/wastewater facilities nationwide. He also lectures on homeland security and health and safety topics throughout the country and teaches water/wastewater operator short courses at Virginia Tech.

Spellman has been cited in more than four hundred publications and written more than fifty books that cover topics in all areas of environmental science and occupational health, including *Chemistry for Nonchemists* (2006), *Biology for Nonbiologists* (2007), *Ecology for Nonecologists* (2008), *Physics for Nonphysicists* (2009), and *Geology for Nongeologists* (2009), all published by Government Institutes.

Breinigsville, PA USA
06 December 2010
250775BV00003B/87-300/P

9 781605 9068